Analytical Microextraction Techniques

Edited by

Miguel Valcárcel

Soleded Cárdenas

Rafael Lucena

Department of Analytical Chemistry Institute of Fine Chemistry and Nanochemistry University of Córdoba, Córdoba, Spain

ANALYTICAL MICROEXTRACTION TECHNIQUES

Editors: M. Valcárcel, S. Cárdenas and R. Lucena

eISBN (Online): 978-1-68108-379-7

ISBN (Print): 978-1-68108-380-3

First published in 2017.

Bentham Science Publishers Ltd.
Executive Suite Y - 2
PO Box 7917, Saif Zone
Sharjah, U.A.E.
Email: subscriptions@benthamscience.org

BENTHAM
SCIENCE

CONTENTS

FOREWORD

Sample preparation involving extraction is normally required prior to analysis by chromatography, electrophoresis, or mass spectrometry. The main purposes of the sample preparation are to clean-up the sample, to make it compatible with the analytical instrumentation, and to enrich the analytes of interest. Sample clean-up is intended to remove major matrix components from the sample, which can interfere with the analyte detection or which can reduce the performance of the analytical instrumentation. Enrichment of the analytes is intended to improve the trace level detectability.

Most sample preparation in routine laboratories today is performed with classical extraction techniques, like solid-phase extraction or liquid-liquid extraction. However, in the scientific literature, a large number of research papers have been published on the development of microextraction techniques. The difference between the classical extraction techniques and the microextraction techniques is the mass / volume of the extraction phase, which has been substantially down-scaled in the latter.

The field of microextraction was essentially initiated in 1990 by Professor Janusz Pawliszyn and his co-workers by the introduction of solid-phase microextraction (SPME). In the years after this, a large number of scientific papers related to SPME emerged, SPME became commercially available, and several other microextraction techniques different from SPME were introduced. In general, a main driving force for all the activities in the field of microextraction has been to reduce the consumption of hazardous organic solvents and other materials required for sample preparation (green chemistry). In addition, the development of different microextraction techniques has been motivated by reduced sample volumes, reduced extraction times, improved analyte enrichments, improved sample clean-up, and improved compatibility with analytical instrumentation.

Definitely microextraction techniques will replace traditional methods in the future, but this will take time. Meanwhile, a lot of progress takes place, and microextraction is currently a very active field of research within analytical chemistry. New techniques and methods based on these are continuously being developed. Some of them are based solid-phase extraction principles, and includes solid-phase microextraction, stir bar sorptive extraction, microextraction by packed sorbent, dispersive micro-solid phase extraction, magnetic solid phase extraction, and thin film extraction. Others are based on liquid-liquid extraction principles, like single-drop microextraction, membrane-based microextraction, hollow-fibre microextraction and dispersive liquid-phase microextraction. In addition to all this, new materials are also entering the field of microextraction, like nanoparticles, ionic liquids and novel solid gels. All this you can read about in this e-book, which is a very comprehensive guide to the most important developments of analytical microextraction techniques. Enjoy the reading!

Stig Pedersen-Bjergaard
School of Pharmacy
University of Oslo
Norway

PREFACE

The miniaturization of the preliminary steps of the chemical measurement process is one of the challenges of Analytical Chemistry as it clearly facilitates the reduction of dimensions of the whole analytical process, being also favorable for the design of portable analyzers and opens up a door for on-site analysis. Among the substeps included in sample treatment, extraction techniques are unavoidable for the majority of the samples, either for analyte preconcentration, interference removal or conditioning prior to instrumental analysis. Therefore, their miniaturization must be systematically considered.

The new microextraction techniques, including the solid and liquid formats have emerged in this context. The advances and innovations in this field affect not only on the units designed for analytes isolation but also on the new materials that are used. They are characterized to be more efficient than classical ones, thus yielding better enrichment factors. This results in an enhanced sensitivity/selectivity of the whole analytical process.

The Anaytical Microextraction Techniques ebook is fully devoted to this subject matter of increasing interest in the last decades. It is divided into three parts. The first one, composed of four chapters, deals with the general aspects of microextraction techniques in the green chemistry context. It also affords the use of new phases, which results in more efficient devices, fully compatible with the green chemistry principles.

The second part of this ebook is fully devoted to micro solid phase extraction. It starts with the reference technique in this context: solid phase microextraction. Next, fully consolidated formats such as stir bar sorptive extraction and microextraction by packed sorbents are afforded. Dispersive solid phase extraction is usually employed for interference removal. However, its applicability for analytes isolation and preconcentration has been recently evaluated. In this ebook, it is considered in two chapters, one of them dealing with the use of magnetic nanoparticles. Their role is crucial to simplify the extraction process avoiding the need for centrifugation and/or filtration. This section concludes with a chapter devoted to thin film microextraction.

The liquid phase microextraction is compiled in section 3 where single drop and membrane-based ones are presented. Single drop microextraction was the first miniaturized technique developed in the liquid-liquid format, reducing the volume of solvent needed to the low microliter level. Since its proposal, several alternatives have been reported aimed at increasing the versatility of the initial approach. The use of porous materials (flat membranes or hollow fiber) as support of the liquid phase, allows the use of large volumes of solvent, which results in higher enrichment factors.

This ebook would serve as a useful reference to readers to provide the current state of the art of this research area as well as to stimulate the developments of innovative approaches on the topic.

Miguel Valcárcel, Soledad Cárdenas and Rafael Lucena
Department of Analytical Chemistry
Institute of Fine Chemistry and Nanochemistry
University of Córdoba
Córdoba
Spain

List of Contributors

Abuzar Kabir	Department of Chemistry and Biochemistry, International Forensic Research Institute, Florida International University, 11200 SW 8th Street, Miami, Florida 33199, USA
Adam Kloskowski	Department of Physical Chemistry, Gdansk University of Technology, Faculty of Chemistry, Narutowicza Str.11/12, Gdansk 80-233, Poland
Ángel Ríos	Department of Analytical Chemistry and Food Technology, University of Castilla-La Mancha, E- 13004, Ciudad Real, Spain
Ana M. Afonso	Departamento de Química, Área de Química Analítica, Universidad de La Laguna (ULL), La Laguna (Tenerife), 38206, Spain
Carmen Molins-Legua	Department of Analytical Chemistry, Faculty of Chemistry, University of Valencia, Dr. Moliner 50, 46100-Burjassot, Valencia, Spain
Carlos Bendicho	Departamento de Química Analítica y Alimentaria, Área de Química Analítica, Facultad de Química, Universidad de Vigo, Campus As Lagoas-Marcosende s/n, 36310 Vigo, Spain
Francisco Pena-Pereira	Departamento de Química Analítica y Alimentaria, Área de Química Analítica, Facultad de Química, Universidad de Vigo, Campus As Lagoas-Marcosende s/n, 36310 Vigo, Spain
Francisco Pena-Pereira	Department of Analytical Chemistry, Gdansk University of Technology, Faculty of Chemistry, Narutowicza Str.11/12, Gdansk 80-233, Poland Analytical and Food Chemistry Department, University of Vigo, Faculty of Chemistry, Campus As Lagoas-Marcosende s/n, 36310 Vigo, Spain
Guillermo Lasarte-Aragonés	Department of Analytical Chemistry, Institute of Fine Chemistry and Nanochemistry, Marie Curie Building (Annex), Campus de Rabanales, University of Córdoba, Córdoba, Spain
Isela Lavilla	Departamento de Química Analítica y Alimentaria, Área de Química Analítica, Facultad de Química, Universidad de Vigo, Campus As Lagoas-Marcosende s/n, 36310 Vigo, Spain
Jacek Namiesnik	Department of Analytical Chemistry, Faculty of Chemistry, Gdansk University of Technology, Narutowwicza Str. 11/12, Gdansk 80-233, Poland
Jan Åke Jönsson	Center for Analysis and Synthesis, Department of Chemistry, Lund University, P.O. Box 124, SE-221 00 Lund, Sweden
Jorge Verdú Andrés	Department of Analytical Chemistry, Faculty of Chemistry, University of Valencia, Dr. Moliner 50, 46100-Burjassot, Valencia, Spain
José Manuel Florêncio Nogueira	Chemistry and Biochemistry Department and Centre of Chemistry and Biochemistry, University of Lisbon, Faculty of Sciences, Campo Grande, Ed. C8, 1749-016 Lisbon, Portugal
Juan H. Ayala	Departamento de Química, Área de Química Analítica, Universidad de La Laguna (ULL), La Laguna (Tenerife), 38206, Spain

Łukasz Marcinkowski	Department of Physical Chemistry, Gdansk University of Technology, Faculty of Chemistry, Narutowicza Str.11/12, Gdansk 80-233, Poland
Miguel Valcárcel	Department of Analytical Chemistry, Institute of Fine Chemistry and Nanochemistry, Marie Curie Building, Campus de Rabanales, University of Córdoba,14071 Córdoba, Spain
Miguel Roldán-Pijuán	Department of Analytical Chemistry, Institute of Fine Chemistry and Nanochemistry, Marie Curie Building, Campus de Rabanales, University of Córdoba,14071 Córdoba, Spain
María J. Trujillo–Rodríguez	Departamento de Química, Área de Química Analítica, Universidad de La Laguna (ULL), La Laguna (Tenerife), 38206, Spain
Miguel de la Guardia	Department of Analytical Chemistry, Research Building, University of Valencia, 50th Dr. Moliner St., E-46100 Burjassot, Valencia, Spain
Miguel Valcárcel	Department of Analytical Chemistry, Institute of Fine Chemistry and Nanochemistry, Marie Curie Building (Annex), Campus de Rabanales, University of Córdoba, Córdoba, Spain
Mohammed Zougagh	Regional Institute for Applied Chemistry Research, IRICA, E-13004, Ciudad Real, Spain Albacete Science and Technology Park, E-02006 Albacete, Spain
Mohamed Abdel-Rehim	Department of Analytical Chemistry, Stockholm University, SE10691 Stockholm, Sweden
Mohammad Mahdi Moein	Department of Analytical Chemistry, Stockholm University, SE10691 Stockholm, Sweden
Pascual Serra-Mora	Department of Analytical Chemistry, Faculty of Chemistry, University of Valencia, Dr. Moliner 50, 46100-Burjassot, Valencia, Spain
Pilar Campíns-Falcó	Department of Analytical Chemistry, Faculty of Chemistry, University of Valencia, Dr. Moliner 50, 46100-Burjassot, Valencia, Spain
Rafael Lucena	Department of Analytical Chemistry, Institute of Fine Chemistry and Nanochemistry, Marie Curie Building (Annex), Campus de Rabanales, University of Córdoba, Córdoba, Spain
Rosa Herráez-Hernández	Department of Analytical Chemistry, Faculty of Chemistry, University of Valencia, Dr. Moliner 50, 46100-Burjassot, Valencia, Spain
Sergio Armenta	Department of Analytical Chemistry, Research Building, University of Valencia, 50th Dr. Moliner St., E-46100 Burjassot, Valencia, Spain
Soledad Cárdenas	Department of Analytical Chemistry, Institute of Fine Chemistry and Nanochemistry, Marie Curie Building (Annex), Campus de Rabanales, University of Córdoba, Córdoba, Spain
Verónica Pino	Departamento de Química, Área de Química Analítica, Universidad de La Laguna (ULL), La Laguna (Tenerife), 38206, Spain
Yolanda Moliner-Martínez	Department of Analytical Chemistry, Faculty of Chemistry, University of Valencia, Dr. Moliner 50, 46100-Burjassot, Valencia, Spain

Analytical Microextraction Techniques

CHAPTER 1

Green Microextraction

Sergio Armenta[1,*], **Miguel de la Guardia**[1] and **Jacek Namiesnik**[2]

[1] *Department of Analytical Chemistry, Research Building, University of Valencia, 50th Dr. Moliner St., E-46100 Burjassot, Valencia, Spain*

[2] *Department of Analytical Chemistry, Faculty of Chemistry, Gdansk University of Technology, Narutowwicza Str. 11/12, Gdansk 80-233, Poland*

Abstract: In the present chapter, the microextraction techniques have been considered from the perspective of the Green Analytical Chemistry and an attempt has been made in order to propose a modified eco-scale suitable to provide a green certification of the extraction steps in the frame of the consideration of reagents and energy consumption and waste generation. Especially important is the evaluation of the intrinsic toxicity and risks associated to the use of reagents and the need of an accurate evaluation of the amounts of reagents and wastes in order to avoid misclassifications. It has been proposed a new eco-scale, called "Green Certificate", based on the application of a color code associated to a letter; from A to G, being A the greenest one, that gives an idea of how much green an analytical methodology is. In short, it is evident that the microextraction enhances the method greenness providing class A steps in front of classical analytical preconcentration and matrix removal which changes the method category to class D.

Keywords: Eco-scale, Green analytical chemistry, Green certificate, Green evaluation, Hollow-fiber liquid phase microextraction, Liquid phase microextraction, Single drop microextraction, Solid phase microextraction, Solvent microextraction, Stir bar sorptive extraction, Stir cake sorptive extraction, Stir rod sorptive extraction.

* **Corresponding author Sergio Armenta:** Department of Analytical Chemistry, Research Building, University of Valencia, 50th Dr. Moliner St., E-46100 Burjassot, Valencia, Spain; Tel/Fax: 0034 96 354 40 04; E-mail:sergio.armenta@uv.es

M. Valcárcel, S. Cárdenas, R. Lucena (Eds.)

1.1. INTRODUCTION

In recent years, the interest and attention of the scientific chemistry community in the development of green practices has increased substantially. From the texts and books edited by Paul Anastas between 1994 and 1998 devoted to Green Chemistry [1 - 4] the activity in the frame of Green Chemistry has considerably been increased.

Analytical methods are used to solve problems. Using, as inputs, preliminary information and knowledge related to a particular problem, solvents, reagents, samples, energy and instruments, the analysts can solve a specific problem. On the other hand, the outputs of those processes are results, data and models, in other words, solutions to the defined problem.

However, on the other side, the outputs of analytical methodologies are harmful wastes that create risks for operators and damage the environment [2]. In 1999 [3], J. Namiesnik proposed the term Green Analytical Chemistry (GAC) to include important parameters such as wastes, amount and toxicity of employed reagents, as important criteria to select the most appropriate methodology to solve a specific problem, thus creating a new frame for the previously named environmentally friendly analytical chemistry [4].

The twelve principles of GAC, proposed by A. Gałuszka *et al.* [5] can be found elsewhere and include:

1. Direct analytical techniques.
2. Minimize sample size and number of samples.
3. *In situ* measurements.
4. Integration of analytical processes and operations to reduce energy and reagents usage.
5. Automation and miniaturization of methods.
6. Avoid derivatization.
7. Avoid waste generation and provide appropriate waste management.
8. Multi-analyte or multi-parameter methods.
9. Minimize the use of energy.

10. Preferable use of reagents obtained from renewable sources.
11. Elimination of toxic reagents.
12. Increase of the safety of the operator.

Taking into consideration the 12 mentioned principles, in-field direct analysis can be considered as the greenest alternative: i) there are no reagents implied, ii) the method do not generate analytical wastes, iii) the risks to operators and the environment are generally, minimal. However, those remote, non invasive and direct analytical techniques can not be used in all the situations and because of that greener analytical method, in many cases, implies the reduction of the energy and reagents consumption, take care about the side effects of sample treatments, mainly isolation/enrichment steps of analytes and minimize the wastes.

Although, in many cases, direct analysis of samples is not possible and a sample preparation step should be included in the analytical method. The evolution of these sample preparation tools from the classical extraction methods such as soxhlet and liquid–liquid extraction (LLE) to the new and innovative solutions in solid phase extraction (SPE) [6] pursues the following objectives: i) miniaturization of the device, ii) multiclass compound extraction and iii) automation and/or high-throughput determination (Fig. **1**).

Analytical microextraction can be defined as a sample preparation step using volumes in the microliter or nanoliter range of extracting phase (solid, semi-solid polymeric or liquid material). In summary, analytical microextraction includes the aforementioned guidelines of sample treatment evolution, especially in the frame of miniaturization, automation, on-site analysis and time efficiency, most of them directly related with the twelve principles of GAC.

1.2. ANALYTICAL STRATEGIES OF MICROEXTRACTION

The extraction procedures can be classified according its exhaustive and non-exhaustive nature. The main objective of the exhaustive techniques is to extract analytes completely from the sample to the extraction phase. The main advantage of these techniques is that they do not require to extract the standards following the extraction procedure, because the analyte is transferred quantitatively to the extraction phase and thus, external standards dissolved in the extracting solvent can be used for calibration. On the other hand, the main drawbacks of complete extraction concern that they are, frequently, labour-intensive and costly procedures.

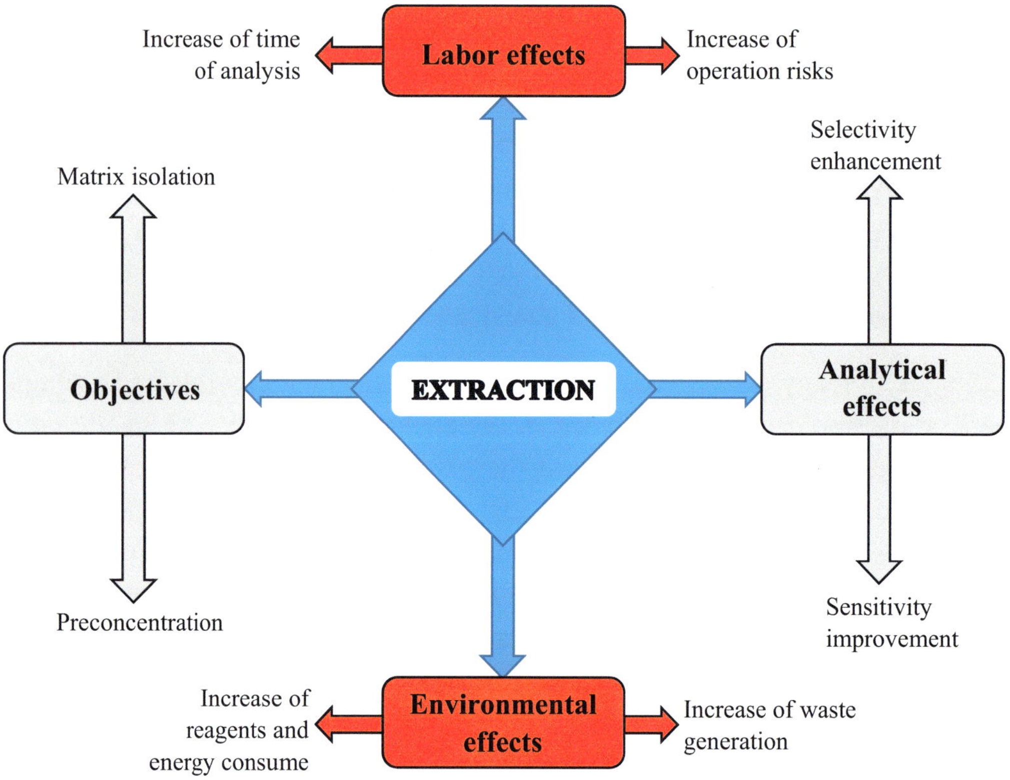

Fig. (1). Extraction steps in the frame of Green Analytical Chemistry.

An improvement of those methodologies includes the replacement of batch equilibrium techniques are frequently replaced by flow-through techniques to reduce volume of solvents and also the analysis time. An example of this is the SPE in which the sample is passed through a cartridge which includes the solid sorbent. The analytes are retained on the bed and extracted using a small solvent volume, resulting in substantial preconcentration. The introduction of flow injection analysis (FIA) has increased the number of samples analyzed per hour [7]. Further miniaturization of FIA technology resulted in the lab in a valve concept in which a single valve is used of to perform all the operations of an analytical procedure [8].

Alternatively, non-exhaustive approaches can be used to extract and preconcentrate analytes from the sample matrix. The non-exhaustive extraction techniques have been developed to reduce solvent consumption and improve performance, fitting, perfectly, with the green analytical principles (Fig. **2**). Those techniques can be used in equilibrium or non-equilibrium conditions. The basis of equilibrium non-exhaustive methods is analogous to equilibrium exhaustive ones. However, the main differences is that the reduced capacity of the extraction phase is not sufficient to extract completely the analytes from the sample. It is due to the use of a volume of extracting phase really smaller than the sample volume in the case of both, solid phase microextraction (SPME) and liquid phase microextraction (LPME) or due to the low sample matrix-extraction phase distribution constant as in the headspace techniques. However, in these cases, standards must be prepared following the same procedure than samples and it reduces the sampling throughput.

The advances in extraction techniques has been directed towards a reduction of the working scale, in other words, a miniaturization of the devices. Currently, there is an increasing interest in the incorporation of sample extraction into miniaturized devices to perform on-site deployment and/or automations, such as the in-micro-needle and on-fibre micro-sampling devices. The main advantages of microextraction procedures are related to the reduction of i) solvent use, ii) extraction time, both in the laboratory and on-site applications and iii) the impact on sampled systems.

In summary, the versatility of microextraction techniques includes the fully automated sample treatment (extraction and injection) when coupled coated fibres for SPME with gas chromatography (GC) or liquid chromatography (LC), the possibility of simultaneous sample treatment using the 96-fibre configuration which increases sample processing throughput and the small size of device possibilities for the *in vivo* analysis.

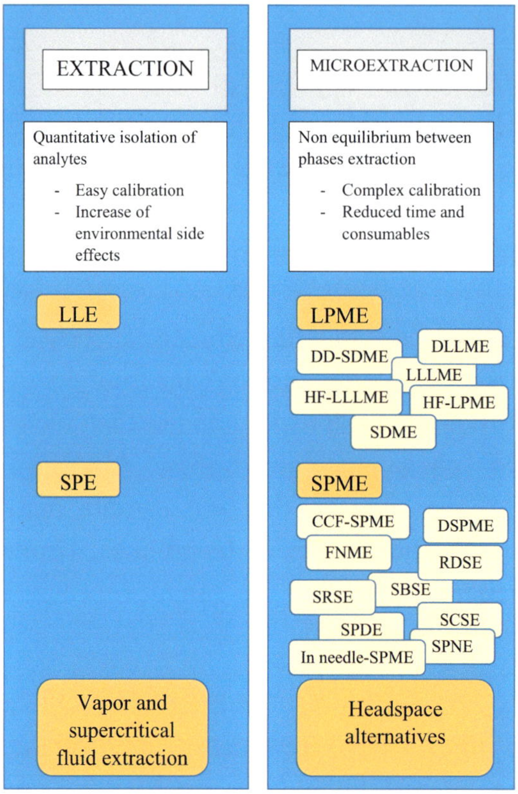

Fig. (2). Extraction *versus* microextraction techniques. Note: Abbreviations are explained in Table **1**.

1.3. TYPES OF MICROEXTRACTION

We would limit our discussion on microextraction techniques to nonexhaustive techniques. In this sense, we can classify the different microextraction formats according the physical state of the extracting phase as: SPME and LPME.

The use of microextraction techniques increased significantly with the commercialization of the SPME in 1993 [9]. In this technique, the sample itself or its headspace is exposed for a well-defined period of time to a small amount of extracting phase dispersed on a fused silica fiber or a metal core. It should be noted that the name of SPME proceeded from the extraction of analytes on solid fused-silica fibers. However, the term SPME was broader used, even when the extraction phase is not a solid [8].

In SPME the objective is not to obtain a concentration in the equilibrium between the analyte in the liquid/gas phase and that in the fiber, because the exposition time in this case would be quite long. The idea is to obtain appropriate matrix removal together with the analyte preconcentration. The main advantages of SPME includes features such as: i) simplicity, ii) versatility, iii) automation iv) thermal desorption, v) relative low cost, vi) short extraction time, vii) elimination of organic solvents and vi) recently, possibility to collect samples *in situ* and *in vivo* [10].

Different geometries of the SPME system has been developed to facilitate speed, convenience of use, and sensitivity and, thus, the SPME technique has derived into several extraction concepts, such as agitation mechanism (stir bar sorptive extraction (SBSE) [11], multi stir bar sorptive extraction (MSBSE) [12], stir rod sorptive extraction (SRSE) [13], stir cake sorptive extraction (SCSE) [14], rotating disk sorptive extraction (RDSE) [15]), vessels, dispersed particles into the solution, also called dispersive solid phase microextraction (DSPME) [16] and needles. In this later case, in-needle SPME [17], solid phase dynamic extraction (SPDE) [18] and accelerated solid phase dynamic extraction (ASPDE) [19], microextraction by packed sorbent (MEPS) [20], microextraction in a packed syringe [21], fiber-packed-needle microextraction (FNME) [22] have been proposed, being also used pipette tips as sorbent-packed in-tip SPME [23], fiber-packed in-tip SPME [24] and disks [25].

Capillaries have been also used as microextraction elements, such as capillary microextraction (CME) [26], fiber in tube solid phase extraction (FIT-SPE) [27] and in-tube solid phase microextraction (in-tube SPME) [28].

The internally coated cooled fiber SPME (CCF-SPME) approach has been developed to enhance the sensitivity of SPME and to preserve the convenience of coupling it to analytical instrumentation. It is a very powerful approach [29] in which a cooled gas or liquid passes through the inside of the coated, hollow fiber during sampling. The main effect is to cool the extraction phase, increasing the percentage of extracted analytes.

Other novel methodological solutions in the field of SPME include the micro-solid phase extraction (μSPE) [30], the adsorptive micro extraction (AμE) [31] and the solid phase nanoextraction (SPNE) [32] which are focused to a drastic reduction of the volume of reagents and samples, thus enhancing the miniaturization.

Membrane based procedures based on the use of a thin sheet of poly (dimethylsiloxane) (PDMS) membrane as an extraction phase have been used as SPME alternatives under the name of thin film microextraction (TFM) [33]. Novel designs for the SPME have been proposed, including the membrane solid-phase microextraction (M-SPME) [34], which implies the use of a membrane to separate the polar extraction medium and the sample. Another membrane based procedure is the polymer coated hollow fiber membrane (PC-HFM) [35]. In this method, the HFM coated with a functional polymer (adsorbent) is introduced in the sample to extract and preconcentrate the analytes. Other extraction method, named membrane extraction with sorbent interface (MESI) [36], consisted in a permeable silicone membrane and an adsorbent trap to preconcentrate organic compounds.

Another SPME technique is that named electrochemically aided SPME [37]. However, its use is not very extended due to its very low extraction efficiency and the difficulty to coupling to a chromatographic system. Therefore a new variant, electrosorption-enhanced SPME (EE-SPME) [38], was proposed in 2007.

The second type of microextraction techniques are called LPME. A definition of LPME includes all modes of sample preparation techniques which used solvent volumes of 100 μL or less for analyte extraction [39]. The number of techniques based on LPME, also called solvent microextraction (SME), is very big. In words of J.M. Kokosa: "SME literature contains a confusing array of more than 100 acronyms – and more keep appearing each year. A rather frustrating fact of academic research is that each experimentalist tends to vary a general method and then apply a new name, along with a new acronym, to their new variation".

The aim of this chapter is not to do a deep discussion about the different LPME modes which have been recently reviewed [40] in terms of green analytical chemistry and, thus, only the most important, used, variations would be described and commented. LPME techniques can be classified in two broad groups named exposed solvent and membrane-protected solvent. The exposed-solvent techniques include single-drop microextraction (SDME) [41], drop to drop-SDME (DD-SDME) [42], liquid-liquid-liquid microextraction (LLLME) [43] and dispersive liquid-liquid microextraction (DLLME) [44]. The protected LPME modes include hollow-fiber LPME (HF-LPME) [45] and hollow-fiber 3-phase liquid microextraction (HF LLLME) [46] and, in all the cases, the main purpose is to dramatically reduce the amount of solvents used in analyte preconcentration and matrix removal.

Table **1** summarizes the different abbreviations of the microextraction techniques to facilitate their use and understanding.

1.4. GREEN CERTIFICATE FOR THE EVALUATION OF ANALYTICAL PROCEDURES

There are different methodologies to evaluate the green character of analytical procedures. The Life Cycle Assessment (LCA) is a tool that includes the impact of all the activities/inputs (resources, energy, emissions, waste) of the whole life of a product or activity. The applicability of this tool, obviously, requires an important knowledge of all the stages of the life-cycle of the product or service involved [47]. The Green Chemistry Institute (GCI) of the American Chemical Society (ACS) developed a second criteria, that was applied to the National Environmental Methods Index (NEMI), a free internet-searchable database of environmental analytical methods [48].

The green character of the database methods was evaluated based on four key terms: PBT (persistent, bioaccumulative and toxic) nature of the used reagents, hazardous nature of processes and products, corrosive character of used reagents and generated waste. A similar system that includes energy as a criterion has been developed [49]. The inclusion of energy as an important parameter is justified because of the use of non-renewable resources for electrical energy production. Moreover, the hazard section was divided into health, safety and environmental hazards.

Table 1. Abbreviations used in microextraction procedures.

Microextraction strategy		
SPME	SPME	Solid phase microextraction
	μSPE	Micro-solid phase extraction
	ASPDE	Accelerated solid phase dynamic extraction
	AμE	Adsorptive microextraction
	CCF-SPME	Coated cooled fiber solid phase microextraction
	CME	Capillary microextraction
	DSPME	Dispersive solid phase microextraction
	EE-SPME	Electrosorption-enhanced solid phase microextraction
	FIT-SPE	Fiber in tube solid phase extraction
	FNME	Fiber-packed-needle microextraction
	In needle-SPME	In-needle solid phase microextraction
	In-tube SPME	In-tube solid phase microextraction
	MEPS	Micorextraction by packed sorbent
	MESI	Membrane extraction with sorbent interface
	MSBSE	Multi stir bar sorptive extraction
	M-SPME	Membrane solid phase microextraction
	PC-HFM	Polymer coated hollow fiber membrane
	RDSE	Rotating disk sorptive extraction
	SBSE	Stir bar sorptive extraction
	SCSE	Stir cake sorptive extraction
	SPNE	Solid phase nanoextraction
	SPDE	Solid phase dynamic extraction
	SRSE	Stir rod sorptive extraction
	TFM	Thin film mocroextraction
LPME	LPME	Liquid phase microextraction
	SDME	Single drop microextraction
	DD-SDME	Drop to drop single drop microextraction
	DLLME	Dispersive liquid-liquid microextraction
	LLLME	Liquid-liquid-liquid microextraction
	HF-LLLME	Hollow fiber 3-phase liquid microextraction
	HF-LPME	Hollow fiber liquid phase microextraction

More recently, a concept based on the introduction of penalty points and an analytical Eco-scale was proposed by A. Gałuszka *et al.* [50]. The basis of this concept is that the ideal green analysis has a score value of 100, and for each of analytical procedure, as a whole or divided in the respective steps, penalty points are assigned if it departs from ideal green analysis. These penalty points concern the amount of reagents, hazards related to the use of reagents and solvents, consumed energy and generated waste. This approach is similar to the Eco-Scale proposed by Van Aken *et al.* [51] for evaluating the greenness of organic synthesis that creates a first approach for the quantification of the green parameters of a method.

In this chapter, we will apply the concept of analytical eco-Scale to perform a green evaluation of the different microextraction techniques. Additionally to the aforementioned eco-scale, what we propose in this chapter is to classify the analytical methodologies, and particularly the microextraction techniques, according to a modified Eco-scale by introducing different categories such as those used in the energy performance certificate employed in the classification of buildings, appliances and so on. The proposed scale, called "Green Certificate", is based on the application of a color code associated to a letter; from A to G, being A the greenest one, that gives an idea of how much green an analytical methodology is (Fig. **3**). A class analytical methods are those with less than 10 penalty points. B class methods sum between 11 and 20 penalty points. C class methods have between 21 and 30 penalty points. D class methods sum between 31 and 45 penalty points, E class methods between 46 and 60 penalty points, F class methods between 61 and 80 penalty points and, finally, G class methods accounts for more than 81 penalty points.

In summary, the penalty points employed to calculate the analytical Eco-scale are assigned according to the use of reagents, concerning both, amount and hazards, the consumed energy, possible occupational hazards and generated wastes. Because of the influence of risks associated to reagents depend on their amount, thus, it was proposed that the calculations of the total penalty points included a factor relative to the amount and hazard of the reactive. The evaluation of physical, environmental and health hazards were made on the basis of the Globally Harmonized System of Classification and Labeling of Chemicals [52]. It was proposed the following criterion in which less than 10 mL (g) of reagent applies for 1 penalty point, between 10 and 100 mL (g) applies for 2 penalty points and more than 100 mL (g) applies for 3 penalty points. These penalty points should be multiplied by the penalty points of hazards: the absence of pictograms concerns 0 penalty points, less severe hazard, described by warning pictogram, concerns 1 penalty point and more severe hazard indicated by danger pictogram involves 2 penalty points.

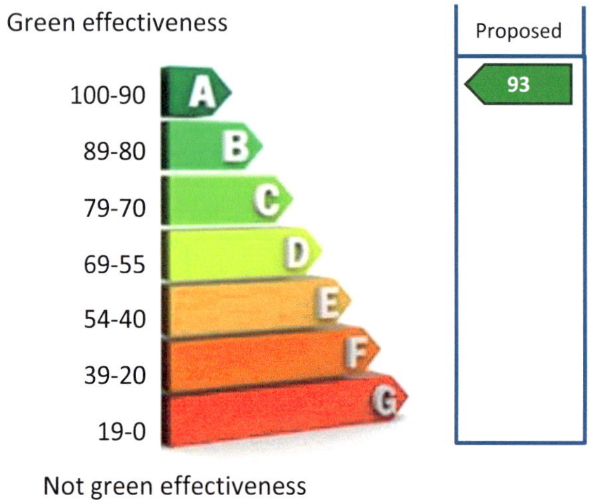

Fig. (3). Green evaluation of analytical methodologies based on the green certificate classification.

Penalty points for energy consumption will be assigned according to the values proposed by Raynie and Driver [53] regarding the power involved in sample analysis. Less or equal than 0.1 kWh per sample which corresponds to 0.04455 Kg CO_2 involves 0 penalty points, between 0.1 and 1.5 kWh per sample (0.04455 to 0.668 Kg CO_2) is 1 penalty points, and more than 1.5 kWh per sample is 2 penalty points.

As an example, different microextraction procedures employed for the measurement of volatile sulfur compounds in beer and beverage by gas chromatography with flame photometric detection [54] can be used to evaluate headspace SDME, direct SDME and HS SPME as microextraction alternatives. The nature and amounts of reagents used in the different procedures developed in [57] are summarized in Table **2**.

According to the aforementioned data, the 1.5 µL n-hexane of the SDME accounts for 8 penalty points, 67% of the penalty points of the microextraction procedure, and seems to be overrated in front of the 8 penalty points that would obtain a procedure using 9.9 mL of n-hexane, thus involving a volume 6600 times higher than that. To avoid this inconsistency we propose to consider an Eco-microscale. It consists of multiply by 0.1 the penalty points when the volume of reagent is from 1 to 9 µL and by 0.5 when the volume is from 10 to 500 µL in order to well discriminate the microextraction from the macroextraction procedures.

Table 2. Penalty points for the proposed microextraction procedure for the determination of volatile sulfur compounds in beer and beverages.

	Penalty points							**Category**
Microextraction method	**Reagents amount**	**Hazard**	**Subtotal**	**Energy**	**Occupational hazard**	**Waste**	**Total**	
HS-SDME	1 (2 g NaCl)	0	0	0	0	3 (10 mL sample)		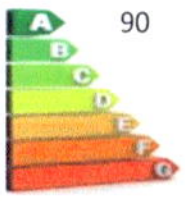
	1 (60 mg EDTA)	1	1					
	1 (2 µL DMF)	6	6					
							10	
SDME	1 (120 mg EDTA)	1	1	0	0	3 (2 mL sample)		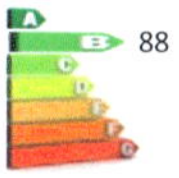
	1 (1.5 µL n-hexane)	8	8					
							12	
HS-SPME	1 (2 g NaCl)	0	0	1	0	3 (10 mL sample)		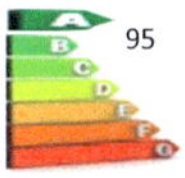
	1 (60 mg EDTA)	1	1					
							5	

However, even applying the Eco-microscale, we will have an inconsistent situation where 0.8 mL of n-hexane would be 8 penalty points, 9.9 mL n-hexane would be equivalent to 8 penalty points and 10 mL of n-hexane (just 0.1 mL more than the previous situation) would be 16 penalty points. Thus, it seems necessary to find an appropriate relationship between the penalty point values and the volume of used reagents or generated wastes, providing a "true" sense of the value of the penalty points on evaluating the greenness of a method.

To do a correct evaluation of the penalty points concerning the use of reagents and solvents, it has been plotted the penalty parameter *versus* the volume of consumed reagents or generated waste and the appropriate relationship between them was established. We used the LAB Fit Curve Fitting Software 7.2.48 for the analysis of experimental data and curve-fitting [55] and the Levenberg-Marquardt algorithm [56] to increase the possibility to have convergence. To find the best model fitting from our data, the sum of squares of the differences between experimental points and theoretical ones were minimized. The program tested two hundred and eight curves with up two parameters, using a maximum number of iterations of 300, and the tolerance criterion for convergence of 10^{-5}.

From the results obtained concerning the equation y=a×xb (Figs. **4** and **5**), it can be seen that those models fits with the experimental data points, minimizing the residuals. y is the penalty point value, x the volume of reagents and a and b two coefficients (a=0.61±0.05 and b=0.31±0.02 for reagents consumption and a=1.50±0.08 and b=0.40±0.02 for waste generation). The aforementioned models provide coefficient of determination (R^2) values of 0.996, in both cases.

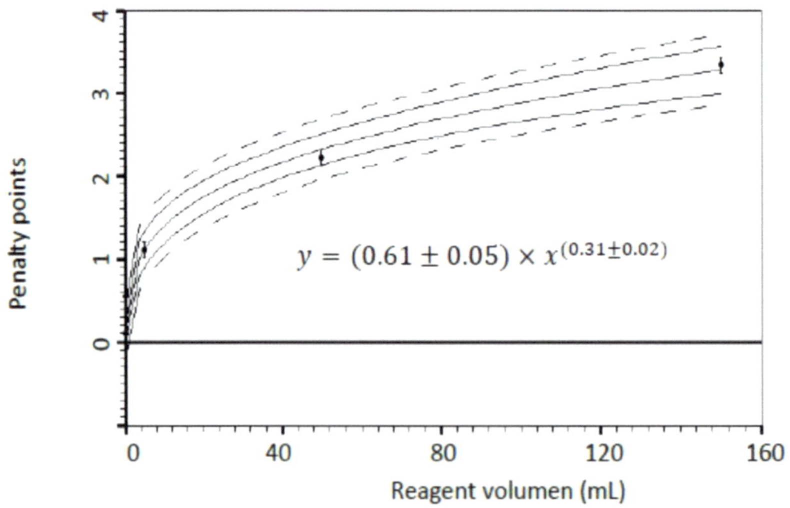

Fig. (4). Curve fitting for the penalty parameter *versus* the volume of consumed reagent.

1.5. GREEN EVALUATION OF MICROEXTRACTION PARAMETERS

In this part of the chapter we will provide, as some practical examples, a green analytical evaluation of microextraction procedures previously published in the literature. It should be noted that we would evaluate only the microextraction step, instead of the whole analytical method.

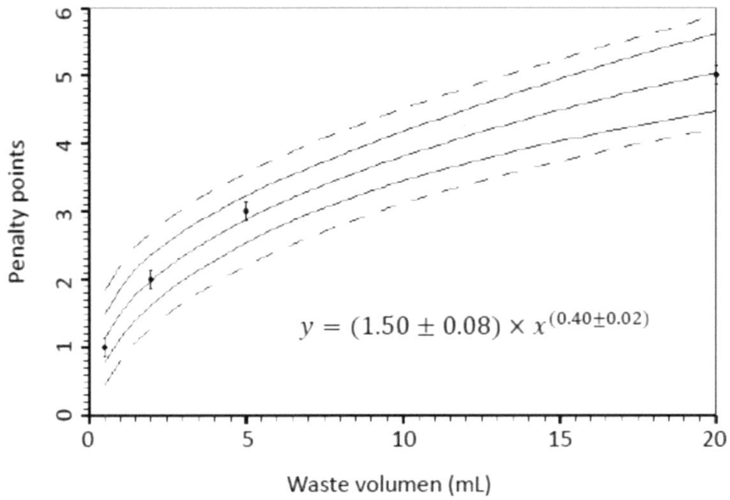

Fig. (5). Plotted the penalty parameter *versus* the volume of generated wastes.

The green certificate of the microextraction procedures suggested for the volatile sulfur compounds determination in beer and Table **3** clearly shows the advantages of using this treatment as compared with the basic Eco-scale applied to the same methods (Table **2**).

The use of HF-LPME has been compared with the SPME for the determination of organophosphorous pesticides in wastewater samples [57]. The reagents and solvents used in this method are summarized in Table **4**. As it can be seen, the green certificate for HF-LPME is comparable to that obtained by SPME.

Another example of comparison of microextraction techniques can be found for the determination of clenbuterol in urine [58] for which HS-SPME, LLLME and SCSE have been proposed (Table **5**) [61].

Table 3. The green certificate eco-sclae employed for the evaluation of the microextraction of sulfur volatile compounds in beer.

	Penalty points							
Microextraction method	**Reagents amount**	**Hazard**	**Subtotal**	**Energy**	**Occupational hazard**	**Waste**	**Total**	
HS-SDME	0.76 (2 g NaCl)	0	0	0	0	3.8 (10 mL sample)		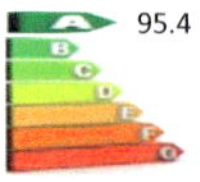
	0.26 (60 mg EDTA)	1	0.26					
	0.09 (2 µL DMF)	6	0.54					
							4.6	
SDME	0.32 (120 mg EDTA)	1	0.32	0	0	2.0 (2 mL sample)		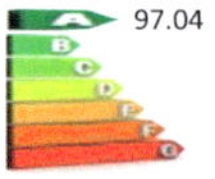
	0.08 (1.5 µL n-hexane)	8	0.64					
							2.96	
HS-SPME	0.76 (2 g NaCl)	0	0	1	0	3.8 (10 mL sample)		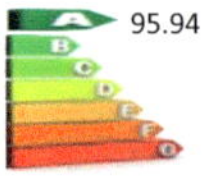
	0.26 (60 mg EDTA)	1	0.26					
							4.06	

Data reported in Table **5** shows that LLLME and HS-SPME are the best alternatives for the clenbuterol microextraction from a green point of view. In this sense, it must be noticed that the heating step involved in HS clearly damages the green eco-scale by a penalty point of 2 points and, thus, it is clear that room temperature methods are preferable to high temperature ones.

In order to compare the proposed analytical Eco-scale results obtained by the microextraction procedures with those based on classical extraction procedures, we have analyzed a LLE, a SPE and a SPME procedures for the GC determination of multiclass priority organic contaminants in wastewater [59, 62] (Table **6**).

Table 4. Evaluation of the green certificate eco-scale for the evaluation of HF-LPME and SPME for the determination of organophosphorous pesticides in wastewater.

	Penalty points							
Microextraction method	**Reagents amount**	**Hazard**	**Subtotal**	**Energy**	**Occupational hazard**	**Waste**	**Total**	
HF-LPME	Membrane PP	0	0	0	0	3.8 (10 mL sample)		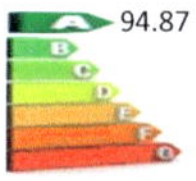
	0.095 (2.5 µL hexane)	8	0.76					
	0.095 (2.5 µL toluene)	6	0.57					
							5.13	
SPME	1 (NaCl)	0	0	1	0	3.8 (10 mL sample)		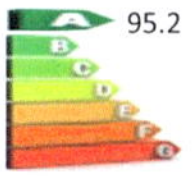
	1 (HCl)	1	1					
							4.8	

Table 5. Evaluation of the green certificate eco-scale for the microextraction procedures used for the determination of clenbuterol.

	Penalty points							
Microextraction method	**Reagents amount**	**Hazard**	**Subtotal**	**Energy**	**Occupational hazard**	**Waste**	**Total**	
HS-SPME	0.30 (100 µL HDMS)	6	1.8	2	0	1.5 (1 mL sample)		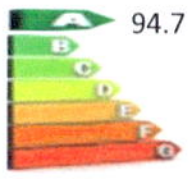
							5.3	
LLLME	1 (NaOH 1 M)	1	1	0	0	3.8 (10 mL sample)		
	0.12 (5 µL H_3PO_4)	1	0.12					
	0.09 (2 µL octanol)	1	0.09					
							5.01	
SCSE	1.2 (8.8 mL MetOH)	6	7.2	0	1	3.8 (10 mL sample)		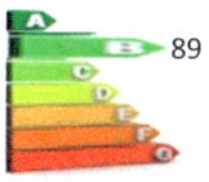
							11	

Table 6. Evaluation of the green certificate eco-scale of extraction procedures and for multiclass priority organic contaminants in wastewaters.

	Penalty points							
Microextraction method	**Reagents amount**	**Hazard**	**Subtotal**	**Energy**	**Occupational hazard**	**Waste**	**Total**	
LLE	1 (H_2SO_4 1M)	1	1	1	0	12.82 (200 mL sample)		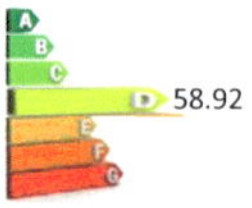
	0.40 (250 mg NaCl)	0	0			8.62 (75 mL n-hexane)		
	2.33 (75 mL n-hexane)	8	18.64					
							41.08	
SPE	1.32 (12 mL methanol)	6	7.92	1	1	12.82 (200 mL sample)		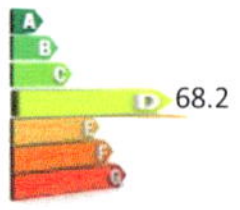
	1.03 (5.5 mL ethylacetate)	4	4.12					
	1.03 (5.5 mL DCM)	2	2.06					
	0.61 (1 mL n-hexane)	8	4.88					
							31.8	
HS-SPME	0.61 (1 mL methanol)	6	3.66	1	0	3.81 (10 mL sample)		
	1 (1 g NaCl)	0	0					
							8.47	

Table **6** clearly shows that classical macroextraction procedures are unable to provide a green certificate of type better than D, thus providing a quantitative evaluation of the microextraction as a green alternative for sample preparation.

In summary, as it can be seen in Tables **3** to **6**, the microextraction procedures account for less than 10 penalty points, with an average of 5.7 penalty points while the classical extraction methods based on LLE and SPE account for approximately 40 and 30 penalty points, respectively. It corresponds to 6.5 to 5 times higher penalty points than those corresponding to microextraction procedures, and the category changes from Eco class A of the microextraction methods to Eco class D of the classic ones.

1.6. FUTURE GREEN PERSPECTIVES

It is clear that the scaling down of the matrix isolation and analyte preconcentration steps in an important way for greening the analytical methods because, as Prof. Kaljurand said, the pollutant capability of a single drop is not a dramatic question. However, additionally than the miniaturization of extraction systems, it must be also taken into account the energy consumption of extraction processes and, from a green point of view, high temperature steps or heating processes must be avoided as possible.

Looking for the future, we hope that the advances in the quantification of green parameters of methods could encourage the researchers and method users to move from classical solutions to new A class methodologies.

Regarding the available methodologies for analyte extraction and microextraction, in the near future we are confident on the development of multianalyte simultaneous extraction methodologies far from the classical approaches based on the preconcentration of a single analyte or a reduced number of chemical substances of the same type and thus, able to be used for the simultaneous preconcentration of different types of analytes in a mixture of solid phase easy to be separated. So, the aforementioned methodologies will be suitable to be employed for in-field preconcentration and stabilization of analytes which can be eluted on-line in the laboratory for reducing the contamination and losses during sample transport and sample storage and to increase the sensitivity and selectivity of the analytical tools.

CONFLICT OF INTEREST

The authors confirm that they have no conflict of interest to declare for this publication.

ACKNOWLEDGEMENTS

Authors gratefully acknowledge the financial support of the Ministerio de Economía y Competitividad and FEDER (Projects CTQ2012-38635) and the

Universitat de Valencia (Project Precompetitiu UV-INV_PRECOMP13-115346).

ABBREVIATIONS

DCM dichloromethane
EDTA Ethylenediaminetetraacetic acid
FIA Flow injection analysis
GAC Green analytical chemistry
GC Gas chromatography
HS Head space
LC Liquid chromatography
LLE Liquid–liquid extraction
PDMS poly(dimethylsiloxane)
SME Solvent microextraction
SPE Solid phase extraction

REFERENCES

[1] Anastas, P.T.; Farris, C.A. Benign by Design: Alternative synthetic design for pollution prevention *ACS Symposium Series N,* **1994**, *577.* [http://dx.doi.org/10.1021/bk-1994-0577.ch014]

[2] de La Guardia, M.; Armenta, S. *Green Analytical Chemistry: Theory and Practice*; Comprehensive Analytical Chemistry, Elsevier B.V.: Oxford, **2011**.

[3] Namiesnik, J. Pro-ecological education: Chemical faculty of the technical university of gdańsk, poland. *Environ. Sci. Pollut. Res. Int.,* **1999**, *6*(4), 243-244. [http://dx.doi.org/10.1007/BF02987339] [PMID: 19005670]

[4] de la Guardia, M.; Ruzicka, J. Towards environmentally conscientious Analytical Chemistry through miniaturization, containment and reagent replacement. *Analyst (Lond.),* **1995**, *120*, 17N. [http://dx.doi.org/10.1039/an995200017n]

[5] Gałuszka, A.; Migaszewski, Z.; Namiesnik, J. The 12 principles of green analytical chemistry and the SIGNIFICANCE mnemonic of green analytical practices, *TrAC Trend. Anal. Chem.,* **2013**, *50*, 78-84.

[6] Pawliszyn, J.; Pedersen-Bjergaard, S. Analytical microextraction: current status and future trends. *J. Chromatogr. Sci.,* **2006**, *44*(6), 291-307. [http://dx.doi.org/10.1093/chromsci/44.6.291] [PMID: 16884584]

[7] Ruzicka, J.; Hansen, E.H. Flow injection analyses. 1. New concept of fast continuous-flow analysis. *Anal. Chim. Acta,* **1975**, *78*, 145-157.

[8] Miró, M.; Hansen, E.H. Recent advances and future prospects of mesofluidic lab-on-a-valve platforms in analytical science critical review. *Anal. Chim. Acta,* **2012**, *750*, 3-15. [http://dx.doi.org/10.1016/j.aca.2012.03.049] [PMID: 23062425]

[9] Arthur, C.L.; Pawliszyn, J. Solid phase microextraction with thermal desorption using fused silica optical fibers. *Anal. Chem.,* **1990**, *62*, 2145-2148. [http://dx.doi.org/10.1021/ac00218a019]

[10] Spietelun, A.; Marcinkowski, Ł.; de la Guardia, M.; Namieśnik, J. Recent developments and future trends in solid phase microextraction techniques towards green analytical chemistry. *J. Chromatogr. A,* **2013**, *1321*, 1-13. [http://dx.doi.org/10.1016/j.chroma.2013.10.030] [PMID: 24238710]

[11] Baltussen, E.; Sandra, P.; David, F.; Cramers, C. Stir bar sorptive extraction (SBSE), a novel extraction technique for aqueous samples: Theory and principles. *J. Microcol.,* **1999**, *11*, 737-747. [http://dx.doi.org/10.1002/(SICI)1520-667X(1999)11:10<737::AID-MCS7>3.0.CO;2-4]

[12] Ochiai, N.; Sasamoto, K.; Ieda, T.; David, F.; Sandra, P. Multi-stir bar sorptive extraction for analysis of odor compounds in aqueous samples. *J. Chromatogr. A,* **2013**, *1315*, 70-79. [http://dx.doi.org/10.1016/j.chroma.2013.09.070] [PMID: 24094753]

[13] Luo, Y.B.; Ma, Q.; Feng, Y.Q. Stir rod sorptive extraction with monolithic polymer as coating and its application to the analysis of fluoroquinolones in honey sample. *J. Chromatogr. A,* **2010**, *1217*(22), 3583-3589. [http://dx.doi.org/10.1016/j.chroma.2010.03.036] [PMID: 20394937]

[14] Huang, X.; Chen, L.; Lin, F.; Yuan, D. Novel extraction approach for liquid samples: stir cake sorptive extraction using monolith. *J. Sep. Sci.,* **2011**, *34*(16-17), 2145-2151. [PMID: 21674796]

[15] Richter, P.; Leiva, C.; Choque, C.; Giordano, A.; Sepúlveda, B. Rotating-disk sorptive extraction of nonylphenol from water samples. *J. Chromatogr. A,* **2009**, *1216*(49), 8598-8602. [http://dx.doi.org/10.1016/j.chroma.2009.10.044] [PMID: 19879588]

[16] Cárdenas, M.S. Dispersive Solid-Phase (Micro)Extraction. In: *Encyclopedia of Analytical Chemistry*; John Wiley & Sons: Chichester, **2010**. [http://dx.doi.org/10.1002/9780470027318.a9167]

[17] Zhang, H.; Lee, H.K. Plunger-in-needle solid-phase microextraction with graphene-based sol-gel coating as sorbent for determination of polybrominated diphenyl ethers. *J. Chromatogr. A,* **2011**, *1218*(28), 4509-4516. [http://dx.doi.org/10.1016/j.chroma.2011.05.016] [PMID: 21636089]

[18] Lipinski, J. Automated solid phase dynamic extraction of organics using a wall coated syringe needle. *Fresenius J. Anal. Chem.,* **2001**, *369*(1), 57-62. [http://dx.doi.org/10.1007/s002160000618] [PMID: 11210231]

[19] Van Durme, J.; Demeestere, K.; Dewulf, J.; Ronsse, F.; Braeckman, L.; Pieters, J.; Van Langenhove, H. Accelerated solid-phase dynamic extraction of toluene from air. *J. Chromatogr. A,* **2007**, *1175*(2), 145-153. [http://dx.doi.org/10.1016/j.chroma.2007.10.043] [PMID: 17980886]

[20] Abdel-Rehim, M. Method and apparatus for sample preparation using solid phase microextraction, PCT, World Intellectual Property Organization. *Int. Pub. Number: WO,* **2003**, *03/019149*, A1.

[21] Abdel-Rehim, M. New trend in sample preparation: on-line microextraction in packed syringe for

liquid and gas chromatography applications. I. Determination of local anaesthetics in human plasma samples using gas chromatography-mass spectrometry. *J. Chromatogr. B Analyt. Technol. Biomed. Life Sci.,* **2004**, *801*(2), 317-321.
[http://dx.doi.org/10.1016/j.jchromb.2003.11.042] [PMID: 14751801]

[22] Saito, Y.; Ueta, I.; Ogawa, M.; Abe, A.; Yogo, K.; Shirai, S.; Jinno, K. Fiber-packed needle-type sample preparation device designed for gas chromatographic analysis. *Anal. Bioanal. Chem.,* **2009**, *393*(3), 861-869.
[http://dx.doi.org/10.1007/s00216-008-2400-6] [PMID: 18818906]

[23] Xie, W. In-Tip Solid Phase Microextraction for High Throughput Drug Analysis. PhD thesis University of Waterloo: Ontario, Canada, **2011**.

[24] Vuckovic, D. High-throughput solid-phase microextraction in multi-well-plate format. *Trends Analyt. Chem.,* **2013**, *45*, 136-153.
[http://dx.doi.org/10.1016/j.trac.2013.01.004]

[25] Lambropoulou, D.; Sakellarides, T.; Albanis, T. Determination of organophosphorus insecticides in natural waters using SPE-disks and SPME followed by GC/FTD and GC/MS. *Fresenius J. Anal. Chem.,* **2000**, *368*(6), 616-623.
[http://dx.doi.org/10.1007/s002160000542] [PMID: 11228713]

[26] Bigham, S.; Medlar, J.; Kabir, A.; Shende, C.; Alli, A.; Malik, A. Sol-gel capillary microextraction. *Anal. Chem.,* **2002**, *74*(4), 752-761.
[http://dx.doi.org/10.1021/ac0109523] [PMID: 11866054]

[27] Saito, Y.; Imaizumi, M.; Takeichi, T.; Jinno, K. Miniaturized fiber-in-tube solid-phase extraction as the sample preconcentration method for microcolumn liquid-phase separations. *Anal. Bioanal. Chem.,* **2002**, *372*(1), 164-168.
[http://dx.doi.org/10.1007/s00216-001-1126-5] [PMID: 11939188]

[28] Eisert, R.; Pawliszyn, J. Automated in-tube solid-phase microextraction coupled to high-performance liquid chromatography. *Anal. Chem.,* **1997**, *69*, 3140-3147.
[http://dx.doi.org/10.1021/ac970319a]

[29] Chen, Y.; Pawliszyn, J. Miniaturization and automation of an internally cooled coated fiber device. *Anal. Chem.,* **2006**, *78*(14), 5222-5226.
[http://dx.doi.org/10.1021/ac060542k] [PMID: 16841953]

[30] Basheer, C.; Alnedhary, A.A.; Rao, B.S.; Valliyaveettil, S.; Lee, H.K. Development and application of porous membrane-protected carbon nanotube micro-solid-phase extraction combined with gas chromatography/mass spectrometry. *Anal. Chem.,* **2006**, *78*(8), 2853-2858.
[http://dx.doi.org/10.1021/ac060240i] [PMID: 16615802]

[31] Neng, N.R.; Silva, A.R.; Nogueira, J.M. Adsorptive micro-extraction techniquesnovel analytical tools for trace levels of polar solutes in aqueous media. *J. Chromatogr. A,* **2010**, *1217*(47), 7303-7310.
[http://dx.doi.org/10.1016/j.chroma.2010.09.048] [PMID: 20950811]

[32] Wang, H.; Campiglia, A.D. Determination of polycyclic aromatic hydrocarbons in drinking water samples by solid-phase nanoextraction and high-performance liquid chromatography. *Anal. Chem.,* **2008**, *80*(21), 8202-8209.
[http://dx.doi.org/10.1021/ac8014824] [PMID: 18834147]

[33] Bruheim, I.; Liu, X.; Pawliszyn, J. Thin-film microextraction. *Anal. Chem.,* **2003**, *75*(4), 1002-1010. [http://dx.doi.org/10.1021/ac026162q] [PMID: 12622398]

[34] Kloskowski, A.; Pilarczyk, M.; Namieśnik, J. Membrane solid-phase microextractiona new concept of sorbent preparation. *Anal. Chem.,* **2009**, *81*(17), 7363-7367. [http://dx.doi.org/10.1021/ac9007978] [PMID: 19653632]

[35] Basheer, C.; Suresh, V.; Renu, R.; Lee, H.K. Development and application of polymer-coated hollow fiber membrane microextraction to the determination of organochlorine pesticides in water. *J. Chromatogr. A,* **2004**, *1033*(2), 213-220. [http://dx.doi.org/10.1016/j.chroma.2004.01.050] [PMID: 15088741]

[36] Luo, Y.Z.; Pawliszyn, J. Calibration of membrane extraction for air analysis. *Anal. Chem.,* **2000**, *72*(5), 1064-1071. [http://dx.doi.org/10.1021/ac990746j] [PMID: 10739213]

[37] Aziz-Zanjani, M.O.; Mehdinia, A. Electrochemically prepared solid-phase microextraction coatings-a review. *Anal. Chim. Acta,* **2013**, *781*, 1-13. [http://dx.doi.org/10.1016/j.aca.2013.03.012] [PMID: 23684460]

[38] Chai, X.; He, Y.; Ying, D.; Jia, J.; Sun, T. Electrosorption-enhanced solid-phase microextraction using activated carbon fiber for determination of aniline in water. *J. Chromatogr. A,* **2007**, *1165*(1-2), 26-31. [http://dx.doi.org/10.1016/j.chroma.2007.07.048] [PMID: 17706235]

[39] Kokosa, J.M. Advances in solvent-microextraction techniques, *TrAC Trend. Anal. Chem.,* **2013**, *43*, 2-13.

[40] Spietelun, A.; Marcinkowski, Ł.; de la Guardia, M.; Namieśnik, J. Green aspects, developments and perspectives of liquid phase microextraction techniques. *Talanta,* **2014**, *119*, 34-45. [http://dx.doi.org/10.1016/j.talanta.2013.10.050] [PMID: 24401382]

[41] Keller, B.O.; Li, L. Nanoliter solvent extraction combined with microspot MALDI TOF mass spectrometry for the analysis of hydrophobic biomolecules. *Anal. Chem.,* **2001**, *73*(13), 2929-2936. [http://dx.doi.org/10.1021/ac001323g] [PMID: 11467537]

[42] Wu, H.F.; Yen, J.H.; Chin, C.C. Combining drop-to-drop solvent microextraction with gas chromatography/mass spectrometry using electronic ionization and self-ion/molecule reaction method to determine methoxyacetophenone isomers in one drop of water. *Anal. Chem.,* **2006**, *78*(5), 1707-1712. [http://dx.doi.org/10.1021/ac052076p] [PMID: 16503626]

[43] Ma, M.; Cantwell, F.F. Solvent microextraction with simultaneous back-extraction for sample clean-up and preconcentration: quantitative extraction. *Anal. Chem.,* **1998**, *70*, 3912-3919. [http://dx.doi.org/10.1021/ac980174n]

[44] Rezaee, M.; Assadi, Y.; Hosseini, M.R.; Aghaee, E.; Ahmadi, F.; Berijani, S. Dispersive liquid-liquid microextraction combined with gas chromatography-flame photometric detection. Very simple, rapid and sensitive method for the determination of organophosphorus pesticides in water. *J. Chromatogr. A,* **2006**, *1116*, 1-9. [http://dx.doi.org/10.1016/j.chroma.2006.03.007] [PMID: 16574135]

[45] Pedersen-Bjergaard, S.; Rasmussen, K.E. Liquid-liquid-liquid microextraction for sample preparation of biological fluids prior to capillary electrophoresis. *Anal. Chem.,* **1999**, *71*(14), 2650-2656. [http://dx.doi.org/10.1021/ac990055n] [PMID: 10424162]

[46] Basheer, C.; Alnedhary, A.A.; Madhava Rao, B.S.; Balasubramanian, R.; Lee, H.K. Ionic liquid supported three-phase liquid-liquid-liquid microextraction as a sample preparation technique for aliphatic and aromatic hydrocarbons prior to gas chromatography-mass spectrometry. *J. Chromatogr. A,* **2008**, *1210*(1), 19-24. [http://dx.doi.org/10.1016/j.chroma.2008.09.040] [PMID: 18823629]

[47] Eionet, European Topic Centre on Sustainable Consumption and Production, http://scp.eionet.europa.eu/themes/lca

[48] National Environmental Methos Index (NEMI) 2002. https://www.nemi.gov/home/

[49] Raynie, D.; Driver, J.L. *13th Green Chemistry and Engineering Conference,* **2009**.Washington, DC

[50] Gałuszka, A.; Konieczka, P.; Migaszewski, Z.M.; Namiesnik, J. Analytical Eco-Scale for assessing the greenness of analytical procedures, *TrAC Trend. Anal. Chem.,* **2012**, *37*, 61-72.

[51] Van Aken, K.; Strekowski, L.; Patiny, L. EcoScale, a semi-quantitative tool to select an organic preparation based on economical and ecological parameters. *Beilstein J. Org. Chem.,* **2006**, *2*(1), 3. [http://dx.doi.org/10.1186/1860-5397-2-3] [PMID: 16542013]

[52] United Nations, Globally Harmonized System of Classification and Labeling of Chemicals (GHS, Rev. 4), United Nations, New York, 2011.

[53] Raynie, D.; Driver, J.L. Green assessment of chemical methods In: *13th Green Chem. Eng. Conf*; Washington DC, **2009**.

[54] Xiao, Q.; Yu, C.; Xing, J.; Hu, B. Comparison of headspace and direct single-drop microextraction and headspace solid-phase microextraction for the measurement of volatile sulfur compounds in beer and beverage by gas chromatography with flame photometric detection. *J. Chromatogr. A,* **2006**, *1125*(1), 133-137. [http://dx.doi.org/10.1016/j.chroma.2006.06.096] [PMID: 16859693]

[55] Silva, W.P.; Silva, C.M. Fit curve fitting software (nonlinear regression and treatment of data program) V 7.2.48, 1999–2011, http://zeus.df.ufcg.edu.br/labfit (accessed May 2014).

[56] Ahearn, T.S.; Staff, R.T.; Redpath, T.W.; Semple, S.I. The use of the Levenberg-Marquardt curve-fitting algorithm in pharmacokinetic modelling of DCE-MRI data. *Phys. Med. Biol.,* **2005**, *50*(9), N85-N92. [http://dx.doi.org/10.1088/0031-9155/50/9/N02] [PMID: 15843726]

[57] Basheer, C.; Alnedhary, A.A.; Rao, B.S.; Lee, H.K. Determination of organophosphorous pesticides in wastewater samples using binary-solvent liquid-phase microextraction and solid-phase microextraction: a comparative study. *Anal. Chim. Acta,* **2007**, *605*(2), 147-152. [http://dx.doi.org/10.1016/j.aca.2007.10.006] [PMID: 18036377]

[58] Melwanki, M.B.; Hsu, W.H.; Huang, S.D. Determination of clenbuterol in urine using headspace solid phase microextraction or liquid–liquid–liquid microextraction. *Anal. Chim. Acta,* **2005**, *552*, 67-75. [http://dx.doi.org/10.1016/j.aca.2005.07.036]

[59] Robles-Molina, J.; Gilbert-López, B.; García-Reyes, J.F.; Molina-Díaz, A. Comparative evaluation of liquid-liquid extraction, solid-phase extraction and solid-phase microextraction for the gas chromatography-mass spectrometry determination of multiclass priority organic contaminants in wastewater. *Talanta,* **2013**, *117*, 382-391.
[http://dx.doi.org/10.1016/j.talanta.2013.09.040] [PMID: 24209357]

CHAPTER 2

Novel Sol-Gel Sorbents in Sorptive Microextraction

Abuzar Kabir* and **Kenneth G. Furton**

International Forensic Research Institute, Department of Chemistry and Biochemistry, Florida International University 11200 SW 8th Street, Miami, Florida 33199, USA

Abstract: Despite remarkable advances in modern state-of-the-art analytical instruments featured with enhanced sensitivity and automated operation in recent years, the success of an analytical investigation still largely depends on the sampling and sample preparation techniques. The growing concerns about environmental pollution and public safety, solvent-free/solvent-minimized sorptive microextraction techniques have gained enormous popularity among practicing scientists over solvent intensive sample preparation techniques. A major share of the increasing popularity of sorptive microextraction techniques definitely goes to sol-gel coating technology. The enormous potential of sol-gel coating technology as a viable approach for creating hybrid organic-inorganic advanced material systems has been intuitively materialized, resulting in a large number of microextraction sorbents possessing unique selectivity, enhanced extraction efficiency, higher thermal, mechanical, chemical and solvent stability. The current chapter explains the basic principle of sol-gel coating technology as well as the step-by-step procedure to fabricate the coating, classifies and describes different sorptive microextraction sorbents, and presents the most recent developments in the field. Selected and representative applications of different sol-gel microextraction sorbents are also tabulated.

Keywords: Fabric phase sorptive extraction (FPSE), Green sample preparation, Inorganic-organic hybrid sorbents, Sample preconcentration, Sampling and sample preparation, Sol-gel coating, Sol-gel extraction media, Solvent-less extraction, Solvent-minimized extraction, Sorptive microextraction.

* **Corresponding author Abuzar Kabir:** International Forensic Research Institute, Department of Chemistry and Biochemistry, Florida International University 11200 SW 8th Street, Miami, Florida 33199, USA; Tel: 305-348-2396; E-mail: akabir@fiu.edu

M. Valcárcel, S. Cárdenas, R. Lucena (Eds.)

2.1. INTRODUCTION

The introduction of solid-phase microextraction (SPME) by Pawliszyn and co-workers [1] in 1987 is considered to be a ground-breaking event in the field of analytical sample preparation that initiated a new era towards solvent-free/solvent-minimized extraction, miniaturization, reduction of multiple time consuming steps in sample preparation, and automation.

Following the inception of SPME, a number of other microextraction techniques including stir bar sorptive extraction (SBSE) [2], thin film microextraction [3], microextraction by packed syringe (MEPS) [4], micro solid-phase extraction (μ-SPE) [5], fabric phase sorptive extraction (FPSE) [6] have emerged over the period of two decades. All these techniques belong to a broader family named as ***sorbent-based sorptive microextraction*** and rely upon the partition equilibrium of the analyte(s) between the microextraction sorbent and the sample matrix. The microextraction sorbents used in sorbent-based sorptive microextraction may be a pseudo liquid absorbent (*e.g.*, polydimethylsiloxane, polyethylene glycol), or a solid adsorbent (*e.g.*, graphite, carbon nanotubes) or a mixture of both (*e.g.*, PDMS/Carboxen). As such, both absorption and adsorption processes may play role in analyte extraction and preconcentration process. Sorbent-based microextraction systems have significantly reduced the usage of toxic and hazardous organic solvents in sample preparation exercises and therefore are considered to be a significant step forward towards green analytical chemistry [7]. In addition to the significant reduction of organic solvent usage, sorbent-based microextraction systems also integrate sampling, extraction and preconcentration of target analyte into a single step, resulting in a simple, fast and convenient sample preparation process.

Considering the high volume of research in the field of sorbent-based sorptive microextraction, a large number of review articles and books have been published covering different aspects and applications of this modern sample preparation technique [8 - 12].

Over the period of last two decades, a number of new formats and modifications of sorbent-based sorptive microextraction systems have emerged, bringing new attributes and unique applications in analytical sample preparation. A classification scheme of different sorptive-microextraction techniques that share the same principle of non-exhaustive equilibrium driven sample extraction and preconcentration is given in Fig. (**1**).

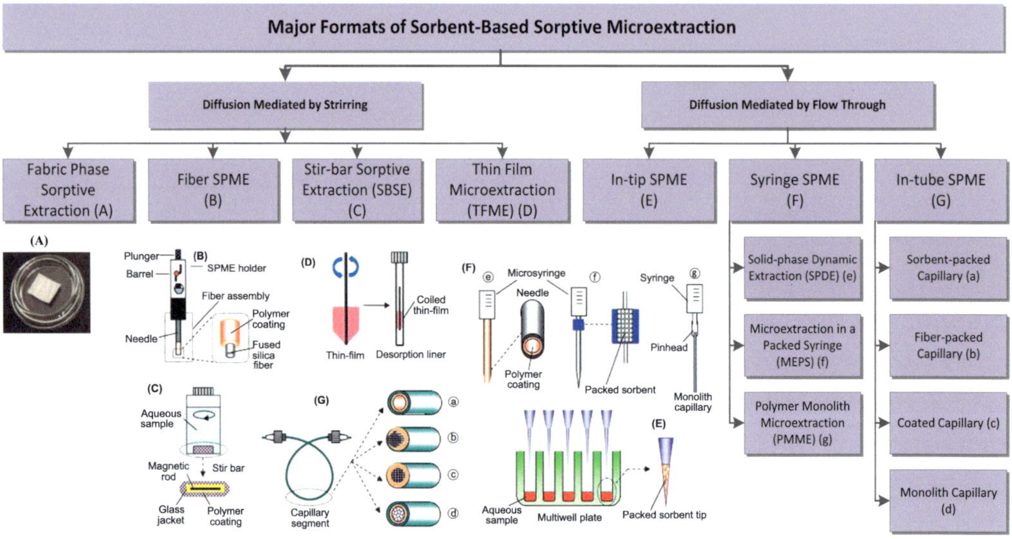

Fig. (1). Classification scheme of major sorption-based micrextraction techniques.

Among all the different formats and modifications of sorbent-based sorptive microextraction, fiber-SPME still dominates among all the developed formats and enjoys high popularity in a broad range of fields including environmental, pharmaceutical, food, toxicological, forensic applications.

2.2. WORKING PRINCIPLE OF SORBENT-BASED SORPTIVE MICROEXTRACTION

All sorbent-based microextraction techniques, regardless of their format, are based on the principle of equilibrium extraction. The extracting sorbents are immobilized either on the outer surface of a substrate (fiber-SPME, TFME, SBSE, FPSE, PSPME) or inside of a fused silica capillary (intube-SPME or CME) or inside of a hollow-fiber (HF-SPME). During the extraction, the microextraction device holding the sorbent may either be directly exposed to the sample matrix containing the target analyte(s) (direct immersion mode), or may be exposed to the headspace of a confined container retaining the sample matrix with the analytes (headspace mode). During the extraction, the polymeric sorbent coating is exposed to the sample matrix containing the target analyte directly or indirectly and time is allowed to establish the equilibrium between the sorbent and the sample matrix. Once the extraction equilibrium is established, further exposure of the sorbent to the sample matrix does not accumulate more analyte(s). Time to establish extraction equilibrium depends on many factors including the nature of the coating, type of the analyte, viscosity of the sample matrix, temperature of the sample matrix, coating thickness of the sorbent and can be expedited by stirring the sample matrix using a magnetic stir bar or by flowing the sample through the capillary. After finishing the extraction, the microextraction device is withdrawn from the sample matrix, and the extracted analytes are rapidly desorbed from the sorbent for introducing into an analytical instrument – generally a gas chromatography (GC), high-performance liquid chromatography (HPLC), capillary electrophoresis (CE), or supercritical fluid chromatography (SFC) system. The extracted analyte can be desorbed either by using thermal energy by introducing the device into a heated GC injection port/thermal desorption unit (thermal desorption) or by rinsing the coated microextraction device with an appropriate solvent (elution/back extraction) followed by injecting an aliquot of the analyte solution into the inlet of the analytical instrument.

Sorbent-based microextraction techniques are generally governed by two principle criteria: (1) thermodynamics; and (2) kinetics [9, 13]. Thermodynamic properties determine the maximum amount of analytes that can be extracted by a given mass of sorbent under a specific set of extraction conditions. Since higher sorbent loading allows a higher amount of target analytes to be accumulated by the sorbent when adequate time is allowed to reach the extraction equilibrium, sorbent loading is directly related to extraction efficiency. On the other hand, kinetics controls the rate of extraction and hence the time required to reach the extraction equilibrium.

2.3. SHORTCOMINGS OF COMMERCIALLY AVAILABLE MICRO-EXTRACTION TECHNIQUES

Despite numerous advantages of sorbent-based sorptive microextraction techniques over conventional sample preparation techniques, they also suffer from significant shortcomings. These limitations generally originate from: (1) physical format of the microextraction unit; and (2) coating technique used to immobilize the organic polymeric sorbents.

Deficiencies due to the physical format of the extraction unit include (1) fragility of SPME fibers; (2) risk of bending the SPME syringe needle during operation; (3) low sorbent loading in SPME fibers often result in poor extraction sensivity; (4) segments of commercial GC column used in in-tube SPME with submicrometer coating thickness does not offer adequate sorbent loading, resulting in poor extraction efficiency; (5) Twister stir bars (stir bar sorptive extraction) are prone to damage the thick sorbent coating when stirring at high speed.

Although several other coating techniques [9] can be used to immobilize the organic polymer on the substrate surface, physical deposition of the organic polymer followed by free radical cross linking is the predominantly used coating technique applied in the majority of the commercially available sorbent-based sorptive microextraction techniques. This coating technique inherently possesses a number of shortcomings including: (1) organic polymer used as the sorbent is the only source that determines the selectivity of the microextraction system; (2) demonstrate low maximum working temperature, poor solvent stability, and limited pH stability due to the physically held sorbent coating on the substrate surface; (3) instability and swelling of the sorbent coating in organic solvents; (4) pronounced run-to-run and batch-to-batch variability; (5) highly viscous and glue-like organic polymer does not allow fast analyte diffusion, resulting in prolonged extraction equilibrium time; (6) absence of chemical bonding between the substrate and the sorbent translates into a short lifetime of the coating; (7) small sorbent volume in SPME fibers limits the extraction sensitivity that promoted the introduction of SBSE with higher sorbent loading. However, slow diffusion of analytes through the viscous organic sorbent coating does not offer significant gain in extraction sensitivity unless prolonged extraction time is allowed to reach extraction equilibrium.

The extensive assessment on the shortcomings of sorbent-based microextraction systems revealed that the sustainability and healthy growth of the microextraction techniques primarily depend on the development of new coating technologies that would offer faster extraction kinetic, better selectivity, superior sensitivity, and higher thermal, solvent and chemical stability. Most of the expected attributes for sorbent-based sorptive microextraction techniques mentioned above were met by the introduction of sol-gel coating technology developed by Malik and co-workers [14].

2.4. SOL-GEL APPROACH IN DEVELOPING MICROEXTRACTION SORBENTS

2.4.1. Fundamentals of Sol-gel Process

The sol-gel process provides a facile, convenient, and economical pathway to create advanced organic-iorganic hybrid material systems with tunable selectivity, adjustable porosity, and remarkably high thermal, solvent, and chemical stability [15]. Sol-gel process involves transition of a liquid colloidal solution (sol) into a three-dimensional network of solid matrix (gel).

Sol-gel coating process consists of following reaction:

(I) Catalytic hydrolysis of sol-gel precursor;

(II) Polycondensation of hydrolyzed precursor leading to a growing sol-gel network;

(III) Random incorporation of sol-gel active organic polymer into the growing sol-gel network;

(IV) Immobilization/chemical incorporation of the growing sol-gel network on the substrate surface *via* polycondensation.

A schematic representation of sol-gel reactions is shown in Fig. (**2**).

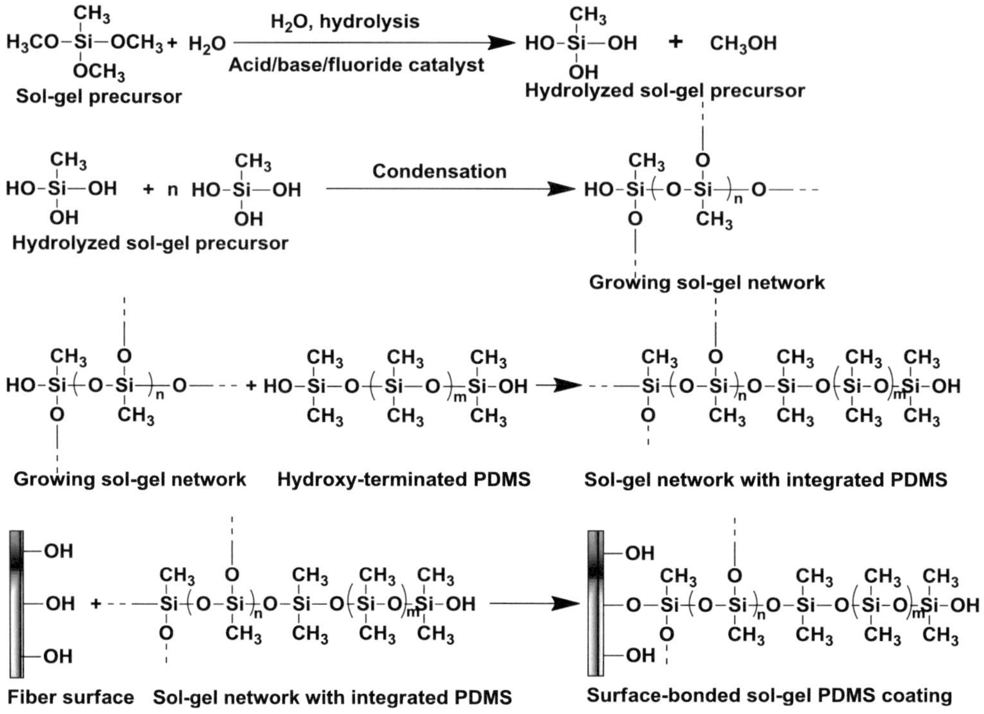

Fig. (2). Schematic representation of sol-gel reactions involve in creating sorbent coatings/monolithic bed. Hydroxy-terminated Poly(di-methyl siloxane) (PDMS) represents a sol-gel active organic polymer.

One of the major advantages of sol-gel process is the ability to control the physicochemical property of sol-gel sorbents using (1) sol solution composition; (2) the relative ratio of different building blocks; (3) reaction conditions; and (4) aging/post-gelation conditions. A number of factors that affect the properties of sol-gel hybrid organic-inorganic sorbents include (a) nature and type of inorganic or organically modified inorganic sol-gel precursor(s); (b) sol-gel precursor/water ratio; (c) catalyst(s) and its concentration (acid, base, fluoride, acid-base dual catalyst); (d) sol solution pH; (e) type of organic solvent(s) and its ratio to the total solution volume; (f) reaction conditions; and (g) aging conditions.

Among all the factors contributing to the properties of sol-gel sorbent, catalyst used in the sol-gel reaction plays key role in determining the final structure of the sol-gel polymeric network. When acid is used as the catalyst, the resulting sorbent is a linear or randomly branched network. On the other hand, base-catalyzed sorbent possesses highly branched and condensed structure.

2.4.2. Sol-gel Technology for Sorbent-based Sorptive Microextraction

Sol-gel technology for creating surface bonded sorbent coating and monolithic bed for sorptive microextraction was developed by Malik and co-workers [14]. Due to the novelty and simplicity of this technology, hundreds of new microextraction sorbents have been emerged during the last 15 years and the surge still continues. Sol-gel coating technology involves several steps:

(1) Surface pretreatment of the substrate to maximize the number of available hydroxyl groups;

(2) Design and preparation of the sol solution;

(3) Creation of the sol-gel coating/monolithic bed;

(4) Conditioning the sorbent.

2.4.2.1. Pretreatment of Surface for sol-gel Coating

Majority of the sol-gel sorbents in the form of coating or monolithic bed have been created either on the outer surface of a fused silica fiber or on the inner surface of a fused silica capillary. However, different substrates *e.g.* stainless steel rod, titanium *etc.* have also been recently reported [16] to impart ruggedness and durability to the coated microextraction fibers. Fused silica fibers are commercially sold with a protective polyimide coating to protect from breaking during handling and offers flexibility to coil on a basket. Prior to the sol-gel coating, this protective polyimide coating is generally removed either by immersing in acetone for an extended period of time or by burning with a flame. Following the removal of this coating, fused silica fibers are treated with NaOH solution to maximize the surface silanol groups. The overall objective of surface pretreatment is to clean the surface of the substrate and to maximize the availability of surface silanol groups so that the sol solution can interact intimately with the substrate surface and integrate the silanol functional groups during polycondensation into the growing three dimensional sol-gel networks, resulting in a surface bonded sol-gel sorbent.

Substrate other than fused silica may require different surface modifications in order to introduce some linkers that may form a bridge between the substrate and the growing sol-gel network.

2.4.2.2. Designing the Sol Solution

One of the most important and challenging tasks in sol-gel coating process is the design of the sol solution which involves selection of the suitable sol-gel precursor(s), organic ligand(s), deactivating reagent, solvent, catalyst(s), porogenic reagents. The ultimate performance of the sol-gel sorbents largely depends on the proper design of the sol solution. A number of important factors *e.g.*, selectivity, porosity, coating thickness, thermal and/or solvent stability play vital roles in designing the sol solution. Although, methyltrimethoxysilane (MTMS) has been found to be the most predominantly used sol-gel precursor [15], other precursors are not unknown. Table **1** presents a list of building blocks that have been used in sol solution design along with their roles in the sol-gel sorbents.

Table 1. Sol-gel building blocks and their roles in creating microextraction sorbents.

Building Block Class	Roles in Sol-gel Reaction	Example
Unchanged sol-gel precursor	Provides inorganic component to the hybrid polymeric network, responsible for mechanical strength of the matrix, offers active hydroxyl group to facilitate chemical bonding to the fiber/capillary surface [17]	Tetraethoxysilane (TEOS) [11], titanium isopropoxide [18-21] zirconium butoxide [22], tetramethoxygermane [21, 24] alumina [25].
Organically modified sol-gel precursor	In addition to forming inorganic backbone of the hybrid material, organic moieties provide intermolecular interactions between analytes and the sorbent [14]	Methyltrimethoxysilane (MTMS) [26], octadecyltrimethoxysilane (C_{18}-TMS) [27], octyltrimethoxysilane (C_8-TMS) [27], 3-(trimethoxysilyl)propylamine (TMSPA) [28], vinyltriethoxysilane (VTEOS) [29], phenyltrimethoxysilane (PTMOS) [30]
Sol-gel active organic polymer	Provides different intermolecular interactions *e.g.*, hydrophobicity, μ-μ interaction, hydrogen bonding between the analyte and the extracting sorbent	OH-terminated polydimethylsiloxane (PDMS) [31], polyethylene glycol (PEG) [32], polytetrahydrofuran (PTHF) [33], polydimethyldiphenylsiloxane (PDMDPS) [22]

(Table 1) contd.....

Building Block Class	Roles in Sol-gel Reaction	Example
Deactivation reagents	Aid in controlling the polarity of the hybrid organic-inorganic coatings by derivatizing free silanol groups in the sol-gel network.	Hexamethyldisilazane (HMDS) [22], polymethylhydrosiloxane (PMHS) [11], trimethylmethoxysilane (TMMS) [34]
Solvent	Facilitates mixing of sol-gel precursors, organic polymers, water, and deactivating reagents into a homogeneous system. In addition, the polarity of the solvent can influence the hydrolysis of the precursor by increasing or decreasing the catalytic activity. Solvents can also reduce reaction rates by diluting the sol solution and consequently reducing the precursor concentration [35]	Methanol, methylene chloride, tetrahydrofuran, ionic liquids, *etc.* [17].
Porogen	Controls and manage porosity of the sol-gel matrix, offers easy access of the target analyte to the sorptive sites and facilitates fast and efficient extraction [28].	Ionic liquid, non-ionic surfactant, PEG, polypropylene glycol (PPG), polytetramethyleneoxide (PTMO) [36], triblock polymer F38 [37]. Various solvents have been used as porogens including cyclohexanol, dodecanol, and granules of salts such as Na_2SO_4 [38].
Catalyst (acid, base, fluoride)	Catalyzes the hydrolysis and condensation reactions and controls the network structure and porosity by branching/linear propagation of the evolving polymeric network. Acid catalyzed sol-gel materials possess weakly branched microporous structures. On the contrary, base catalyzed sol-gel materials have highly branched particulate structures with large pore sizes [17].	Trifluoroacetic acid (TFA), glacial acetic acid (GAA), ammonium hydroxide (NH_4OH), hydrochloric acid (HCl), hydrofluoric acid (HF), sodium hydroxide (NaOH), nitric acid (HNO_3), basic amino acids [39], citric acid [40], transition metal complexes [41]
Water	Hydrolyzes the sol-gel precursor to create reactive species for condensation reactions leading to the formation of sol-gel sorbents [17].	H_2O

2.4.2.3. Creation of Sol-gel Coating

A rational design of the sol solution is generally followed by the coating process. The sol solution ingredients are added to each other and are thoroughly mixed together to ensure the formation of a homogeneous sol solution. Vortexing and ultrasonication are often employed at this step. The homogenous sol solution is then subjected to centrifugation in order to remove particulates from the sol solution. The particle free supernatant sol solution is then removed from the container and is immediately applied to the surface to create the sol-gel coating.

2.4.2.4. Conditioning of Sol-gel Coating

Both the sol-gel coating and monolithic bed are subjected to post-gelation conditioning. The objective of this step is to complete the condensation reaction and to remove solvents and unreacted residuals from the sol-gel matrix. A well designed post-gelation conditioning ensures a clean, surface bonded sol-gel sorbent free of structural deformation and internal stress. Conditioning of sol-gel coatings/monolithic beds is often carried by heating the sol-gel matrix very slowly under continuous inert gas flow. The maximum temperature for conditioning is generally determined by the end use of the sol-gel sorbent. For solvent-mediated desorption, high temperature conditioning is not necessary. On the other hand, for thermal desorption, the sol-gel coatings are exposed to very high temperature during this conditioning step.

2.5. DIFFERENT FORMATS OF SOL-GEL BASED SORPTIVE MICROEXTRACTION

Since the sol-gel coating process begins with the liquid sol solution, the sorbent can be coated conveniently *in-situ* inside or outside surface of the substrate. The simplicity in the coating process has prompted the development of a large number of new sorbents on different substrates in many different formats. Fig. (**3**) demonstrates the most common formats sol-gel sorptive microextraction techniques. Similar to commercial SPME fibers, fiber SPME format dominates in sol-gel coated microextraction systems.

2.6. NOVEL SOL-GEL MICROEXTRACTION SORBENTS

Although most of the sol-gel sorbents developed to date are silica based, a number of new sorbents have been developed using titania, zirconia, alumina, germania based precursors. Non-silica based sol-gel sorbents have offered new properties that have not been offered by silica based sol-gel sorbents *e.g.*, high pH stability, ability to extract elemental metals, higher chemical and mechanical stability.

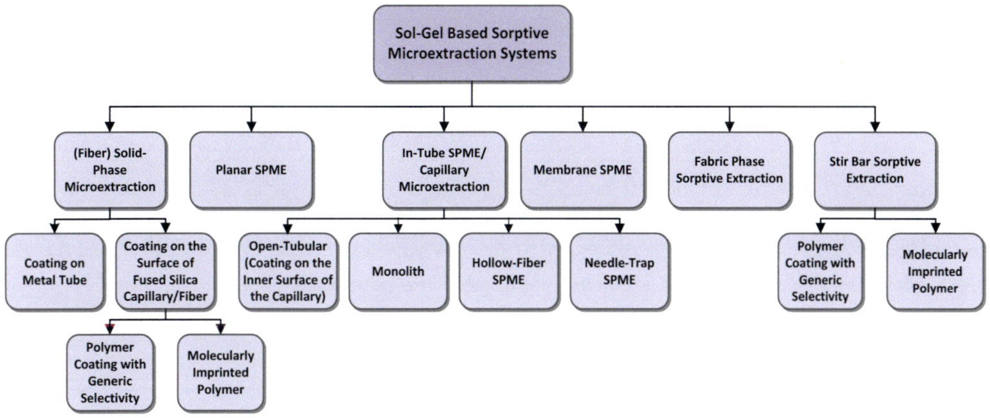

Fig. (3). Most common formats of sol-gel based sorptive microextraction systems.

2.6.1. Silica Based Microextraction Sorbents

The availability of a large number of highly pure and relatively inexpensive silica based precursors, in-depth knowledge in silica chemistry, availability of a large number of silica-based organic polymers and deactivating reagents have resulted in the development of hundreds of new sol-gel microextraction sorbents. Following the high popularity of PDMS coatings in SPME and SBSE, a majority of sol-gel coatings have been developed using PDMS as one of the premier organic ligands.

2.6.1.1. Sol-gel Active Organic Polymers and Ligands

A wide variety of new sol-gel hybrid sorbents with various organic ligands chemically integrated in their hybrid organic-inorganic structure have been reported over the last decade. Most common organic polymers include poly(dimethylsiloxane) (PDMS), poly(ethylene glycol) (PEG) and poly(tetrahydrofuran) (PTHF). Fig. (**4**) provides a list of major organic polymers along with different additives used to manipulate the selectivity and sensitivity of microextraction sorbents.

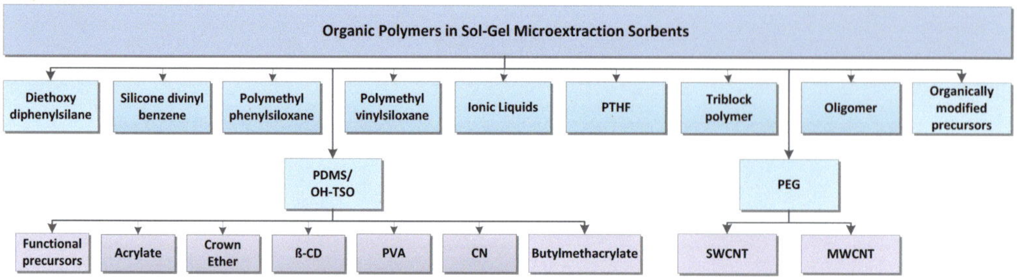

Fig. (4). Major organic polymers and additives used in sol-gel sorptive micrextraction.

2.6.1.1.1. Poly(dimethylsiloxane) (PDMS) and its Modifiers

Owing to its favorable physicochemical properties, poly(dimethylsiloxane) (PDMS), a silicone-based polymer, has been extensively used as the sorbent in sorptive microextraction techniques. In commercial SPME fibers, PDMS is one of the most popular sorbents. Malik and co-workers [14] first reported the use of hydroxy-terminated poly(dimethylsiloxane) (PDMS) to create surface-bonded organic-inorganic hybrid coating (~10 µm) for SPME. Sol-gel PDMS coatings provided extraction selectivity which differs from that of commercially available PDMS phases used on SPME fibers. As such, the sol-gel PDMS coated SPME fibers were found to be efficient in extracting both polar and non-polar analytes from aqueous samples. The chemical bonding of the sol-gel PDMS coating to the substrate surface provided significantly higher thermal stability (≥ 320°C) compared to commercial PDMS fibers (~ 250 °C). Due to the open, porous structure of sol-gel PDMS coating, a thinner sol-gel coating can provide comparable or better extraction efficiency with a shorter equilibrium time.

Malik and coworkers [42] also reported the use of sol-gel PDMS and sol-gel polyethylene glycol phases in in-tube SPME. For the first time, they coined the term capillary microextraction (CME) to describe a microextraction system containing chemically bonded sol-gel coating/monolithic bed inside a fused silica capillary. Sol-gel PDMS and sol-gel PEG capillaries were interfaced with the GC using a two-way press-fit fused silica connector. The extraction efficiency of sol-gel PDMS coating was evaluated by polycyclic aromatic hydrocarbons (PAHs), aldehydes, and ketones, whereas alcohols, phenols, and amines were used to assess the CME performances of sol-gel PEG coated capillaries in combination with GC-FID.

Liu *et al.* [43] exploited sol-gel PDMS coated fiber for trace analysis of anti-estrogens in biological matrices. Due to the high affinity of sol-gel PDMS coated fibers toward the selected analyte, no derivatization was required. The method was validated using rat liver sample spiked with selected anti-estrogens and was found to offer good recovery (78.3-113.1%) as well as good accuracy and precision (RSD 5.1-8.7%).

One major limitation of PDMS sorbent is its limited selectivity towards medium polarity and high polarity analytes and is generally recommended for non-poar analytes. Sol-gel coating technology has enabled to modify the selectivity of PDMS sorbent by using different modifiers *e.g.*, β-cyclodextrin [44], crown ethers [45], poly(vinylalcohol) [34], calixarenes [46], hydroxyfullerene [47], carbon nanotube [31], metal-organic frameworks [48], ionic liquids [49] *etc.*

2.6.1.1.2. Sol-gel Microextraction Sorbents Based on Polyethylene Glycol (PEG) and its Derivatives

Although commercial solid-phase microextraction fibers have enjoyed an exponential growth in recent years with continuously expanding application areas, most applications are still limited to nonpolar or medium polar analytes. Extraction of highly polar analytes from aqueous samples is still considered to be a daunting task for separation scientists and often requires analyte derivatization, pH adjustment and/or salting out procedures [50]. Commercial SPME fibers coated with polyethylelene glycol (PEG) are recommended for highly polar analytes. However, low thermal stability (200-240°C) and poor solvent stability of PEG coated SPME fibers due to the absence of chemical bonding between the coating and fused silica substrate limit practical application of this polar sorbent to a great extent. Fortunately, sol-gel technology innovatively addressed this problem and has made tremendous contribution in developing a large number of highly polar sorbents using PEG [51, 52], PEG derivatives [53], and grafted PEGs [54 - 56].

Wang *et al.* [51] were first to introduce sol-gel PEG coated SPME fibers using superox-4 (PEG with average molar mass 4 megagram/mol) as the organic polymer. The sol-gel PEG sorbents were highly porous resulting in faster mass transfer. In addition, the sol-gel PEG sorbents demonstrated high extraction efficiency for both polar and nonpolar analytes, superior thermal stability (300°C) and extended life span (>150 extractions).

Among different chain-length of PEG polymers, Carbowax 20M (molar mass of 14-16 kg/mol) is frequently used as GC stationary phase as well as in SPME sorbent preparations. Augusto and co-workers developed sol-gel PEG SPME sorbent using Carbowax 20M [57]. A sponge-like porous coating with a thickness of 8±1 μm of sol-gel PEG demonstrated 230% and 540% enhancement in headspace extraction of o-xylene compared with 100-μm PDMS and 65-μm Carbowax-DVB commercial SPME fibers, respectively.

Jia *et al.* [53] recently developed three functional coatings, β-cyclodextrin, β-cyclodextrin-co-poly(ethylenepropylene glycol) and poly(ethylene glycol) (PEG) based sorbents coated on a titanium wire. In order to achieve the chemical anchorage between the sorbent and the substrate, nanotubular film was prepared first *in situ* on titanium wire. The extraction performance of these coatings was evaluated using a diverse range of compounds including polar aromatic compounds, alcohols and ketones. The coated fibers demonstrated good mechanical strength, excellent thermal stability, high sample capacity and selectivity. The fibers were used in combination with gas chromatography-flame ionization detector. Among the three sorbents, sol-gel PEG was found superior for extracting ephedrine and methylephedrine. Repeating an inherent trend of sol-gel coated microextraction fibers, sol-gel PEG coated SPME fibers also demonstrated remarkable repeatability and reproducibility.

The advantages of poly(ethylene glycol) and graphene were exploited in poly(ethylene glycol)-grafted SPME fibers. Seven volatile aromatic compounds (VAC) in water were extracted. The sol-gel coating was found highly porous consisting of nano-sized particles. The headspace SPME-GC-FID method for VAC provided LOD values ranging from 1.0-6.0 ng/L [58]. The fibers were used more than 200 times without any noticeable loss of extraction performance.

Sarafraz-Yazdi and his research group utilized PEG and PEG/MWCANT coatings to extract monocyclic aromatic amines in water and urine samples using headspace microextraction coupled with GC-FID. Extraction efficiency of PEG/MWCNT was found better compared to PEG coating [32].

2.6.1.1.3. Poly(tetrahydrofuran) (poly-THF)

Poly(tetrahydrofuran), also known as polytetramethylene oxide (PTMO), is a hydroxyl-terminated medium polar polymer. Kabir *et al.* [59] first utilized this polymer to develop sol-gel poly-THF coating for in-tube SPME. This medium polarity sorbent was found equally effective for both polar, medium polar, and nonpolar analytes without any derivatization. Recently, Kumar *et al.* [33] used sol-gel poly-THF coated fabric phase sorptive extraction media for efficient extraction of selected endocrine disruptors from environmental water and urine samples with very high enrichment factors.

The applicability of this polymer as microextraction sorbent has been greatly enhanced by incorporating it with titania network [60]. Due to the strong resistance of titania against highly acidic and basic environment, sol-gel titania poly-THF coating demonstrates stable extraction performance even after prolonged exposure to pH 0 and pH 14.

2.6.1.1.4. Dendrimer

Regardless of the sorbent immobilization process, both commercial SPME fibers/Twister stir bar and sol-gel coated SPME fibers rely upon organic polymers which are inherently polydisperse (wide range of molecular weight distribution). The polydispersity of organic polymers is among many factors that contribute to poor batch-to-batch reproducibility. One approach to resolve that problem is to use dendrimer macromolecules as the replacement of organic polymers. Dendrimers are highly branched macromolecules that allow tailoring their properties by attaching desired functional groups to the core, the peripheral surface, the branching unit or at multiple sites within the dendrimer. Contrary to the typical organic polymers, dendrimers are uniquely monodisperse *i.e.*, behaves like a molecule with fixed molecular weight. Kabir *et al.* [61] were first to report the use of dendrimer as microextraction sorbent. A novel benzyl-terminated Dendron-based sol-gel coating was created for capillary microextraction (CME).

A group of polar and nonpolar analytes including alcohols, phenols, ketones, aldehydes, and polycyclic aromatic hydrocarbons (PAHs) was successfully extracted by this coating followed by gas chromatography-flame ionization detection. The sol-gel dendrimer coating demonstrated remarkable run-to-run peak area reproducibility with RSD values ranging from 2.04 to 9.20%. The tree-like branched architecture of benzyl-terminated dendrimer incorporated into sol-gel coating provided enhanced analyte preconcentration leading to low LODs ranging from 1.0 to 11.2 ppt for alcohols, 1.9 to 44.3 ppt for ketones, 3.3 to 19.4 ppt for aldehydes, and 2.1 to 3.6 ppt for PAHs. In addition, strong covalent bonding between the sol-gel dendrimer coatings to the fused silica capillary surface provided high thermal stability (300 °C) of the sorbent.

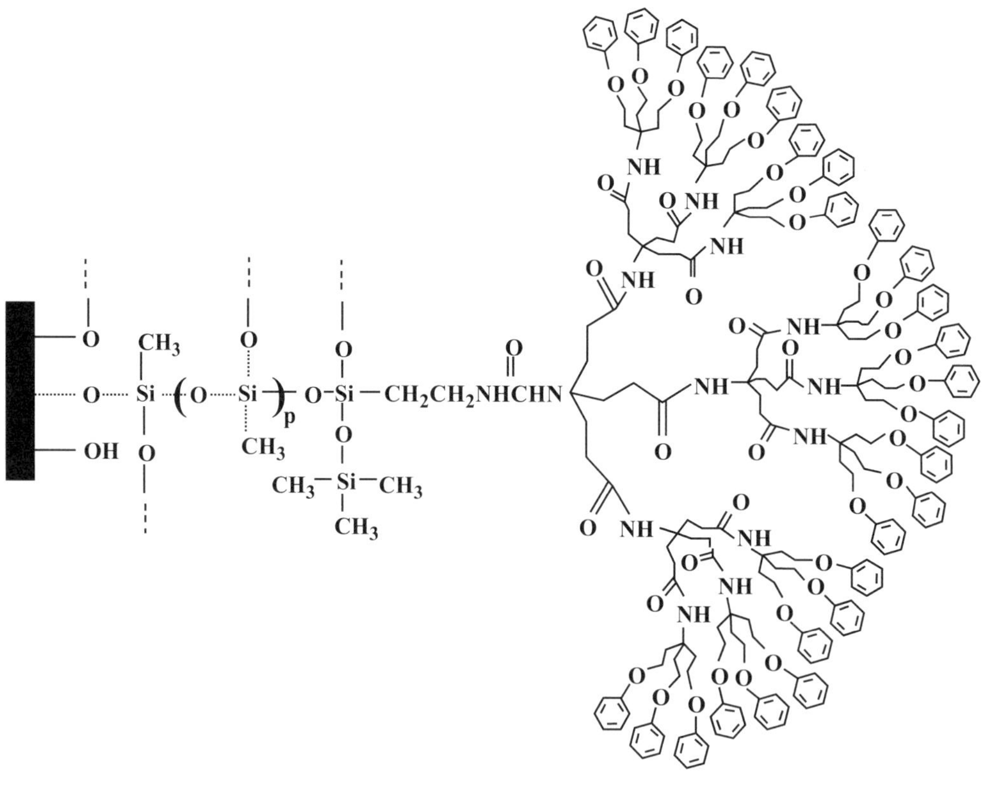

Fig. (5). Schematic representation of sol-gel dendrimer coating inside a fused silica capillary.

2.6.1.1.5. Polypyrrole

Polypyrrole (PPY) is a conducting polymer. Since PPY is formed in its oxidized state carrying positive charges, the counter ions are incorporated into the PPY film. The counter ions provided a convenient way to introduce functional groups into the film and thus the extraction properties of the film can be modified by selecting an appropriate counterion. Due to the inherent multifunctional properties of PPY, a number of intermolecular interactions *e.g.*, μ-μ interactions, acid-base interactions, hydrogen bonding, dipole-dipole interaction *etc.* can be exploited when PPY is used as the microextraction sorbent [62]. A number of PPY based microextraction sorbents have been developed lately [63, 64].

2.6.1.1.6. Ionic Liquids

Ionic liquids (ILs) are organic salts possessing melting points at or below 100 °C. Ionic liquid is composed of an organic cation and an iorganic/organic anion. A number of unique attributes of ILs including high thermal stability, negligible vapor pressure, non-flammability, ability to tailor selectivity by introducing different substituents into IL structure, tunable viscosity, excellent solvating property have earned tremendous interests from researchers working in developing novel microextraction sorbents. As a result, within a short period of time, a large number of sol-gel based sorbents containing ionic liquids have emerged [65 - 70, 49].

Considering the advantageous features that potentially benefit sol-gel microextraction media, ILs have been used as solvents, pore templates, drying control chemical additives, and catalysts. ILs also reportedly reduce cracking and shrinking of coatings [71]. Shearrow *et al.* [66, 68], were first to report IL-mediated sol-gel hybrid organic-inorganic materials for capillary microextraction of nonpolar and moderately polar analytes. The ILs were found to play key roles as porogens and co-solvents; however, they did not contribute to extraction efficiency as the selected ILs decomposed or were rinsed out of the sorbent during the post-gelation processing of sol-gel coatings (thermal conditioning, rinsing). The IL-mediated sol-gel coatings prepared with organic polymers containing Si-OH termini provided up to 28 times higher extraction efficiencies compared with the same coatings prepared without ILs. On the other hand, IL-mediated sol-gel coatings prepared in conjunction with organic polymers containing C-OH termini provided lower efficiencies compared to their IL-free counterparts. This strongly suggests that IL-generated porous morphology alone is not sufficient to provide affinity toward target analyte for effective extraction. The authors concluded that lack of effective chemical bonding between C-OH termini of the organic polymer and Si-OH functionality of the sol-gel network during polycondensation reaction of the sol-gel process may have contributed to the lower extraction efficiencies and emphasized that a careful selection of organic polymers and sol-gel precursors with similar sol-gel reactivity is needed for successful sorbent development for microextraction.

2.6.1.1.7. Functional Coatings

It is a common practice to use functional/organically modified inorganic silica precursors in the sol solution to fine tune the polarity/selectivity of the microextraction sorbent in addition to the sol-gel active organic polymer. However, several researchers used functional precursor along with other network-forming precursors in the sol solution and totally eliminated the organic polymer.

Li *et al.* [72] discussed the reaction of 3-mercaptopropyl trimethoxysilane (MPTS) with silver wire to produce MPTS-Ag wire. The MPTS-Ag wire was then used as a substrate to incorporate C_{12} functionality using dedecyltrimethoxysilane in the sol solution. The extraction performance of C_{12}-Ag wire was evaluated by extracting three benzophenone UV filters from river water. This approach demonstrated an alternate unbreakable substrate to replace fragile fused silica fibers.

Wang *et al.* [73] utilized octyl-functionalized mesoporous SBA-15 as a precursor to create SPME fibers. 8 polycyclic aromatic hydrocarbons (PAHs) were used as the probe analytes to evaluate the performance of C_8-SBA-15 coated fibers. The C_8-SBA-15 coated fibers outperformed commercial SPME fibers coated with 30 μM-PDMS and 100 μM-PDMS fibers. In addition, the sol-gel sorbent demonstrated excellent fiber-to-fiber and batch-to-batch reproducibility.

Other functional coatings include dual precursors, phenyltrimethoxysilane

(PTMOS) and methyltrimethoxysilane (MTMS) [27, 30, 74], octadecyl-functionalized nanotubular TiO_2/Ti wire [75].

2.6.2. Non-silica Microextraction Sorbents

Although silica-based sorbents still dominate among all the available microextraction sorbents due to the comprehensive understanding of silica chemistry, availability of a large number of highly pure and inexpensive silica-based inorganic precursors as well as organically modified silica precursors, they suffer from poor pH stability (pH 2-8) and are consequently limited to applications where matrix pH values do not need to adjust beyond 2-8 pH values. As a result, extraction of highly acidic and basic analytes using matrix pH adjustment is not feasible when silica based sorbents are used. In order to increase the pH stability of sol-gel sorbents, recently a few sol-gel hybrid organic-inorganic sorbents have been developed based on titania [20, 76], zirconia [22], Germania [24], alumina [25] chemistry. Compared to silica-based sorbents, these sorbents demonstrate remarkable resistance against highly acidic and basic environment.

Titania has long been known for its high pH stability, good physical strength and excellent mechanical stability. As such, titania-based sorbents should be ideal microextraction sorbents. Titania possesses anion-exhange and cation-exchange properties at acidic pH and alkaline pH, respectively. Therefore, titania-based sorbents can be conveniently used for metal extraction. Titania-based hybrid materials show pronounced affinity towards Bronsted bases and Lewis bases [77].

Germania is an isothermal analogue of silica [78] and therefore possesses material properties similar to silica. As a result, Germania is considered to be an excellent alternative to silica. Malik and co-workers have recently presented novel sorbents based on Germania chemistry [23, 24, 79].

Zirconia possesses very high resistance against basic pH compared to other metal oxides. It is practically insoluble within the whole pH range (1-14). In addition to distinct pH stability, zirconia demonstrates high thermal stability, chemical inertness as well as superior mechanical strength-all these attributes are favourable for a microextraction sorbent [22].

2.6.3. Advances in Stir Bar Sorptive Extraction Coatings

Stir-bar sorptive extraction (SBSE) was developed by Pat Sandra and coworkers [2] in 1999 as a new format for solvent-free microextraction. SBSE, like SPME, is also an equilibrium-based non-exhaustive sample preparation technique. In addition to the high volume of sorbent loading (24 µL in SBSE *vs.* 0.5 µL in SPME), SBSE has integrated a sample diffusion mechanism into its design (sorbent coating on a magnetic stir-bar). High sorbent loading and self stirring capability of SBSE should offer high sample preconcentration factor for the target analytes in a relatively short period of time. However, due to some inherent shortcomings *e.g.* limited number of commercially available sorbent coating, absence of chemical bonding between the coated polymer sorbent and the substrate, likelihood of the sorbent being washed away if improper solvent is used, low diffusion of analyte through highly viscous polymeric sorbents, high cost of coated stir-bar have undermined the universal appeal of this highly potential sample preparation technique.

Similar to other formats of sorbent based sorptive microextraction system, sol-gel technology has generously contributed to the growth of SBSE, resulting in a large number of novel sorbents [80 - 82].

2.6.4. Advances in Sol-gel Molecularly Imprinted Coatings

Sol-gel technology offers a unique synthesis pathway to create molecularly imprinted polymers (MIPs) both in the form of surface coating as well as in the form of a monolithic bed. In presence of a template molecule (real or dummy) in the sol solution, a self assembly process of sol-gel precursor(s) begins under optimum reaction conditions, resulting in a three dimensional network of sol-gel sorbent with entrapped template. The template can be removed from the sol-gel sorbent using different approaches that leave a solid material imprinted with the shape, size and functional selectivity complementary to the template. The artificially created multidimensional selectivity has been found to be very effective in extracting the target analytes from different matrices with high efficiency. MIPs are specially used for targeted analyte extraction from complicated sample matrix.

Du *et al.* imprinted clenbuterol (CLB) on silica particles using sol-gel process [83]. The imprinted silica was then packed in a syringe (SMIPs-MEPS). The resulting extraction system was evaluated by extracting CLB from pork samples. The SMIPs-MEPS method combined with HPLC provided the limit of quantitation (LOQ) and limit of Detection (LOD) for CLB 0.02 and 0.009 μg/kg, respectively. The sample pretreatment was very fast, completing within 2 min.

An interesting comparison between grafting *via* surface attached monomer and grafting *via* sol-gel was carried out by Tiwari *et al.* [84] for developing molecularly imprinted μ-solid phase extraction fiber. Grafting *via* sol-gel approach was found inferior in terms of the sensitivity of insulin detection.

Chaves *et al.* [85] developed a protein imprinted sol-gel polymer inside the inner surface of a fused silica capillary to form in-tube SPME fiber. The device was applied to extract α2a in plasma sample. Detection of the extracted protein was carried out in high-performance liquid chromatography with fluorescence detection (HPLC-FLD). The sol-gel synthesis of MIP provided fast extraction and short extraction equilibrium time. Higher precision, accuracy and low sample volume requirement (50 μL) clearly distinguished the new method from other methods.

Prasad *et al.* [86] presented a very interesting study designed for dihydro pyrimidine dehydrogenase (DPD) deficient patients to analyze uracil and 5-fluorouracil (5FU) biomarkers for inborn errors of metabolism. Uracil and 5FU templated MIP were developed individually on SPME fibers and used in combination with voltammetric sensor. The unique combination of sol-gel MIP–voltammetry provided as high as 10- and 8.4- fold preconcentrations of uracil and 5FU respectively, sufficient to diagnose uracil disorders and fluoropyrimidine toxicity. The proposed method demonstrated LOD of uracil and 5FU in plasma samples at 0.0245 and 0.484 ng/mL, respectively and reduced the chance of non-specific false positives and cross reactivities which often complicates the analysis of biological samples.

Caffeine imprinted sol-gel coated SPME fibers have been presented in a recent publication [87] using vinyl trimethoxysilane as sol-gel precursor and methacrylic acid as functional polymer. Coupling caffeine imprinted SPME fiber with GC/MS allowed analysis of caffeine in human serum. The sol-gel imprinted SPME fiber displayed excellent specific affinity toward caffeine with high recovery.

2.6.5. Sol-gel Sorbents for Metals and Organometallics

Metal oxides are well known for their ability to efficiently extract metal ions from different matrices.

Due to the lack of volatility, metal ions and organometallics are analyzed preferentially in liquid phase *e.g.*, HPLC-ICP-MS. Commercial SPME fibers demonstrate poor solvent stability when exposed to organic solvents. Sol-gel coating technology successfully addressed the weakness of coating by incorporating organic polymers into inorganic network and by anchoring the coating to the substrate *via* strong covalent bonding. As a result, the potential of sol-gel derived sorbents can be easily exploited to extract metal ions and organometallics from different sample matrices.

Boyaci *et al.* [88] synthesized amine functionalized SPME fibers using both electrospinning and dip coating approach. The amine functionalized SPME fibers were employed to extract dimethyl arsenic acid (DMA), monomethylarsinic acid (MMA) and arsenate (AsIV) from aqueous solution using direct immersion mode followed by analysis in HPLC-ICP-MS. Electrospinning was found to be a better coating approach than dip coating. Furthermore, electrospinning resulted in homogeneous coating, easily accessible amine functional groups, and higher extraction sensitivity for arsenicals. Also, electrospinning coating was faster than dip coating process.

Gbatu [89] reported the first sol-gel coated SPME fibers capable of extracting organometallics from aqueous solution followed by HPLC analysis. Several organometallics with known detrimental impact on human health *e.g.*, triphenylarsine (Ph_3AS), diphenyl mercury (Ph_2Hg) and trimethyl tin (TMPhT) were used to evaluate the extraction performance of sol-gel SPME fibers. Limit of detection values range from 80 μg/L to 647 μg/L in HPLC-UV.

Elemental Cu, Zn, Ni, Hg and Cd were also successfully extracted by SPME fibers coated with $-NH-CH_2-CH_2-NH_2-$ functional group [90].

Ordered mesoporous titania film was exploited in in-tube SPME for extracting V, Cr, and Cu from aqueous solution [91]. The sol-gel coated CME fibers demonstrated excellent recoveries (89.2-105%) of V, Cr and Cu from human urine and lake water.

2.6.6. Monoliths as Sol-gel Microextraction Sorbent

Although sol-gel sorbent coating and monolithic bed utilize the same sol-gel technology to create the three dimensional network of hybrid organic-inorganic material, unlike sorbent coating, the sol-solution used for creating the monolithic bed is allowed to stay inside the fused silica capillary until the *sol* solution transforms into solid *gel* containing entrapped solvent and other unreacted sol solution ingredients. After aging for a predetermined period, the monolith is slowly heated to expel all the trapped unreacted entities including the solvent. Finally, the monolith is rinsed with a suitable solvent to further clean the sorbent bed before it is ready to use for extraction. The monolith, like its counterpart surface coating, is chemically bonded to the substrate which eventually imparts mechanical stability to the monolithic bed. Monolith behaves like a highly porous sponge-like cylindrical rod possessing easily accessible functional groups ready for analyte-sorbent interaction. Monolithic beds are inherently porous and thus easily permeable. The porosity of the monolithic bed can be controlled by optimizing sol-gel reaction conditions and by introducing different porogenic agents to the sol solution. As a result, a combination of flow-through channels, micropores and mesopores form throughout the monolithic bed that allows rapid interaction between the analyte(s) and sorbent resulting in faster extraction equilibrium with higher analyte recovery.

Unlike surface coating, sol-gel technology for creating monolithic bed is still evolving. Generation of a surface bonded and ctrack-free monolithic bed is very challenging. As a consequence, the huge potential and advantages of monolithic bed is yet to be exploited.

One successful monolithic bed was created by Zheng *et al.* [92] The monolith was created *in situ* within a 250-μm I.d. fused silica capillary using an acid catalyst for hydrolysis and a basic catalyst for polycondensation. The monolithic bed was coupled with μ-HPLC-UV. When the bed was used as microextraction sorbent, a high preconcentration factors ranging from 299-465 and LOD values ranging from 2.4-8.1 ng/mL for selected polycyclic aromatic hydrocarbons (PAHs) were achieved.

2.6.7. Fabric Phase Sorptive Extraction (FPSE)

As discussed before, majority of the shortcomings pertaining to commercially available microextraction techniques (SPME, SBSE) can be classified into two groups: (1) coating technology used for immobilizing organic polymers onto the substrate surface; (2) the physical format of the extraction system that determines the primary contact surface area (PCSA) of the device. The sample matrix containing the analyte(s) makes direct contact with this surface to interact with the sorbent for extraction. Sol-gel coating technology [14] has made phenomenal contribution to sorbent-based microextraction techniques by providing a facile synthesis pathway to creating a surface bonded hybrid inorganic-organic sorbent systems. The technology introduced organically modified inorganic precursors into the overall selectivity equation, resulting in a tunable and more favorable selectivity parameter. The selectivity of a pristine organic polymer is different than a sol-gel composite containing the same polymer. The inherent porosity of the sol-gel matrix offers unrestrained access to numerous functional groups for fast analyte-sorbent interaction. As a result, relatively thin sol-gel sorbent coating translates into similar or higher sensivity than commercial SPME/SBSE coatings. Nonetheless, sol-gel coating technology followed the same foot steps of commercially available microextraction techniques by limiting itself in the same geometric formats. Similar to the commercially available microextraction techniques, most of the sol-gel coatings reported so far utilized a fused silica fiber or a metal rod as the substrate (fiber-SPME), or fused silica capillary inner surface (intube-SPME or CME) or a magnet covered with glass tube (stir bar). As a result, problem with limited sorbent loading arised by the substrate geometry remains unsolved. It is imperative that, if a microextraction technique is to be highly sensitive as well as fast, both the sorbent volume and the primary contact area have to be augmented.

In addition to increasing the sorbent volume and enhancing the primary contact surface area of the microextraction device, some other factors which may potentially raise the quality of sample preparation to a new level need further consideration: (1) the ability to preconcentrate target analytes directly from the unmodified samples without any clean-up exercises; (2) resistance to harsh chemical environments (*i.e.*, highly acidic and basic) so that matrix pH can be adjusted to wider pH values; (3) the ability to use any organic solvent to elute the extracted analytes so that the final solution can be injected into an analytical instrument of choice; (4) equal effectiveness in field sampling and sample preparation to eliminate the burden of sample collection, transportation, storage, and sample preparation in the laboratory; (5) the ability to achieve a high preconcentration factor during the extraction so that solvent evaporation and sample reconstitution are unnecessary (a frequent practice in LLE and SPE); (6) the ability to reach extraction equilibrium quickly so that field sampling and sample preparation do not become an inconvenient task; and (7) the ability to analyze the same sample using both gas chromatography (GC), high performance liquid chromatography (HPLC), and/or capillary electrophoresis (CE) to obtain complementary information depending on the analytical need.

Taking all these challenges into consideration, a new sorbent-based sorptive microextraction, fabric phase sorptive extraction (FPSE) was developed by Kabir and Furton [6] that effectively incorporated majority of the aforementioned attributes. In addition to the advanced material properties of sol-gel derived hybrid organic-inorganic sorbents, FPSE has successfully utilized flexibility, permeability, and the rich surface chemistry of natural fabric substrates resulting in a microextraction sorbent chemically bonded to the substrate with a very high, readily accessible active extraction surface for fast and high-efficiency analyte(s) extraction. FPSE has introduced major advantages in solvent-less/solvent-minimized sample preparation techniques including an extraordinarily high primary contact surface area (PCSA) available for direct sorbent-analyte interaction, and the ability to directly preconcentrate the target analyte(s) even from excessively complicated sample matrices containing debris, cells, proteins, particulates *etc*.

There are some additional features of this technique which make it more efficient and useful. Some of the features include:

- It is simple, inexpensive, fast and solvent-free sample preparation;
- FPSE media can be introduced directly into the container holding the original sample for sample preparation;
- Sample containing excessive matrix interferents including particulates, cells, debris, fat, protein do not need any sample pretreatment *e.g.*, filtration, centrifugation *etc.*;
- Diffusion of analytes into the liquid sample and the FPSE media can be augmented by magnetic stirring, sonication *etc.*;
- Any organic solvent can be used for the back extraction. Due to strong covalent bonding between the surface and sorbent coating, there is no restriction in solvent selection for back extraction;
- Large number of substrates can be used with their own unique surface properties, *e.g.*, hydrophilicity, hydrophobicity;
- Hundreds of novel sol-gel coatings can be used as the sorbent;
- Extremely high analyte pre-concentration factor;
- Totally eliminates the necessity of solvent evaporation and sample reconstitution.

Fig. (**6**) compares the primary contact surface area of SPME, SBSE and FPSE. It is obvious that extraordinarily high surface area of FPSE allows high sorbent loading in the form of thin or ultra-thin coating, resulting in high sample capacity and fast extraction equilibrium.

Fabric phase sorptive extraction utilizes flexible fabric surface (cotton, polyester, mixed) as the substrate. Unlike other sorbent-based microextraction systems, FPSE substrates are not inert. Depending on the substrate used, it may be hydrophilic (cotton) or hydrophobic (polyester) and exerts significant impact on the overall selectivity/polarity of the FPSE media. As such, if the target analytes are nonpolar, a nonpolar polymer (*e.g.*, polydimethyldiphenyl siloxane) coated on a polyester substrate is recommended. On the other hand, for polar analytes, a polar polymer (*e.g.*, polyethylene glycol) coated on a cellulose substrate is suggested. The impact of the substrates on extraction performances is demonstrated in Fig. (**7**). As seen in the graph, same polymer (polytetrahydrofuran) coated on three different substrates (hydrophobic, medium hydrophilic, and hydrophobic) shows different affinity towards different analytes.

Hydrophilic substrate extracted polar analytes *e.g.*, 4,5-dimethyl phenol, 2-nitrophenol, benzophenol with higher efficiency. Hydrophobic substrate demonstrated preferential affinity towards nonpolar fluoranthene.

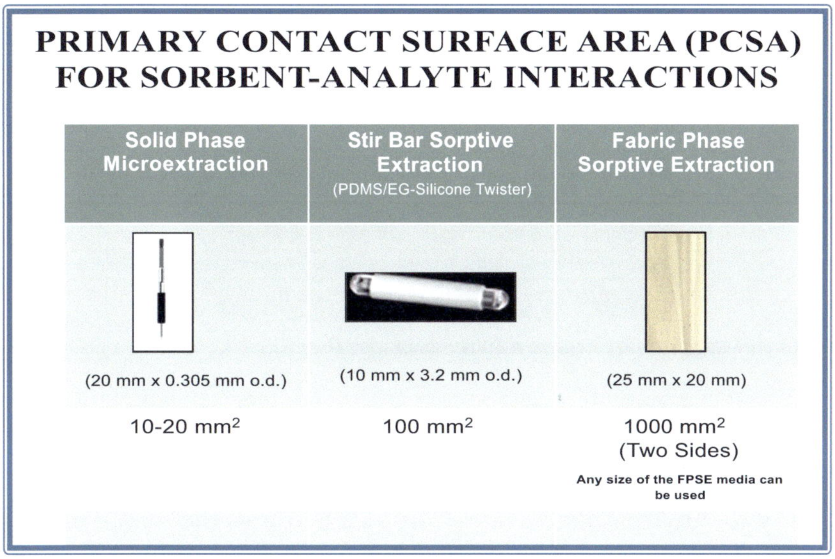

Fig. (6). A comparison between primary contact surface areas (PCSA) of major microextraction devices.

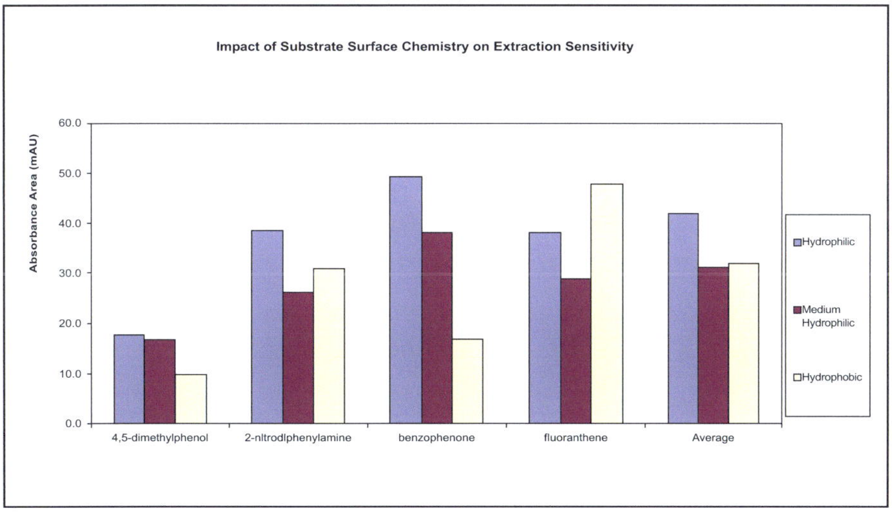

Fig. (7). Impact of substrate surface chemistry on extraction sensitivity.

Fabric phase sorptive extraction is not only very sensitive but also simple and user friendly. It does not require any specialized training. Fig. (**8**) demonstrates a photographic image of FPSE process.

Fig. (8). Photographic image of fabric phase sorptive extraction (FPSE).

Due to the inherent simplicity, fabric phase sorptive extraction is equally effective in both the laboratory environment and in the field applications. This opens up a new possibility for environmental sampling. FPSE allows the whole sampling and sample preparation *in situ* in the field. At the end, using FPSE as the sample preparation technique a highly concentrated target analyte solution in a suitable organic solvent in a sample vial ready for instrumental analysis can be obtained. Fig. (**9**) shows a comparison between laboratory and field sample preparation. Statistical analysis of triplicate data obtained from the field and the laboratory revealed no significance differences between the two sample preparation places.

The enormous potential and advantages of fabric phase sorptive extraction have drawn encouraging attention from separation scientists working with food, forensics, environmental, toxicological, pharmaceutical problems. A number of research projects [33, 93, 94] have already showcased the application benefits and performance superiority of this emerging sorbent-based sorptive microextraction technique. Taking one step forward, Valcarcel and co-workers have successfully integrated fabric phase sorptive extraction with magnetic stirring mechanism in their new microextraction technique known as stir fabric phase sorptive extraction SFPSE) [95].

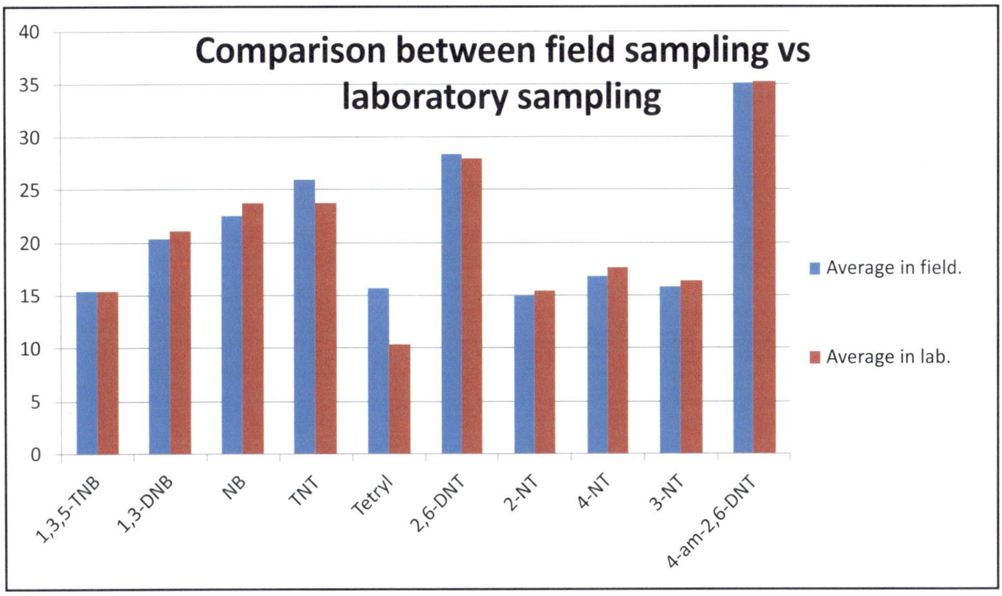

Fig. (9). Comparison of extraction performances of FPSE between laboratory and in the field.

2.7. APPLICATIONS OF SOL-GEL SORPTIVE MICROEXTRACTION SORBENTS

The tunable selectivity, adjustable porosity, convenient synthesis pathway, excellent thermal, mechanical, chemical, and solvent stability, availability of a large number of inorganic precursors and a wide variety of organic ligands have catalyzed the development of novel sol-gel sorptive microextraction sorbents. To date, hundreds of unique microextraction sorbents have emerged, aiming at a wide variety of application areas including environment, food, pharmaceutical, natural products, toxicology, forensics, metallics and organometallics. Applications of only selected sol-gel microextraction sorbents are presented in Table **2**. In order to avoid potential contamination of the microextraction system by the matrix interferents, majority of the analysis were carried out in headspace extraction mode. This may pose a limitation in extraction efficiency to particularly polar analytes. However, fabric phase sorptive extraction has eloquently addressed this issue and demonstrated the capability of extracting both polar and nonpolar analytes directly from the sample matrix containing high amount of interferents.

Table 2. Applications of selected sol-gel hybrid organic-inorganic microextraction sorbents.

Microextraction Format/ Instrument	Inorganic Network	Organic Ligand	Application	Reference
DI-SPME[1]/GC-FID[2]	Silica	Poly(dimethylsiloxane)	Polycyclic aromatic hydrocarbons, Dimethyl phenols, Anilines, Alcohols, Alkanes	[14]
HS-SPME[3]/GC-FID	Silica	Poly(dimethylsiloxane)/Poly(vinyl alcohol)	o-Xylene, Naphthalene, Ethyl caprate, p-Chloro toluene, PCBs	[34]
CME[4]/GC-FID	Silica	Poly(dimethylsiloxane), Poly(ethylene glycol)	Polycyclic aromatic hydrocarbons, Aldehydes, Ketones, Phenols, Alcohols, Amines	[42]
HS-SPME/GC-FID	Silica	Poly(dimethylsiloxane)/β-Cyclodextrin	Phenols, Amines	[44]
HS-SPME/GC-FID	Silica	Hydroxy terminated silicone oil/Dibenzo-18-crown-6	Aliphatic amines	[45]
HS-SPME/GC-FID	Silica	Hydroxy terminated silicone oil/Hydroxyfullerene	Polychlorinated biphenyls, Polycyclic aromatic hydrocarbons, Aromatic amines	[47]
SBSE[4]/HPLC-UV[5]	Silica	Poly(dimethylsiloxane)/Metal organic frameworks	Estrogens	[48]
DI-SPME/GC-FID	Silica	Poly(ethylene glycol) grafted multi-walled carbon nanotubes	Non-steroidal anti-inflammatory drugs	[56]
HS-SPME/GC-FID	Silica	Poly(ethylene glycol)-modified ormosils	BTEX	[57]
HS-SPME/GC-FID	Silica	Ionic liquid mediated poly(dimethylsiloxane) -grafted carbon nanotubes	Methyl-tert-butyl ether	[65]
DI-SPME/GC-NPD[6]	Silica	Polypyrrole	Organophosphorous pesticides	[63]
DI-FPSE[7]/HPLC-UV	Silica	Poly(tetrahydrofuran)	Estrogens	[33]
DI-FPSE/HPLC-UV	Silica	Short chain Poly(ethylene glycol)	Amphenicols	[94]
DI-SFPSE/HPLC-UV	Silica	Poly(tetrahydrofuran)	Triazine	[95]

(Table 2) contd.....

Microextraction Format/ Instrument	Inorganic Network	Organic Ligand	Application	Reference
CME/HPLC-UV	Silica	Methyl/Octyl/Octadecyl functional ligands	Polycyclic aromatic hydrocarbons, Ketones, Amines, Phenols	[27]
DI-SPME/HPLC-UV	Silica	Octyl functional ligand mesoporous SBA-15	Polycyclic aromatic hydrocarbons	[73]
DI-SPME/HPLC-UV	Silica	Octyl functionalized ligand	Organometals triphenylarsine, diphenylmercury, trimethylphenyltin	[89]
CME/ICP-AES	Silica	3-mercaptopropyl trimethoxysilane modified network	Cu, Hg. Pb	[90]
DI-SPME/GC-MS[8]	Silica	Caffeine imprinted SPME coating	Caffeine	[87]
In-Tube SPME/FD	Silica	Interferon alpha 2a imprinted capillary tube	Interferon alpha 2a	[85]
HS-SPME/GC-FID	Titania	Poly(ethylene glycol)	BTEX	[21]
CME/GC-FID	Zirconia	Poly(dimethyldiphenylsiloxane)	Polycyclic aromatic hydrocarbons, Aldehydes, Ketones	[22]
CME/HPLC-UV	Germania	PEO-PPO-PEO	Polycyclic aromatic hydrocarbons, Amines, Alcohols, Phenols	[24]
HS-SPME/GC-ECD[9]	Alumina/ Titania	Poly(ethylene glycol)	Chlorinated organic solvents	[25]

[1]DI-SPME Direct Immersion-Solid Phase Microextraction; [2]GC-FID Gas Chromatography-Flame Ionization Detector;
[3]HS-SPME Headspace-Solid Phase Microextraction; [4]CME Capillary Microextraction; [5]SBSE Stir Bar Sorptive Extraction;
[6]GC-NPD Gas Chromatography-Nitrogen Phosphorous Detector; [7]DI-FPSE Direct Immersion-Fabric Phase Sorptive Extraction; [8]GC-MS Gas Chromatography-Mass Spectrometry; [9]GC-ECD Gas Chromatography-Electron Conductivity Detector.

CONCLUSION

Sol-gel technology for creating surface coating and monolithic bed of hybrid organic-inorganic composite sorbents has proven itself as an inevitable tool for the growth and sustainance of sorben-based microextraction techniques and has catalyzed the explosive growth of sol-gel microextraction sorbents in recent years.

Controllable porosity, tunable selectivity, adjustable coating thickness and morphology as well as other physico-chemical attributes of sol-gel microextraction sorbents are now within the reach of researchers working in the field. High volume of research directed towards sol-gel based material systems has already started paying dividends in not only in the field of separation science but also in other important fields including catalysts, sensors, drug delivery, corrosion resistance *etc.*

Sol-gel coatings and monolithic beds are often chemically anchored to the substrate surface by strong covalent bonding. This strong adhesion of the sorbent material to the substrate surface offers high thermal and solvent stability and allows the sorbents to expose to high temperature and any organic solvent of choice for thermal desorption and solvent mediated elution, respectively. High chemical stability of sol-gel sorbents permits exposure to harsh chemical environments *e.g.*, highly acidic and basic solutions which can be exploited to extract ionizable compounds (organic acids and bases) by adjusting the pH value of the sample matrix to bring the compounds in their neutral form without enduring performance loss of the sorbent.

Fabric phase sorptive extraction (FPSE) has elevated sorptive microextraction to a new level. High sorbent loading in the form of ultra-thin coating chemically anchored to the permeable and flexible fabric support in FPSE allows near-exhaustive extraction of the target analyte(s) from unmodified sample matrix in a very short period of time. Due to the high preconcentration factor achieved in FPSE, no post-sample preparation treatment *e.g.*, solvent evaporation, sample reconstitution is required.

Looking forward, further developments and innovations in novel sol-gel microextraction sorbents as well as new substrates are expected to surface in near future.

DISCLOSURE

Part of this chapter has been previously published in Innovations in sol-gel microextraction phases for solvent-free sample preparation in Analytical Chemistry. TrAC Trends in Analytical Chemistry. Volume 45, April 2013, Pages

197-218.

CONFLICT OF INTEREST

The authors confirm that they have no conflict of interest to declare for this publication.

ACKNOWLEDGEMENTS

The authors gratefully acknowledge the help of Mr. Rodolfo Mesa, Department of Chemistry and Biochemistry at Florida International University, Miami, FL, USA for collecting data on fabric phase sorptive extraction presented in the current chapter.

REFERENCES

[1] Pawliszyn, J.; Liu, S. Sample introduction for capillary gas chromatography with laser desorption and optical fibers. *Anal. Chem.,* **1987**, *59*(10), 1475-1478. [http://dx.doi.org/10.1021/ac00137a024]

[2] Baltussen, E.; Sandra, P.; David, F.; Cramers, C. Stir bar sorptive extraction (SBSE), a novel extraction technique for aqueous samples: Theory and principles. *J. Microcolumn Sep.,* **1999**, *11*(10), 737-747. [http://dx.doi.org/10.1002/(SICI)1520-667X(1999)11:10<737::AID-MCS7>3.0.CO;2-4]

[3] Bruheim, I.; Liu, X.; Pawliszyn, J. Thin-film microextraction. *Anal. Chem.,* **2003**, *75*(4), 1002-1010. [http://dx.doi.org/10.1021/ac026162q] [PMID: 12622398]

[4] Abdel-Rehim, M. New trend in sample preparation: on-line microextraction in packed syringe for liquid and gas chromatography applications. I. Determination of local anaesthetics in human plasma samples using gas chromatography-mass spectrometry. *J. Chromatogr. B Analyt. Technol. Biomed. Life Sci.,* **2004**, *801*(2), 317-321. [http://dx.doi.org/10.1016/j.jchromb.2003.11.042] [PMID: 14751801]

[5] Lu, Z.L.; Mallet, C.; Diehl, D.M.; Cheng, Y.F. Micro scale solid-phase extraction: Decrease your sample size by 50% while increasing sensitivity by 10X without matrix suppression. *LC GC Eur.,* **2004**, 44-45.

[6] Kabir, A.; Furton, K.G. Fabric phase sorptive extractors. US, 9, 283, 544, B2. , **2014**.

[7] Spietelun, A.; Marcinkowski, Ł.; de la Guardia, M.; Namieśnik, J. Recent developments and future trends in solid phase microextraction techniques towards green analytical chemistry. *J. Chromatogr. A,* **2013**, *1321*, 1-13. [http://dx.doi.org/10.1016/j.chroma.2013.10.030] [PMID: 24238710]

[8] Pawliszyn, J.; Lord, H.L. *Handbook of Sample Preparation*; Wiley, **2012**.

[9] Pawliszyn, J. *Handbook of Solid Phase Microextraction*; Elsevier, **2011**.

[10] Risticevic, S.; DeEll, J.R.; Pawliszyn, J. Solid phase microextraction coupled with comprehensive two-dimensional gas chromatography-time-of-flight mass spectrometry for high-resolution metabolite profiling in apples: implementation of structured separations for optimization of sample preparation procedure in complex samples. *J. Chromatogr. A,* **2012**, *1251*, 208-218. [http://dx.doi.org/10.1016/j.chroma.2012.06.052] [PMID: 22771060]

[11] Dietz, C.; Sanz, J.; Cámara, C. Recent developments in solid-phase microextraction coatings and related techniques. *J. Chromatogr. A,* **2006**, *1103*(2), 183-192. [http://dx.doi.org/10.1016/j.chroma.2005.11.041] [PMID: 16337213]

[12] Kataoka, H. Recent advances in solid-phase microextraction and related techniques for pharmaceutical and biomedical analysis. *Curr. Pharm. Anal.,* **2005**, *1*(1), 65-84. [http://dx.doi.org/10.2174/1573412052953373]

[13] Lucena, R. Extraction and stirring integrated techniques: examples and recent advances. *Anal. Bioanal. Chem.,* **2012**, *403*(8), 2213-2223. [http://dx.doi.org/10.1007/s00216-012-5826-9] [PMID: 22354573]

[14] Chong, S.L.; Wang, D.; Hayes, J.D.; Wilhite, B.W.; Malik, A. Sol-gel coating technology for the preparation of solid-phase microextraction fibers of enhanced thermal stability. *Anal. Chem.,* **1997**, *69*(19), 3889-3898. [http://dx.doi.org/10.1021/ac9703360] [PMID: 21639208]

[15] Kabir, A.; Furton, K.G.; Malik, A. Innovations in sol-gel microextraction phases for solvent-free sample preparation in analytical chemistry. *TrAC. Trends Analyt. Chem.,* **2013**, *45*(0), 197-218. [http://dx.doi.org/10.1016/j.trac.2012.11.014]

[16] Bagheri, H.; Piri-Moghadam, H.; Naderi, M. Towards greater mechanical, thermal and chemical stability in solid-phase microextraction. *Trac-. Trends Analyt. Chem.,* **2012**, *34*, 126-139. [http://dx.doi.org/10.1016/j.trac.2011.11.004]

[17] Brinker, C. J. S.; Scherer, G. W. Sol-Gel science. Academic Press, Inc.: San Diego, California, **1990**.

[18] Fontanals, N.; Marcé, R.M.; Borrull, F. New materials in sorptive extraction techniques for polar compounds. *J. Chromatogr. A,* **2007**, *1152*(1-2), 14-31. [http://dx.doi.org/10.1016/j.chroma.2006.11.077] [PMID: 17187808]

[19] Su, C.J.; Srimurugan, S.; Chen, C.; Shu, H.C. Sol-gel titania-coated needles for solid phase dynamic extraction-GC/MS analysis of desomorphine and desocodeine. *Anal. Sci.,* **2011**, *27*(11), 1107-1113. [http://dx.doi.org/10.2116/analsci.27.1107] [PMID: 22076337]

[20] Maleki, R.; Farhadi, K.; Tahmasebi, R. Preparation of a Sol-Gel Titania Based Coating for HS-SPME of Aliphatic Alcohols from Non-Alcoholic Beer Samples. *Chromatographia,* **2009**, *69*(7-8), 775-778. [http://dx.doi.org/10.1365/s10337-009-0964-8]

[21] Farhadi, K.; Tahmasebi, R.; Maleki, R. Preparation and application of the titania sol-gel coated anodized aluminum fibers for headspace solid phase microextraction of aromatic hydrocarbons from water samples. *Talanta,* **2009**, *77*(4), 1285-1289. [http://dx.doi.org/10.1016/j.talanta.2008.09.009] [PMID: 19084636]

[22] Alhooshani, K.; Kim, T.Y.; Kabir, A.; Malik, A. Sol-gel approach to *in situ* creation of high pH-resistant surface-bonded organic-inorganic hybrid zirconia coating for capillary microextraction (in-

tube SPME). *J. Chromatogr. A,* **2005**, *1062*(1), 1-14.
[http://dx.doi.org/10.1016/j.chroma.2004.10.103] [PMID: 15679137]

[23] Segro, S.S.; Malik, A. High-temperature solvent stability of sol-gel germania triblock polymer coatings in capillary microextraction on-line coupled to high-performance liquid chromatography. *J. Chromatogr. A,* **2010**, *1217*(37), 5746-5752.
[http://dx.doi.org/10.1016/j.chroma.2010.07.038] [PMID: 20691975]

[24] Segro, S.S.; Triplett, J.; Malik, A. Sol-gel germania triblock polymer coatings of exceptional pH stability in capillary microextraction online-coupled to high-performance liquid chromatography. *Anal. Chem.,* **2010**, *82*(10), 4107-4113.
[http://dx.doi.org/10.1021/ac100209p] [PMID: 20405898]

[25] Farhadi, K.; Maleki, R.; Tahmasebi, R. Preparation of Al2O3/TiO2 composite sol-gel fiber for headspace solid-phase microextraction of chlorinated organic solvents from urine. *J. Sep. Sci.,* **2011**, *34*(14), 1669-1674.
[http://dx.doi.org/10.1002/jssc.201000924] [PMID: 21695681]

[26] Duan, C.F.; Shen, Z.; Wu, D.P.; Guan, Y.F. Recent developments in solid- phase microextraction for on-site sampling and sample preparation. *TrAC-. Trends Analyt. Chem.,* **2011**, *30*(10), 1568-1574.
[http://dx.doi.org/10.1016/j.trac.2011.08.005]

[27] Segro, S.S.; Malik, A. Sol-gel coatings with covalently attached methyl, octyl, and octadecyl ligands for capillary microextraction. Effects of alkyl chain length and sol-gel precursor concentration on extraction behavior. *J. Chromatogr. A,* **2009**, *1216*(45), 7677-7686.
[http://dx.doi.org/10.1016/j.chroma.2009.09.030] [PMID: 19782372]

[28] Bagheri, H.; Ghanbarnejad, H.; Khalilian, F. Immersed sol-gel based amino-functionalized SPME fiber and HPLC combined with post-column photochemically induced fluorimetry derivatization and fluorescence detection of pyrethroid insecticides from water samples. *J. Sep. Sci.,* **2009**, *32*(17), 2912-2918.
[http://dx.doi.org/10.1002/jssc.200900044] [PMID: 19655323]

[29] Yun, L. High extraction efficiency solid-phase microextraction fibers coated with open crown ether stationary phase using sol-gel technique. *Anal. Chim. Acta,* **2003**, *486*(1), 63-72.
[http://dx.doi.org/10.1016/S0003-2670(03)00491-4]

[30] Azenha, M.; Malheiro, C.; Silva, A.F. Ultrathin phenyl-functionalized solid phase microextraction fiber coating developed by sol-gel deposition. *J. Chromatogr. A,* **2005**, *1069*(2), 163-172.
[http://dx.doi.org/10.1016/j.chroma.2005.02.013] [PMID: 15830942]

[31] Zhu, S.; Tong, T.; Zhang, W.; Dai, W.; He, S.; Chang, Z.; Gao, X. Preparation of multiwalled carbon nanotubes/hydroxyl-terminated silicone oil fiber and its application to analysis of crude oils. *Scientific World J.,* **2014**, *2014*, 758043.
[http://dx.doi.org/10.1155/2014/758043] [PMID: 24578659]

[32] Sarafraz-Yazdi, A.; Ardaki, M.S.; Amiri, A. Determination of monocyclic aromatic amines using headspace solid-phase microextraction based on sol-gel technique prior to GC. *J. Sep. Sci.,* **2013**, *36*(9-10), 1629-1635.
[http://dx.doi.org/10.1002/jssc.201200940] [PMID: 23418161]

[33] Kumar, R.; Gaurav, H.; Malik, A.K.; Kabir, A.; Furton, K.G. Efficient analysis of selected estrogens

using fabric phase sorptive extraction and high performance liquid chromatography-fluorescence detection. *J. Chromatogr. A,* **2014**, *1359*(0), 16-25.
[http://dx.doi.org/10.1016/j.chroma.2014.07.013] [PMID: 25082524]

[34] Lopes, A.L.; Augusto, F. Preparation and characterization of polydimethylsiloxane/poly(vinylalcohol) coated solid phase microextraction fibers using sol-gel technology. *J. Chromatogr. A,* **2004**, *1056*(1-2), 13-19.
[http://dx.doi.org/10.1016/j.chroma.2004.05.047] [PMID: 15595527]

[35] McDonagh, C.; Bowe, P.; Mongey, K.; MacCraith, B.D. Characterisation of porosity and sensor response times of sol-gel-derived thin films for oxygen sensor applications. *J. Non-Cryst. Solids,* **2002**, *306*(2), 138-148.
[http://dx.doi.org/10.1016/S0022-3093(02)01154-7]

[36] Higuchi, T.; Kurumada, K.; Nagamine, S.; Lothongkum, A.W.; Tanigaki, M. Effect of addition of polymeric species with ether moieties on porous structure of silica prepared by sol-gel method. *J. Mater. Sci.,* **2000**, *35*(13), 3237-3243.
[http://dx.doi.org/10.1023/A:1004811019300]

[37] He, C.; Oka, T.; Kobayashi, Y.; Oshima, N.; Ohdaira, T.; Kinomura, A.; Suzuki, R. Evolution of pores in mesoporous silica films: Porogen loading effect. *Appl. Surf. Sci.,* **2008**, *255*(1), 183-186.
[http://dx.doi.org/10.1016/j.apsusc.2008.05.302]

[38] Zhang, M.; Sun, Y. Cooperation of solid granule and solvent as porogenic agents. Novel porogenic mode of biporous media for protein chromatography. *J. Chromatogr. A,* **2001**, *922*(1-2), 77-86.
[http://dx.doi.org/10.1016/S0021-9673(01)00935-9] [PMID: 11486893]

[39] Yokoi, T.; Karouji, T.; Ohta, S.; Kondo, J.N.; Tatsumi, T. Synthesis of mesoporous silica nanospheres promoted by basic amino acids and their catalytic application. *Chem. Mater.,* **2010**, *22*(13), 3900-3908.
[http://dx.doi.org/10.1021/cm9037846]

[40] Wagh, P.B.; Rao, A.V.; Haranath, D. Influence of Catalyst (Citric Acid) Concentration on the Physical Properties of TEOS Silica Aerogels. *J. Porous Mater.,* **1997**, *4*(4), 295-301.
[http://dx.doi.org/10.1023/A:1009633524487]

[41] Mayo, E.I.; Pooré, D.D.; Stiegman, A.E. Catalysis of the silica sol-gel process by divalent transition metal bis(acetylacetonate) complexes. *Inorg. Chem.,* **2000**, *39*(5), 899-905.
[http://dx.doi.org/10.1021/ic990577l] [PMID: 12526367]

[42] Bigham, S.; Medlar, J.; Kabir, A.; Shende, C.; Alli, A.; Malik, A. Sol-gel capillary microextraction. *Anal. Chem.,* **2002**, *74*(4), 752-761.
[http://dx.doi.org/10.1021/ac0109523] [PMID: 11866054]

[43] Liu, W.; Zhang, L.; Chen, S.; Duan, H.; Chen, X.; Wei, Z.; Chen, G. A method by homemade OH/TSO-PMHS fibre solid-phase microextraction coupling with gas chromatography-mass spectrometry for analysis of antiestrogens in biological matrices. *Anal. Chim. Acta,* **2009**, *631*(1), 47-53.
[http://dx.doi.org/10.1016/j.aca.2008.10.012] [PMID: 19046677]

[44] Fu, Y.L.; Hu, Y.L.; Zheng, Y.J.; Li, G.K. Preparation and application of poly(dimethylsiloxane)/beta-cyclodextrin solid-phase microextraction fibers. *J. Sep. Sci.,* **2006**, *29*(17), 2684-2691.

[http://dx.doi.org/10.1002/jssc.200600116] [PMID: 17313110]

[45] Cai, L.S.; Zhao, Y.Y.; Gong, S.L.; Dong, L.; Wu, C.Y. Use of a novel sol-gel dibenxo-18-crown-6 solid-phase microcxtraction fiber and a new derivatizing reagent for determination of aliphatic amines in lake water and human urine. *Chromatographia,* **2003**, *58*(9-10), 615-621.

[46] Bianchi, F.; Bisceglie, F.; Careri, M.; Di Berardino, S.; Mangia, A.; Musci, M. Innovative sol-gel coatings for solid-phase microextraction. Development of fibers for the determination of polycyclic aromatic hydrocarbons at trace level in water. *J. Chromatogr. A,* **2008**, *1196-1197*, 15-22. [http://dx.doi.org/10.1016/j.chroma.2008.01.036] [PMID: 18258246]

[47] Yu, J.; Dong, L.; Wu, C.; Wu, L.; Xing, J. Hydroxyfullerene as a novel coating for solid-phase microextraction fiber with sol-gel technology. *J. Chromatogr. A,* **2002**, *978*(1-2), 37-48. [http://dx.doi.org/10.1016/S0021-9673(02)01347-X] [PMID: 12458943]

[48] Hu, C.; He, M.; Chen, B.; Zhong, C.; Hu, B. Polydimethylsiloxane/metal-organic frameworks coated stir bar sorptive extraction coupled to high performance liquid chromatography-ultraviolet detector for the determination of estrogens in environmental water samples. *J. Chromatogr. A,* **2013**, *1310*, 21-30. [http://dx.doi.org/10.1016/j.chroma.2013.08.047] [PMID: 23992883]

[49] Sarafraz-Yazdi, A.; Vatani, H. A solid phase microextraction coating based on ionic liquid sol-gel technique for determination of benzene, toluene, ethylbenzene and o-xylene in water samples using gas chromatography flame ionization detector. *J. Chromatogr. A,* **2013**, *1300*, 104-111. [http://dx.doi.org/10.1016/j.chroma.2013.03.039] [PMID: 23582769]

[50] Quintana, J.B.; Rodríguez, I. Strategies for the microextraction of polar organic contaminants in water samples. *Anal. Bioanal. Chem.,* **2006**, *384*(7-8), 1447-1461. [http://dx.doi.org/10.1007/s00216-005-0242-z] [PMID: 16496135]

[51] Wang, Z.; Xiao, C.; Wu, C.; Han, H. High-performance polyethylene glycol-coated solid-phase microextraction fibers using sol-gel technology. *J. Chromatogr. A,* **2000**, *893*(1), 157-168. [http://dx.doi.org/10.1016/S0021-9673(00)00692-0] [PMID: 11043596]

[52] Sungkaew, S.; Thammakhet, C.; Thavarungkul, P.; Kanatharana, P. A new polyethylene glycol fiber prepared by coating porous zinc electrodeposited onto silver for solid-phase microextraction of styrene. *Anal. Chim. Acta,* **2010**, *664*(1), 49-55. [http://dx.doi.org/10.1016/j.aca.2010.02.004] [PMID: 20226931]

[53] Jia, J.; Xu, L.; Wang, S.; Wang, L.; Liu, X. Sol-gel approach for fabrication of coated anodized titanium wire for solid-phase microextraction: highly efficient adsorbents for enrichment of trace polar analytes. *Anal. Bioanal. Chem.,* **2014**, *406*(13), 3209-3217. [http://dx.doi.org/10.1007/s00216-014-7727-6] [PMID: 24682230]

[54] Sarafraz-Yazdi, A.; Amiri, A.; Rounaghi, G.; Hosseini, H.E. A novel solid-phase microextraction using coated fiber based sol-gel technique using poly(ethylene glycol) grafted multi-walled carbon nanotubes for determination of benzene, toluene, ethylbenzene and o-xylene in water samples with gas chromatography-flame ionization detector. *J. Chromatogr. A,* **2011**, *1218*(34), 5757-5764. [http://dx.doi.org/10.1016/j.chroma.2011.06.099] [PMID: 21782185]

[55] Sarafraz-Yazdi, A.; Sepehr, S.; Es'haghi, Z.; Moghadam, H.P. Application of sol-gel based poly(ethylene glycol)/multiwalled carbon nanotubes coated fiber for SPME of methyl tert-butyl ether in environmental water samples. *Chromatographia,* **2010**, *72*(9-10), 923-931.

[http://dx.doi.org/10.1365/s10337-010-1740-5]

[56] Sarafraz-Yazdi, A.; Amiri, A.; Rounaghi, G.; Eshtiagh-Hosseini, H. Determination of non-steroidal anti-inflammatory drugs in water samples by solid-phase microextraction based sol-gel technique using poly(ethylene glycol) grafted multi-walled carbon nanotubes coated fiber. *Anal. Chim. Acta,* **2012**, *720*, 134-141. [http://dx.doi.org/10.1016/j.aca.2012.01.021] [PMID: 22365131]

[57] da Costa Silva, R.G.; Augusto, F. Highly porous solid-phase microextraction fiber coating based on poly(ethylene glycol)-modified ormosils synthesized by sol-gel technology. *J. Chromatogr. A,* **2005**, *1072*(1), 7-12. [http://dx.doi.org/10.1016/j.chroma.2004.11.027] [PMID: 15881453]

[58] Li, Z.; Ma, R.; Bai, S.; Wang, C.; Wang, Z. A solid phase microextraction fiber coated with graphene-poly(ethylene glycol) composite for the extraction of volatile aromatic compounds from water samples. *Talanta,* **2014**, *119*, 498-504. [http://dx.doi.org/10.1016/j.talanta.2013.11.068] [PMID: 24401447]

[59] Kabir, A.; Hamlet, C.; Malik, A. Parts per quadrillion level ultra-trace determination of polar and nonpolar compounds *via* solvent-free capillary microextraction on surface-bonded sol-gel polytetrahydrofuran coating and gas chromatography-flame ionization detection. *J. Chromatogr. A,* **2004**, *1047*(1), 1-13. [http://dx.doi.org/10.1016/j.chroma.2004.06.078] [PMID: 15481455]

[60] Segro, S.S.; Cabezas, Y.; Malik, A. Ultra-high-stability, pH-resistant sol-gel titania poly(tetrahydrofuran) coating for capillary microextraction on-line coupled to high-performance liquid chromatography. *J. Chromatogr. A,* **2009**, *1216*(20), 4329-4338. [http://dx.doi.org/10.1016/j.chroma.2009.03.036] [PMID: 19344910]

[61] Kabir, A.; Hamlet, C.; Yoo, K. S.; Newkome, G. R.; Malik, A. Capillary microextraction on sol-gel dendrimer coatings. *Journal of Chromatography A.,* **2004**, *1034*(1-2), 1-11. [http://dx.doi.org/10.1016/j.chroma.2004.02.019]

[62] Pragst, F. Application of solid-phase microextraction in analytical toxicology. *Anal. Bioanal. Chem.,* **2007**, *388*(7), 1393-1414. [http://dx.doi.org/10.1007/s00216-007-1289-9] [PMID: 17476482]

[63] Saraji, M.; Rezaei, B.; Boroujeni, M.K.; Bidgoli, A.A. Polypyrrole/sol-gel composite as a solid-phase microextraction fiber coating for the determination of organophosphorus pesticides in water and vegetable samples. *J. Chromatogr. A,* **2013**, *1279*, 20-26. [http://dx.doi.org/10.1016/j.chroma.2013.01.017] [PMID: 23357752]

[64] Zhang, Z.; Zhu, L.; Ma, Y.; Huang, Y.; Li, G. Preparation of polypyrrole composite solid-phase microextraction fiber coatings by sol-gel technique for the trace analysis of polar biological volatile organic compounds. *Analyst (Lond.),* **2013**, *138*(4), 1156-1166. [http://dx.doi.org/10.1039/c2an36231g] [PMID: 23282483]

[65] Vatani, H.; Yazdi, A.S. Ionic-liquid-mediated poly(dimethylsiloxane)- grafted carbon nanotube fiber prepared by the sol-gel technique for the head space solid-phase microextraction of methyl tert-butyl ether using GC. *J. Sep. Sci.,* **2014**, *37*(1-2), 127-134. [http://dx.doi.org/10.1002/jssc.201300796] [PMID: 24285506]

[66] Shearrow, A.M.; Bhansali, S.; Malik, A. Ionic liquid-mediated bis[(3-methyldimethoxysilyl)propyl] polypropylene oxide-based polar sol-gel coatings for capillary microextraction. *J. Chromatogr. A,* **2009**, *1216*(36), 6349-6355. [http://dx.doi.org/10.1016/j.chroma.2009.07.028] [PMID: 19643422]

[67] Graham, C.M.; Meng, Y.; Ho, T.; Anderson, J.L. Sorbent coatings for solid-phase microextraction based on mixtures of polymeric ionic liquids. *J. Sep. Sci.,* **2011**, *34*(3), 340-346. [http://dx.doi.org/10.1002/jssc.201000367] [PMID: 21268258]

[68] Shearrow, A.M.; Harris, G.A.; Fang, L.; Sekhar, P.K.; Nguyen, L.T.; Turner, E.B.; Bhansali, S.; Malik, A. Ionic liquid-mediated sol-gel coatings for capillary microextraction. *J. Chromatogr. A,* **2009**, *1216*(29), 5449-5458. [http://dx.doi.org/10.1016/j.chroma.2009.04.093] [PMID: 19515375]

[69] Liu, M.; Zhou, X.; Chen, Y.; Liu, H.; Feng, X.; Qiu, G.; Liu, F.; Zeng, Z. Innovative chemically bonded ionic liquids-based sol-gel coatings as highly porous, stable and selective stationary phases for solid phase microextraction. *Anal. Chim. Acta,* **2010**, *683*(1), 96-106. [http://dx.doi.org/10.1016/j.aca.2010.10.004] [PMID: 21094387]

[70] Tian, M.; Cheng, R.; Ye, J.; Liu, X.; Jia, Q. Preparation and evaluation of ionic liquid-calixarene solid-phase microextraction fibres for the determination of triazines in fruit and vegetable samples. *Food Chem.,* **2014**, *145*, 28-33. [http://dx.doi.org/10.1016/j.foodchem.2013.08.029] [PMID: 24128445]

[71] Vickackaite, V.; Padarauskas, A. Ionic liquids in microextraction techniques. *Cent. Eur. J. Chem.,* **2012**, *10*(3), 652-674.

[72] Li, J.; Ma, L.; Tang, M.; Xu, L. C12-Ag wire as solid-phase microextraction fiber for determination of benzophenone ultraviolet filters in river water. *J. Chromatogr. A,* **2013**, *1298*, 1-8. [http://dx.doi.org/10.1016/j.chroma.2013.05.010] [PMID: 23721808]

[73] Wang, X.; Rao, H.; Lu, X.; Du, X. Application of sol-gel based octyl-functionalized mesoporous materials coated fiber for solid-phase microextraction. *Talanta,* **2013**, *105*, 204-210. [http://dx.doi.org/10.1016/j.talanta.2012.11.074] [PMID: 23598009]

[74] Segro, S.S.; Malik, A. Sol-gel methyl coating in capillary microextraction hyphenated on-line with high-performance liquid chromatography counterintuitive extraction behavior for polar analytes. *J. Chromatogr. A,* **2008**, *1200*(1), 62-71. [http://dx.doi.org/10.1016/j.chroma.2008.03.087] [PMID: 18420215]

[75] Chen, C.; Yang, S.; Pan, D.; Long, Y.; Yan, Z.; Cai, Q.; Yao, S. Development of octadecyl-functionalized-nanotubular TiO_2/Ti wire solid-phase microextraction fiber. *Analyst (Lond.),* **2013**, *138*(2), 569-575. [http://dx.doi.org/10.1039/C2AN36315A] [PMID: 23172420]

[76] Bagheri, H.; Piri-Moghadam, H.; Rastegar, S.; Taheri, N. Electrospun titania sol-gel-based ceramic composite nanofibers for online micro- solid-phase extraction with high-performance liquid chromatography. *J. Sep. Sci.,* **2014**, *37*(15), 1982-1988. [http://dx.doi.org/10.1002/jssc.201400252] [PMID: 24825073]

[77] Li, X.; Gao, J.; Zeng, Z. pH-resistant titania hybrid organic-inorganic sol-gel coating for solid-phase

microextraction of polar compounds. *Anal. Chim. Acta,* **2007**, *590*(1), 26-33.
[http://dx.doi.org/10.1016/j.aca.2007.03.009] [PMID: 17416219]

[78] Armelao, L.; Fabrizio, M.; Gross, S.; Martucci, A.; Tondello, E. Molecularly interconnected SiO2-GeO2 thin films: sol-gel synthesis and characterization. *J. Mater. Chem.,* **2000**, *10*(5), 1147-1150.
[http://dx.doi.org/10.1039/a907247k]

[79] Fang, L.; Kulkarni, S.; Alhooshani, K.; Malik, A. Germania-based, sol-gel hybrid organic-inorganic coatings for capillary microextraction and gas chromatography. *Anal. Chem.,* **2007**, *79*(24), 9441-9451.
[http://dx.doi.org/10.1021/ac071056f] [PMID: 17994707]

[80] Abdulra'uf, L.B.; Tan, G.H. Review of SBSE technique for the analysis of pesticide residues in fruits and vegetables. *Chromatographia,* **2014**, *77*(1-2), 15-24.
[http://dx.doi.org/10.1007/s10337-013-2566-8]

[81] Kabir, A.; Furton, K.G.; Malik, A. Innovations in sol-gel microextraction phases for solvent-free sample preparation in analytical chemistry. *Trac-. Trends Analyt. Chem.,* **2013**, *45*, 197-218.
[http://dx.doi.org/10.1016/j.trac.2012.11.014]

[82] Gilart, N.; Marce, R.M.; Borrull, F.; Fontanals, N. New coatings for stir-bar sorptive extraction of polar emerging organic contaminants. *TrAC. Trends Analyt. Chem.,* **2014**, *54*, 11-23.
[http://dx.doi.org/10.1016/j.trac.2013.10.010]

[83] Du, W.; Lei, C.; Zhang, S.; Bai, G.; Zhou, H.; Sun, M.; Fu, Q.; Chang, C. Determination of clenbuterol from pork samples using surface molecularly imprinted polymers as the selective sorbents for microextraction in packed syringe. *J. Pharm. Biomed. Anal.,* **2014**, *91*, 160-168.
[http://dx.doi.org/10.1016/j.jpba.2013.12.022] [PMID: 24463040]

[84] Tiwari, B.K.; Donnell, C.P.; Muthukumarappan, K.; Cullen, P.J. Effect of low temperature sonication on orange juice quality parameters using response surface methodology. *Food Bioprocess Tech.,* **2009**, *2*(1), 109-114.
[http://dx.doi.org/10.1007/s11947-008-0156-9]

[85] Rodrigues Chaves, A.; Costa Queiroz, M.E. In-tube solid-phase microextraction with molecularly imprinted polymer to determine interferon alpha 2a in plasma sample by high performance liquid chromatography. *J. Chromatogr. A,* **2013**, *1318*, 43-48.
[http://dx.doi.org/10.1016/j.chroma.2013.10.018] [PMID: 24354000]

[86] Prasad, B.B.; Tiwari, K.; Singh, M.; Sharma, P.S.; Patel, A.K.; Srivastava, S. Ultratrace analysis of uracil and 5-fluorouracil by molecularly imprinted polymer brushes grafted to silylated solid-phase microextraction fiber in combination with complementary molecularly imprinted polymer-based sensor. *Biomed. Chromatogr.,* **2009**, *23*(5), 499-509.
[http://dx.doi.org/10.1002/bmc.1145] [PMID: 19101922]

[87] Rajabi Khorrami, A.; Rashidpur, A. Development of a fiber coating based on molecular sol-gel imprinting technology for selective solid-phase micro extraction of caffeine from human serum and determination by gas chromatography/mass spectrometry. *Anal. Chim. Acta,* **2012**, *727*, 20-25.
[http://dx.doi.org/10.1016/j.aca.2012.03.048] [PMID: 22541818]

[88] Boyaci, E.; Horzum, N.; Cagir, A.; Demir, M.M.; Eroglu, A.E. Electrospun amino-functionalized PDMS as a novel SPME sorbent for the speciation of inorganic and organometallic arsenic species.

RSC Advances, **2013**, *3*(44), 22261-22268.
[http://dx.doi.org/10.1039/c3ra43622e]

[89] Gbatu, T.P.; Sutton, K.L.; Caruso, J.A. Development of new SPME fibers by sol-gel technology for SPME-HPLC determination of organometals. *Anal. Chim. Acta,* **1999**, *402*(1-2), 67-79.
[http://dx.doi.org/10.1016/S0003-2670(99)00532-2]

[90] Zheng, F.; Hu, B. MPTS-silica coated capillary microextraction on line hyphenated with inductively coupled plasma atomic emission spectrometry for the determination of Cu, Hg and Pb in biological samples. *Talanta,* **2007**, *73*(2), 372-379.
[http://dx.doi.org/10.1016/j.talanta.2007.03.060] [PMID: 19073042]

[91] Wu, Y.; Hu, B.; Hu, W.; Jiang, Z.; Li, B. A novel capillary microextraction on ordered mesoporous titania coating combined with electrothermal vaporization inductively coupled plasma mass spectrometry for the determination of V, Cr and Cu in environmental and biological samples. *J. Mass Spectrom.,* **2007**, *42*(4), 467-475.
[http://dx.doi.org/10.1002/jms.1179] [PMID: 17256811]

[92] Zheng, M.M.; Lin, B.; Feng, Y.Q. Hybrid organic-inorganic octyl monolithic column for in-tube solid-phase microextraction coupled to capillary high-performance liquid chromatography. *J. Chromatogr. A,* **2007**, *1164*(1-2), 48-55.
[http://dx.doi.org/10.1016/j.chroma.2007.07.020] [PMID: 17675044]

[93] Kumar, R.; Guarav, ; Kabir, A.; Furton, K.G; Malik, A. K. Development of a fabric phase sorptive extraction with high-performance liquid chromatography and ultraviolet detection method for the analyses of alkyl phenols in environmental samples. *J. Sep. Sci.,* **2015**, *0*, 1-11.

[94] Samanidou, V.; Galanopoulos, L.-D.; Kabir, A.; Furton, K.G Fast extraction of amphenicol residues from raw milk using novel fabric phase sorptive extraction followed by high-performance liquid chromatography-diode array detection. *Anal. Chim. Acta,* **2015**, *855*, 41-50.

[95] Roldan-Pijuan, M.; Lucena, R.; Cardenas, S.; Valcarcel, M.; Kabir, A.; Furton, K.G Stir fabric phase sorptive extraction for the determination of triazine herbicides in environmental water by liquid chromatography. *J. Chromatogr. A,* **2015**, *1376*, 35-45.

CHAPTER 3

Ionic Liquids in the Microextraction Context

María J. Trujillo-Rodríguez, Verónica Pino*, Juan H. Ayala and **Ana M. Afonso**

Departamento de Química, Área de Química Analítica, Universidad de La Laguna (ULL), La Laguna (Tenerife), 38206, Spain

Abstract: The structural versatility of ionic liquids (ILs), the wide range of interest properties that can present: low to negligible vapor pressure, from water-soluble to water-insoluble, from medium viscosity to high viscosity…, their synthetic tuneability, together with their impressive solvation abilities for analytes of quite different nature, make their use in microextraction techniques an obvious approach of great interest.

Thus, ILs and a group of interesting derivatives, such as polymeric ionic liquids (PILs) and IL-based surfactants, pure, mixed, or combined with other materials forming hybrid sorbents, have been used in all variants of liquid-phase and solid-based microextraction strategies.

This chapter will give an exhaustive overview of existing approaches that combine ILs and derivatives in liquid-phase microextraction (LPME) modes such as single-drop microextraction (SDME), hollow fiber liquid-phase microextraction (HF-LPME) and dispersive liquid-liquid microextraction (DLLME), including all their variants; and also in solid-based microextraction methods, such as micro-solid-phase extraction (μ-SPE), solid-phase microextraction (SPME), stir-bar sorptive extraction (SBSE), and stir-cake sorptive extraction (SCSE).

Keywords: Dispersive liquid-liquid microextraction, Environmental applications, Food applications, Hollow-fiber liquid phase-microextraction, Ionic liquids, Ionic liquid-based surfactants, Liquid-phase microextraction, Micro-solid-phase extraction, Polymeric ionic liquids, Single-drop microextraction, Solid-phase extr-

* **Corresponding author Verónica Pino:** Departamento de Química, Área de Química Analítica, Universidad de La Laguna (ULL), La Laguna (Tenerife), 38206, Spain; Tel: (+34) 922318990; Fax: (+34) 922318005; E-mail: veropino@ull.edu.es

M. Valcárcel, S. Cárdenas, R. Lucena (Eds.)

action, Solid-phase microextraction, Stir-bar sorptive extraction, Stir-cake sorptive extraction.

3.1. INTRODUCTION

3.1.1. Ionic Liquids

Ionic liquids (ILs) constitute a group of non-molecular solvents characterized for presenting melting points below 100 ºC [1 - 3] and a number of outstanding properties [4], such as negligible vapor pressure at room temperature, high chemical and electrochemical stabilities, and capability to interact with a variety of analytes through different solvation mechanisms, among others. They are mainly formed by large and asymmetric organic cations, which contain a nitrogen or phosphorus atom, associated with an inorganic or organic anion. Fig. (**1**) shows a scheme of common structures of ILs and derivatives. Those IL which retain their liquid nature at room temperature are designated as room temperature ILs (RTILs).

Relatively simple modifications of the IL structure (change in the nature of the IL anion, incorporation of a certain functional group in the IL cation…) are accompanied by dramatic modifications of their properties: from water soluble to water insoluble, from medium viscosity to high viscosity, and so on [4]. This can clearly be observed from Table **1** [5 - 7], in which the modifications of IL properties with a simple change of the anion nature are shown. As a result, ILs can be tailored for specific applications in a quite simple manner. The incorporation of a polar functional group can promote dipolar interactions between the IL and polar analytes. In the same manner, the incorporation of long aliphatic alkyl chains to an IL can enhance its extraction ability with nonpolar analytes [8].

Table 1. Physicochemical properties of a group of ILs containing the 1-butyl-3-methylimidazolium (C_4MIm^+) cation at 25 ºC (or at the specified temperature) [5 - 7].

IL anion	Abbreviation	Melting point (ºC)	Density (g·mL^{-1})	Viscosity (cP)
hexafluorophosphate	C_4MIm-PF_6	10	1.373	450

(Table 1) contd.....

IL anion	Abbreviation	Melting point (°C)	Density (g·mL^{-1})	Viscosity (cP)
tetrafluoroborate	C_4MIm-BF_4	-81	1.208	219
chloride	C_4MIm-Cl	53	-	1.7 (30 °C)
bromide	C_4MIm-Br	79	-	0.8 (30 °C)

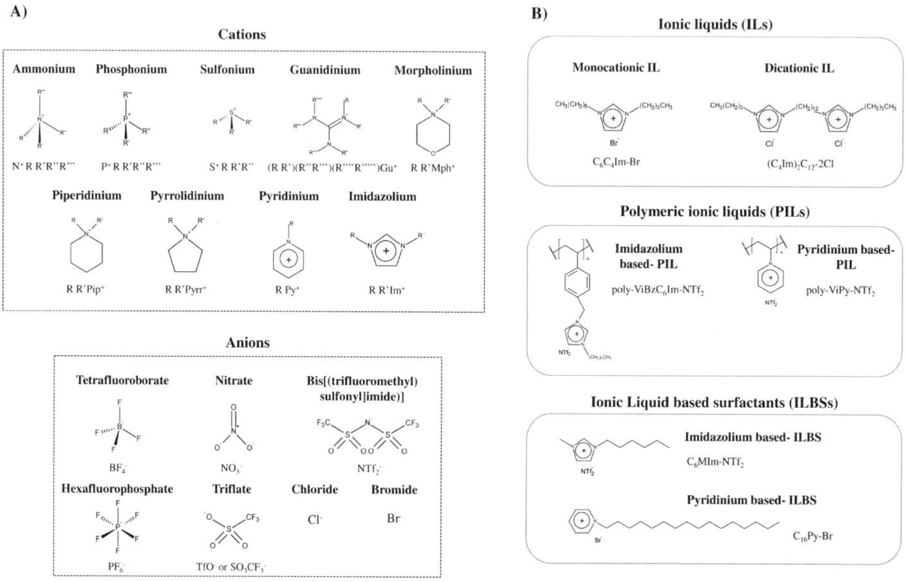

Fig. (1). A) Main cations and anions constituting the structures of ionic liquids and derivatives, and **B)** Examples of structures for ionic liquids, polymeric ionic liquids, and ionic liquids based surfactants.

It is also important to mention the impressive solvation ability of ILs, useful for polar to non-polar compounds. This is due to the electrostatic interactions associated with the cation and anion moieties comprised within the ILs, as well as their ability to undergo unique intermolecular interactions [8]. Their solvation ability has been explored with different models, being particularly successful the Abraham solvation parameter model [9].

ILs have also been subject of topic in Green Analytical Chemistry, given that they do not generate volatile organic compounds (VOCs), therefore being excellent candidates for substituting toxic chlorinated organic solvents [10]. Nevertheless, recent studies have pointed out the toxicity of certain ILs [11]. For this reason, the environmental implications of ILs have been reviewed [12 - 14]. As an important trend to decrease the possible toxicity of ILs, it has been pointed out that the structural manipulation of the pyridinium or morpholinium skeleton is the key to obtain biodegradable ILs [12].

This group of properties mentioned early and the high number of available ILs justify the enormous interest that surrounds them and, particularly, the multiple applications described for these "designer solvents" in Analytical Chemistry [8, 15 - 18].

The abbreviations used for the ILs in this chapter will follow the common trend. Thus, alkyl substituents of the IL cation will be first written showing their length *i.e.*: C_8 for octyl, C_2 for ethyl, or M for methyl; followed by the terms Im for imidazolium, Pyrr for pyrrolidinium, and Py for pyridinium; and finalizing by the anion: -PF_6 for hexafluorophosphate, -Br for bromide, -NTf_2 for bis[(trifluoromethane)sulfonyl]imide, and so on. In the case of imidazolium-based ILs, the substituent located in position 1 will be first written, followed by the substituent in position 3.

Apart from neat ILs, several ILs derivatives need to be highlighted in Analytical Chemistry, and their uses will also be described in this chapter. Polymeric ionic liquids (PILs) are defined as polymers synthesized from IL monomers as opposed to polyelectrolytes, which are synthesized from solid salt monomers [19]. IL-based surfactants are defined as surfactants derived from ILs [20], and not necessarily keep the liquid nature at temperatures below 100 ºC.

3.1.2. Microextraction Methods in Analytical Chemistry

The utilization of microextraction strategies for sample preparation in Analytical Chemistry has been one of the hottest topics in the last twenty-twenty five years [21 - 30]. The idea is simple: the utilization of low quantities of extraction solvents and/or sorbents (micro-amounts) in the analytical extraction stage, while keeping adequate analytical performances (in terms of efficiency and reproducibility) than those achieved with conventional extraction methods (which usually require macro-amounts of solvents and/or sorbents). This trend of minimization of reagents and wastes in the analytical laboratory, or the utilization of less harmful solvents and reagents, follows the green analytical chemistry rules with the aim of developing environmental-friendly methods [31 - 35].

From a simplistic point of view, those strategies that utilize liquid extracting solvents in microextraction procedures are included in the group of liquid-phase microextraction techniques, while those involving a solid sorbent are grouped as solid-based microextraction techniques.

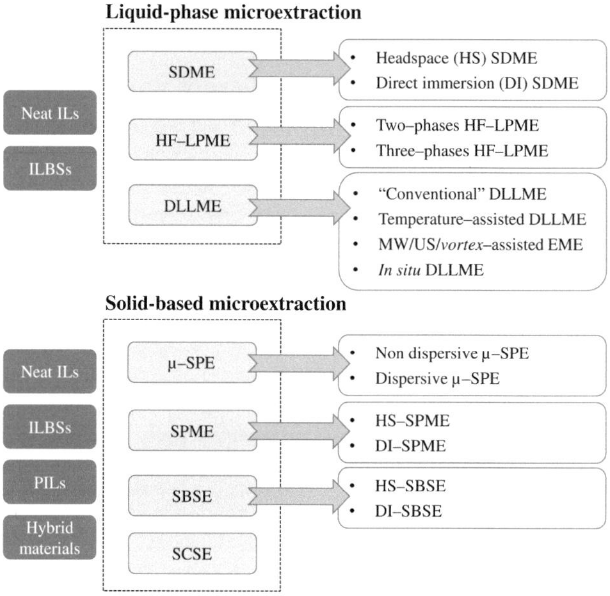

Fig. (2). Applications of ionic liquids and derivatives in different microextraction procedures: single-drop microextraction (SDME), hollow-fiber liquid-phase microextraction (HF-LPME), dispersive liquid-liquid microextraction (DLLME), micro solid-phase extraction (μ-SPE), solid-phase microextraction (SPME), stir-bar sorptive extraction (SBSE), and stir-cake sorptive extraction (SCSE). The following abbreviations have also been used in the figure: ILs for ionic liquids, PILs for polymeric ionic liquids, MW for microwaves, and US for ultrasounds.

Considering the liquid nature of ILs at mild temperatures, it results obvious their link to all liquid-phase microextraction techniques [36 - 39]. However, ILs are also found an important application area in solid-based microextraction strategies. Thus, ILs can be used as solids if working at temperatures lower than their melting points. However, it is important to highlight that IL derivatives, rather than neat ILs, are the preferred solid sorbents used in solid-based microextraction methods. Among IL derivatives, PILs altogether with ILs constituting hybrid materials (*i.e.* ILs linked to silica) need to be highlighted [40, 41]. This chapter intends to give an overview of the main applications of ILs and derivatives in microextraction methods, as it is summarized in Fig. (**2**).

3.2. ILS IN LIQUID-PHASE MICROEXTRACTION

Liquid-phase microextraction methods go beyond a simple miniaturization of the classical liquid-liquid extraction (LLE) procedure. In LLE, large amounts of both the aqueous sample (~0.5–2 L) and the extractant solvent (~200–500 mL, which is water-immiscible) are required to ensure an efficient partition of analytes, which are initially present into the aqueous sample, to the extractant solvent (normally an organic solvent). LLE also requires strong mixing between both solvents, and a timing to ensure adequate phase separation after mixing. Furthermore, cautions must be undertaken to avoid the formation of emulsions. Once the organic solvent contains the extracted analytes, it can be further subjected to stages of clean-up and/or solvent-exchange. A *priori*, it might look like that the utilization of lower amounts of both, aqueous sample and extractant solvent, following the same extraction procedure as LLE, is the basis of LPME. In fact, only one of the liquid-phase microextraction (LPME) modes, the so-called liquid-liquid microextraction (LLME), is really a micro-LLE. All remaining modes of LPME, namely single-drop microextraction (SDME), hollow-fiber liquid phase microextraction (HF-LPME), and dispersive liquid-liquid microextraction (DLLME), as main variants, do not exactly match the concept of a simple micro-LLE. As a matter of fact, all extraction methods which require micro-liters of liquid extracting solvents to accomplish the extraction strategy are included under the umbrella of LPME [23, 42, 43].

ILs have been used in all LPME modes, as it will be summarized in these sections, as a function of the microextraction mode. As a general rule when using ILs in LPME, it is important to take into account that the wavelength must be carefully selected to ensure a high absorptivity for the analytes while minimizing the absorptivity of the IL, if UV–Vis detection is utilized in combination with high–performance liquid chromatography (HPLC) or capillary electrophoresis (CE). If such selection is not possible, adequate chromatographic or electrophoretic resolution between the IL and the analytes is required. This requirement needs to be satisfied in all LPME modes involving ILs and UV detection.

3.2.1. ILs in Single Drop Microextraction

Single-drop microextraction is an LPME mode in which a single drop of extractant solvent (few microliters, ~1-15 μL) is required to accomplish the microextraction step. The technique was mainly developed by Dasgupta [44 - 47]. The simplest SDME mode utilizes the needle of a syringe to support the droplet, which is exposed to the sample solution containing analytes for a certain period of time. It can be performed in the headspace (HS-SDME) or directly in the bulk solution (DI-SDME). Once analytes suffer partition into the droplet, it is retracted in the syringe and further subjected to analytical determination. If there is not adequate compatibility between the microdroplet solvent and the analytical instrumentation, it is possible to incorporate a solvent-exchange as clean-up step. Fig. (**3A**) includes a summarizing scheme of SDME.

The main advantages of this microextraction method are: simplicity, inexistence of memory effect (a new droplet is used in every extraction), and the possibility of automation [48, 49]. Among the disadvantages are relatively long extraction times (~20-90 min), difficulties in keeping the droplet attached to the syringe during long extraction times, necessity of low stirring rates to avoid the falling of the droplet, really low droplet volumes (and thus limited extraction efficiency), and impossibility of using volatile solvents (because they can evaporate before the extraction is completed). Modifications of this conventional SDME mode imply the use of the droplet directly suspended in the sample solution, in the so-called directly suspended droplet microextraction (DSDME) [50, 51], followed in several cases by further freezing of the droplet to facilitate its sampling, the so-called solidification of the floating organic (SFO) [52].

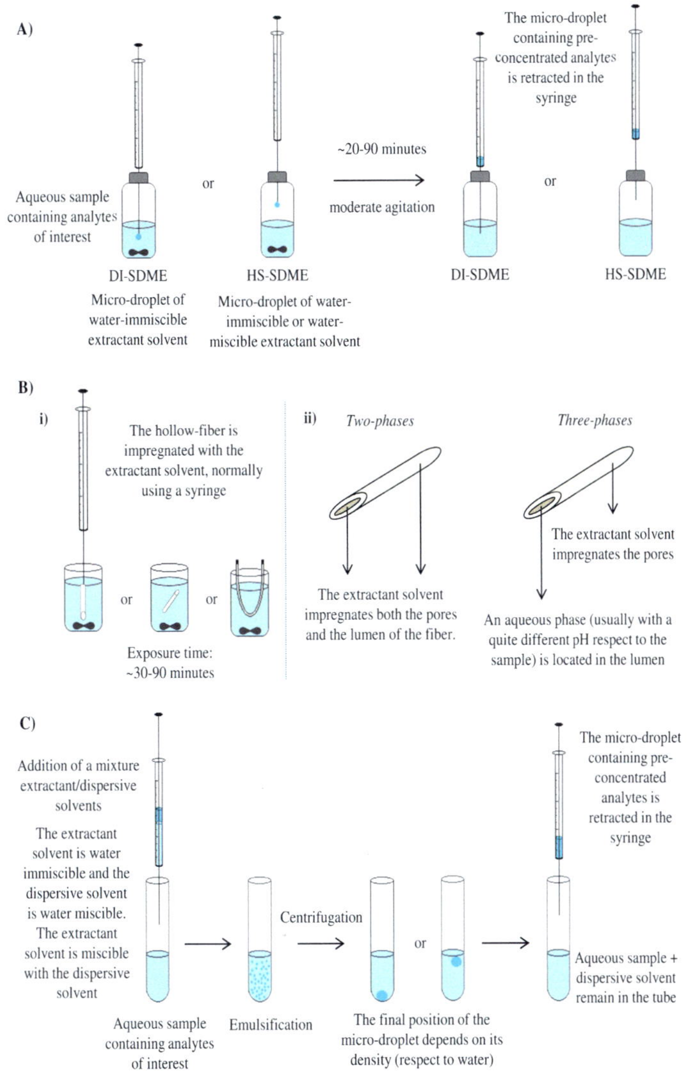

Fig. (3). Main liquid-phase microextraction modes: **A**) single-drop microextraction (SDME); **B**) hollow-fiber liquid-phase microextraction (HF-LPME) showing i) different configurations for the extraction with the HF, and ii) the fundamentals of the two-phases and three-phases HF-LPME; and **C**) conventional dispersive liquid-liquid microextraction (DLLME).

ILs have been used in practically all SDME variants, with approximately the same number of works in DI mode than in HS.

In DI-SDME, all reported applications evidently utilize water-immiscible ILs, mainly C_4MIm-PF_6 [53 - 64], but also C_6MIm-PF_6 [65 - 68] and C_8MIm-PF_6 [69, 70]. In some cases, specific highly hydrophobic ILs have been synthetized, as those containing the tris(pentafluoroethyl)trifluorophosphate (FAP) anion [71]. Recent applications describe the use of mixtures of C_4MIm-PF_6 with Pt nanoparticles (NPs) [56] or even mixtures of C_4MIm-PF_6 with nanocellulose and carboxylated multiwalled carbon nanotubes (c-MWCNTs) [53]. Microdroplet IL volumes usually range from 2 μL [56] to 20 μL [57]. Larger volumes in DI- (from 6 to 20 μL) are normally facilitated with the use of a PTFE tube as support, and using low stirring rates to avoid the falling of the droplet. Common variables used to increase the extraction efficiency in DI-SDME with ILs are the extraction time, varying from 2 min [55] to 120 min [71]; the pH if ionizable analytes are studied, and the ionic strength of the donor solution. The temperature is rarely controlled in DI-SDME with ILs [58].

The reported DI-SDME methods with ILs have been used in combination with spectrophotometry [57], atomic absorption spectrometry (AAS) [54, 61, 62, 72, 73], inductively-coupled plasma (ICP) [63], matrix-assisted laser desorption ionization (MALDI) [56], HPLC [58, 64 - 71], and CE [53, 59, 74]. Recently, a microfluidic device has also been proposed to favor the extraction efficiency using sonicated waves [55]. Interestingly, automation of the DI-SDME with ILs has been possible without any alteration on the CE instrument, utilizing IL droplet volumes of 2.40 nL [59, 60]. In this case, it is important to ensure the total solubility of the IL in the buffer solution within a short time [59, 60]. Generally, the compatibility of the final IL microdroplet containing the preconcentrated analytes with the further analytical instrument is important to be ensured. Proper dissolution of the IL droplet with certain solvents is accomplished prior to the analytical determination when such compatibility is not possible [53, 54, 57, 73].

Applications of IL-DI-SDME include the determination of organic compounds (aromatic amines, sulfonamides, anilines, phenols, carbonyl compounds, hydrocarbons...) [53, 55, 58 - 60, 64 - 66, 68, 71, 75] and metals (Cd, Cu, Co, Pb, Hg, and Mn) utilizing proper chelating agents [54, 57, 61, 62, 65, 67, 72, 73]; in a variety of aqueous samples [53, 54, 58 - 67, 69 - 73], with sample volumes varying from 1 to 20 mL.

In HS-SDME, the utilization of ILs as alternatives to organic solvents is clearly interesting not only for their low toxicity and outstanding solvation abilities, but mainly because it is minimized (if not totally avoided) the losses of extractant solvent due to evaporation. This is totally linked to IL properties, particularly their low to negligible vapor pressures and their thermal stability. Some applications can utilize water-immiscible ILs, mainly C_4MIm-PF_6 [76 - 82], C_6MIm-PF_6 [83 - 85], C_6MIm-NTf_2 [86, 87], or C_8MIm-PF_6 [74, 88 - 95]; as well as water-miscible ILs such as C_6MIm-BF_4 [75] or C_4MIm-Cl [96]. The mixture of C_6MIm-PF_6 with CdSe/ZnS quantum dots has also been shown [83]. Micelles of conventional surfactants such as sodium dodecyl sulfate (SDS) or IL-based surfactants such as C_{10}MIm-Br have also been used as micellar components (~pseudo-stationary phase) in a microdroplet of C_4MIm-Cl, showing improved extraction behavior in HS-SDME [96]. Microdroplet IL volumes usually range from 1 μL [89, 90, 92, 94, 97] to 20 μL [83]. Common variables used to increase the extraction efficiency in HS-SDME with ILs are: the extraction time, varying from 2 min [83] to 45 min [89, 94], and being an average of ~30 min; the pH in case that ionizable analytes are studied, the ionic strength of the donor solution, which is particularly helpful to decrease the solubility of analytes in the water solution forcing them to the HS; and also the temperature, varying from room temperature [84, 96] to 85 ºC [79], and being an average of ~50 ºC. The generation of volatile analytes, or the increase of their content in the headspace is sometimes favored by the dynamic nebulization extraction [74], ultrasonic nebulization [75], hydrodistillation [88], or microwaves [85].

The reported HS-SDME methods with ILs are combined with a variety of analytical techniques, such as HPLC [77, 78, 80, 81, 85, 96], AAS [79, 98], ion mobility spectrometry (IMS) [86, 87], but mainly GC [74 - 76, 82, 84, 88 - 95, 97]. Automation of the overall method has been described in several GC works [89, 90]. In any case, it is really important to ensure that the IL does not fall into the GC system, and thus several interfaces have been developed with this purpose [74, 76, 82, 84, 86 - 95]. In this way, it has been described an interface that contains a removable unit, in which a stainless steel (SS) tube packed with cotton can trap the IL while allowing desorption of the analytes. In other cases, long liners containing cotton or glass wool are utilized, altogether with a guard column.

The IL droplet is retracted back in the syringe after desorption of analytes. In any case, replacement of these systems is needed after every 5 to 15 injections.

The majority of applications of IL-HS-SDME cover the extraction of organic compounds: essential oils, personal care products, flavors, musk, drugs, amines, esters, phenols and derivatives, aromatic hydrocarbons, carbonyl compounds, chlorobenzenes, trihalomethanes... The variety of samples is much higher than that of DI-SDME applications, and covers aqueous samples [76, 80 - 82, 84, 85, 91 - 95, 98] and more complex samples such as wastewaters, fruits, fruit juices, seeds, sewages sludge, fish, wine, gasoline... The HS-SDME can be directly conducted in the samples, and in some cases in the resulting extract of very complex samples, for example after application of pressurized liquid extraction (PLE) [89].

3.2.2. ILs in Hollow-fiber Liquid-phase Microextraction

This LPME mode was mainly developed by Pedersen-Bjergaard and Rassmussen in 1999 [99], but the original idea was published by Audunsson in 1986 [100] and later improved by Thordarson in 1996 [101]. In HF-LPME, a hollow porous membrane is used as inert material to support the extractant solvent onto the sample. The hollow-fiber, normally made of polypropylene, is commonly attached to a syringe but not necessarily. Fig. (**3B**) includes a summarizing scheme of HF-LPME. It can be carried out in the HS- or in DI- mode, and also in static- or dynamic-mode.

Some important advantages of this microextraction procedure can be mentioned: the amount of extractant solvent is much higher than in SDME because the membrane is able to retain efficiently higher volumes of solvent, and thus higher extraction efficiencies can be obtained. In addition to this, HF-LPME is quite adequate for the extraction of complex samples because the membrane can act as a protective barrier *versus* deleterious sample components. Automation in HF-LPME is also possible, as it was described by Pawliszyn *et al.* in 2007 [102]. As in SDME, the main disadvantage of this method is the long extraction time required (~30-60 min).

Interestingly, HF-LPME can also be classified as two-phases HF-LPME or three-phases HF-LPME [103]. The simpler working mode is the so-called two-phases HF-LPME, in which the extractant solvent (acceptor phase) impregnates both the walls and the lumen of the HF. However, in the three-phases HF-LPME mode (also called HF-liquid-liquid-liquid microextraction), the extractant solvent is placed in the walls, and an aqueous phase is placed in the lumen. Such aqueous phase is forced to have a pH value totally different to that of the aqueous sample, to generate a gradient of pH across the supported liquid membrane. Thus, an aqueous acidic solution in the lumen would be beneficial for basic analytes from the sample, and *vice versa*, because the gradient of pH is the driving force of the transference. Evidently, the three-phase HF-LPME mode is limited to compounds presenting ionizable groups. There is also a further variant, the electromembrane extraction (EME) [104], which incorporates two electrodes, one in the sample and one in the lumen, and an external current source. This variant decreases incredibly the extraction time, because the external potential generates an electrokinetic migration.

The number of reported applications of ILs in HF-LPME is significantly lower than in SDME, despite the *a priori* advantages of HF-LPME *versus* SDME. All reported applications of ILs in HF-LPME are performed in DI-mode, and therefore the ILs impregnating the HF pores are water immiscible. Among the most utilized ILs, C_4MIm-PF_6 [105 - 110], C_6MIm-PF_6 [111 - 113] and C_8MIm-PF_6 [114 - 117] need to be highlighted. The IL C_4MIm-Cl has also been described in HF-LPME [105], but as additive of the acceptor phase (located in the lumen of the HF), being in this case 1-octylalcohol the solvent impregnating the pores of the HF.

Several applications use ILs in the two-phases HF-LPME mode [106, 108, 111, 113, 117, 118]. In these cases, HF sizes varied between ~3 cm [111, 118] and 15 cm [108], with IL acceptor phase volumes of 8 μL on average [106, 108, 111, 118], and donor (sample) volumes between 3 mL [111] to 100 mL [108]. As it happened with SDME, extraction times are long, varying from ~15 to 55 min, and even up to 8 h [117].

The applications involving ILs are HF-LPME based on the three-phases approach [105, 107, 109, 110, 112, 114, 115, 119] have utilized HF sizes varying from ~1 cm [107] to 12 cm [115, 119], internal acceptor phases volumes varying from 2.5 μL [117] to 25 μL [115], and sample volumes between 4 mL [115] and 50 mL [109]. Extraction times range between 15 min [105] and 80 min [110], and up to 8 h [115]. The pH of the acceptor phase solution in these applications has always been basic, between 10 and 13, because the organic analytes studied were of acidic nature.

The automation of the method has recently been described by Nitiyanontakit *et al.* [119] using a specific device. Authors employ the IL $MC_8C_8C_8N$-Cl in the three-phases mode, a HF of ~12 cm (~34 μL of acceptor phase), and 1,5-diphenylcarbazide as chromogenic probe for the determination of Cr (VI). Their system clearly decreases the extraction time to only 5 min.

Basheer *et al.* have also described the use of a HF to support the IL $C_4MIm-PF_6$ but the resulting method resembles more a liquid-liquid-liquid microextraction approach than a conventional HF-LPME method in the three-phases mode. Authors require toluene (8 μL) for the back extraction, which is a disadvantage, given the toxicity [107].

The electromembrane extraction using ILs have also been reported [116], specifically for the determination of one dye followed by off-line elution assisted by ultrasounds, and spectrophotometric determination.

The reported HF-LPME methods with ILs have been utilized in combination with AAS [108, 109, 111, 112] and ICP for the determination of metals [113], with GC for the determination of BTEX [106] and hydrocarbons [107], and mainly with HPLC for the determination of UV filters [118], chlorophenols [114], sulfonamides [115], hydrocarbons [105], bisphenol-A and other endocrine disrupting chemicals [117], and dyes [110]. It should be taken into account that the determination of metals always requires the use of chelating reagents [108, 109, 111 - 113], and in some cases EDTA in the acceptor phase when working in the three-phases mode [109, 112]. Proper elution with solvents of the final acceptor phase (IL or not) should be ensured before carrying out the analytical determination.

3.2.3. ILs in Dispersive Liquid-liquid Microextraction

Rezaee *et al.* [120] were the first to develop the DLLME method. This distinct LPME mode performs with relatively higher amounts of extractant solvent (20-250 μL) if compared to SDME and HF-LPME. It is based on the dispersion of the extractant solvent into the aqueous solution (~4-20 mL) containing analytes. In the most conventional mode, such dispersion is possible with the aid of a dispersive solvent (~0.1-2 mL). Thus, the ternary mixing of components: water (containing analytes), water miscible dispersive solvent, and water immiscible extractant solvent (but soluble in the dispersive solvent) is accompanied by an emulsion, in which multitude of microdroplets of extractant solvent are formed in the bulk aqueous solution. The procedure is then accompanied by centrifugation to separate properly the microdroplet of extractant solvent containing the preconcentrated analytes from the aqueous solution. The high surface area between the donor phase (the aqueous sample) and the acceptor phase (the extractant solvent) ensure a high extraction efficiency of analytes. Fig. (**3C**) summarizes the fundamentals of DLLME.

The method is quite fast (~5 min for the overall procedure), particularly if compared to SDME or HF-LPME. Recent works also support the possibility of automation [121]. As a disadvantage of conventional DLLME, the dispersive solvent can partially solubilize the extractant solvent into the aqueous sample, and also can increase the solubility of the analytes in water, which in turn can decrease the overall extraction efficiency. Recent modifications of the DLLME method include the substitution of the dispersive solvent by other approaches to ensure the dispersion of the extractant solvent, such as the utilization of vortex, in the so-called vortex-assisted emulsification microextraction (VAEME) [122].

DLLME is undoubtedly the LPME mode with higher number of applications and variations when incorporating ILs as extractant solvents, with an overall number of literature works of ~150. The main uses of ILs and derivatives in combination with DLLME have been quite reviewed recently by Trujillo-Rodríguez *et al.* [39]. Following the classification established in such review, it is possible to distinguish four main categories: 1) Conventional IL-DLLME; 2) Temperature-assisted IL-DLLME; 3) Ultrasound-assisted, Microwave-assisted, or Vortex-assisted IL emulsification microextraction; and 4) *In-situ* IL-DLLME.

3.2.3.1. Conventional IL-DLLME

This variant is the one that more closely resembles the conventional DLLME using organic solvents. It does not need any energy requirements other than the simple ternary mixture of components: aqueous sample containing analytes, water immiscible IL as extractant, and a dispersive solvent. Centrifugation is applied afterwards to separate the IL microdroplet containing the extracted and preconcentrated analytes.

Since the original work of Liu *et al.* [122], a high number of works have followed this approach for the determination of organic compounds [39, 124 - 128] and metals [39, 129 - 131].

Interestingly, the centrifugation step, which is *a priori* necessary in this mode, has been even eliminated with an in-syringe setup [132].

3.2.3.2. Temperature-Assisted IL-DLLME

This variant takes advantage of the dramatic influence that the temperature exerts on the solubility and viscosity of ILs. Thus, heating (~50-90 ºC) is used to favor the dispersion of the water immiscible IL on the sample solution (aiding or not such dispersion with a dispersive solvent), followed by cooling to ensure an increase of the viscosity and insolubility of the IL, and therefore facilitating its further separation and handling. The main disadvantage of this extraction mode is that it requires long extraction times to ensure adequate dispersion of the IL when heating, sometimes even up to 30 min. This is some way against the most interesting feature of DLLME: its rapidity compared with other LPME modes.

The first report of the technique was by Zhou *et al.* [133] for the determination of organic compounds, and by Baghdadi and Shemirani [134] for the determination of metals. Since that, a variety of works have appeared in the literature regarding its use for organic compounds [39, 135 - 137] and metals [39, 138 - 140].

3.2.3.3. Ultrasound-Assisted, Microwave-Assisted, or Vortex-Assisted IL Emulsification Microextraction

This particular mode utilizes ultrasounds, microwaves, or vortex as an auxiliary energy for ensuring the dispersion of the water immiscible IL through the sample solution (aiding or not such dispersion with a dispersive solvent, but normally requiring lower amounts of them if compared with conventional IL-DLLME). If the application of such auxiliary energy is accompanied by an increase of the solution temperature, the IL not only suffers dispersion but also certain solubilization. Afterwards, the solution can be subjected to centrifugation or cooling to separate the IL droplet containing the preconcentrated analytes.

The utilization of ultrasounds in combination with IL-DLLME was first described by Zhou *et al.* [141] for the determination of aromatic amines in water, and by Li *et al.* [142] for the determination of Cd in water, in 2009. The ultrasound-assisted IL-DLLME has been widely utilized for determination of both organic compounds [39, 143 - 147] and metals [39, 148 - 150]. In any case, cautions must be taken because the utilization of ultrasounds can cause certain degradation of the analytes due to the influence of the ultrasonic energy [21].

The utilization of microwaves was first reported by Xu *et al.* [151] and further applied in a relatively low number of reports [152, 153], probably due to the inherent complication of the instrument while only the dispersion of the IL is intended.

The utilization of vortex-assisted IL-DLLME is more recent, being reported almost at the same time by Asensio-Ramos *et al.* [154] and Ye *et al.* [155], and since that detailed in a high number of applications [39, 156 - 159]. Among the advantages of the application of vortex it must be highlighted the simplicity of the technique while avoiding the reaching of high temperatures, as it can easily happen with ultrasounds or microwaves.

Ku *et al.* have also reported the use of an up-and down-shaker as a stirring device [160], as a simpler alternative to microwaves, ultrasound, or even vortex. Wu *et al.* have reported the use of manual shaking (30 times in 20 s) to avoid other agitation methods [161].

3.2.3.4. In-situ IL-DLLME

This mode requires a hydrophilic IL or IL-based surfactants as extractant solvent, which is obviously soluble in the aqueous sample. Once analytes have interacted with the IL, an anion-exchange reagent is added to promote a metathesis reaction, and the hydrophilic IL (or IL-based surfactant) is transformed into a hydrophobic IL (or IL-based surfactant), which settles due to insolubilization (as a microdroplet containing the analytes). The method generally requires the utilization of vortex, microwaves, ultrasound, or shaking to improve the kinetics of the metathesis reaction. It also requires further centrifugation or even cooling to separate the droplet containing the preconcentrated analytes.

The first description of the method was by Baghdadi and Shemirani for metals in [162] and by Yao and Anderson for organics [163] in 2009. Applications of this variant in the literature have increased since that for a variety of analytes and samples [39, 164 - 168].

3.2.3.5. Trends and Interest of IL-DLLME Methods

As some interesting features of ILs in all DLLME modes, it is important to mention that hydrophilic ILs have also been used as dispersive solvents, and thus organic solvents are completely eliminated in the overall procedure [144, 169 - 172]. In other cases, surfactants have been employed with those purposes [153, 173].

As it has been mentioned early for all LPME modes, proper compatibility of the IL containing analytes and the analytical technique need to be also ensured when using ILs (or IL-based surfactant) in DLLME.

An important characteristic of the IL-DLLME method is the fact that it can be applied directly to aqueous liquid samples, but also to solid samples if they have been previously extracted and then reconstituted with water (water which can contain minor amounts of alcohols or acetonitrile). This evidently highly increases the applicability of the technique to a wide variety of samples, because solid samples can be treated with other methods, and the solid extract can afterwards experience a preconcentration with IL-DLLME [174]. Furthermore, complex solid samples can be also extracted with aqueous solutions of IL-based surfactants, with the aid of microwaves or ultrasounds, followed by the *in situ* IL-DLLME mode in the IL-based surfactant extract. The overall method can therefore performs in complete absence of organic solvents [20, 175, 176], and at the same time being able to deal with solid samples.

Another interesting advantage is the fact that the IL-DLLME method can be performed with derivatization reactions (even simultaneously) [127, 151 - 153, 177, 178].

Main features of the majority of reported IL-DLME methods have been covered in detail by Trujillo-Rodríguez *et al.* [39]. Consequently, the compatibility of these methods with further analytical techniques, the analytical performance of the reported applications (in terms of efficiency, reproducibility and sensitivity), the nature of the specific applications, the nature and amount of IL required in any IL-DLLME sub-mode, the extraction time and so on, will not be described in this chapter.

3.3. ILs IN SOLID-BASED MICROEXTRACTION

Solid-based microextraction methods also go beyond a simple miniaturization of the solid-phase extraction (SPE) procedure, which would be the μ-SPE mode [179]. Other interesting solid-based microextraction approaches include solid-phase microextraction (SPME) [180, 181], stir-bar sorptive extraction (SBSE) [27], and stir-cake sorptive extraction (SCSE) [182]. The performance of each solid based microextraction mode is summarized in Fig. (**4**).

All these techniques are assisted by the aid of a solid stationary phase: the sorbent material. The following sections will describe the applications of ILs as solid sorbent materials (working at temperatures below their respective melting points), and also ILs derivatives like PILs, in all solid-based microextraction modes: μ-SPE, SPME, SBSE and SCSE.

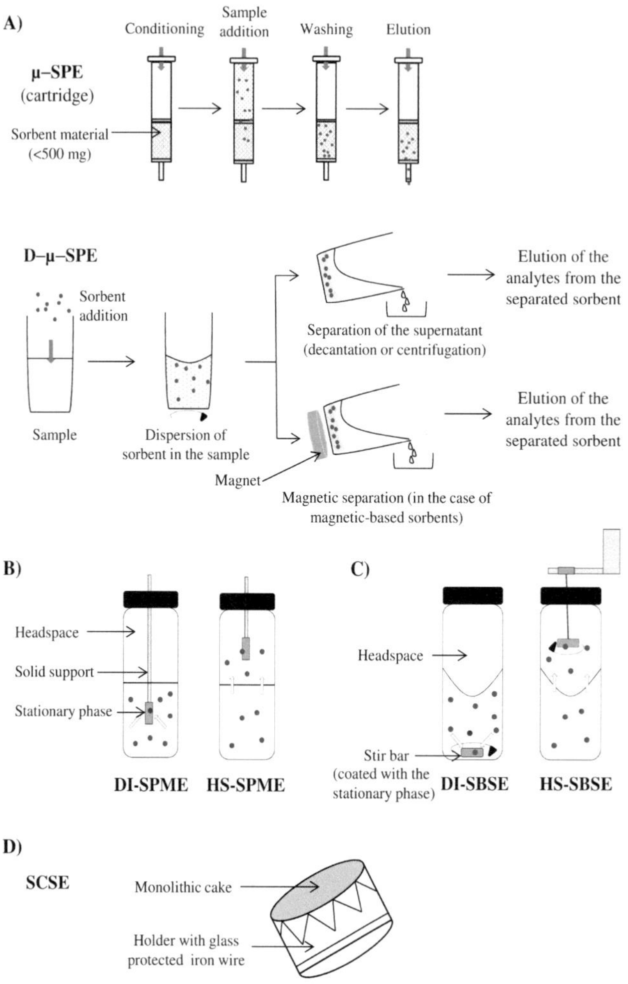

Fig. (4). Main solid-based microextraction modes: **A**) micro-solid-phase extraction (μ-SPE), **B**) solid-phase microextraction (SPME), **C**) stir-bar sorptive extraction (SBSE), and **D**) stir-cake sorptive extraction (SCSE).

3.3.1. ILs in Micro-solid-phase Extraction

Micro-solid-phase extraction (μ-SPE) is a method that utilizes exactly the same principles of SPE: the partitioning of the analytes between the sample and a stationary phase which is a solid sorbent. The main difference respect to the classical SPE is that μ-SPE minimizes both the solid sorbent and sample amount requirements during the extraction. Thus, common SPE procedures are carried out using 3 mL cartridges, which allow ~500 mg of stationary phase and ~100 mg of matrix sample [183]. µ-SPE uses lower amounts than those mentioned before. In this sense, it is a simpler technique, less costly and less time-consuming than conventional SPE [180]. Besides, as only a relative small amount of sorbent is used, less volume of solvent will also be needed to desorb the analytes.

Two main modes can be distinguished in µ-SPE: dispersive (D-µ-SPE) and non-dispersive, as it is shown in Fig. (**4A**). In the non-dispersive µ-SPE mode, the sorbent material remains static during the procedure whereas in D-µ-SPE, the solid sorbent is dispersed into the matrix sample. Such dispersion can occur with the aid of ultrasounds, vortex or any stirring method.

In order to extend the use of µ-SPE, new materials have been developed recently to increase the selectivity and affinity for different types of analytes [184]. ILs and its derivatives have also been studied extensively as new sorbent materials in µ-SPE, alone or in combination with other materials.

It is important to emphasize that not all the authors utilize the term µ-SPE. However, in this chapter this term will be used for all SPE applications with ILs (or derivatives) and ILs forming hybrid materials that require lower amounts that those of conventional SPE.

3.3.1.1. ILs in Non-dispersive µ-SPE

This µ-SPE procedure is exactly a miniaturized version of the conventional SPE method, using a static solid sorbent as extractant material. Classical materials employed as solid sorbents are silica or polymers [184]. ILs and its derivatives have been utilized to modify these classical materials with the purpose of increasing both the selectivity and the efficiency of the overall method. Thus, two types of IL-based sorbent materials are described in the literature: (1) IL-modified silica sorbents, and (2) IL-modified polymers. Many works on this topic have been summarized in a review of Vidal *et al.* in 2012 [184].

Tian *et al.* were the first to utilize IL-modified silica sorbents in µ-SPE for the extraction of tanshinones from a medicinal plant in 2009 [185]. The preparation of the sorbent needed the prior activation of the commercial silica to enhance the content of silanol groups on its surface, followed by the functionalization with a trimethoxysilane IL derivate. The new sorbent provided higher recovery values than traditional silica-based SPE cartridges.

Many works have utilized IL-modified silica sorbents since this first application [184, 186 - 192]. The applications generally utilized imidazolium-based ILs [184 - 191], but quaternary ammonium-based ILs have also been tested [192].

Nevertheless, the utilization of silica has some disadvantages such as limited pH stability. Polymers can be adequate substitutes of silica to solve these problems. Fontanals *et al.* were the first to report the modification of polymers with ILs in µ-SPE, in 2009 [193]. Authors integrated the IL MIm-CF_3COO inside the structure of the polymer divinylbenzene (DVB), and evaluated the performance of the novel sorbent in diverse modes. The IL-modified polymer was suitable to be utilized in strong anion exchange SPE mode.

Since then, a number of works has developed new IL-modified polymers for µ-SPE [184, 194 - 199]. The majority of the ILs employed for the preparation of the sorbent contains imidazolium cations with different substituents [184, 193 - 197, 199], or they belong to the pyridinium family [198, 200]. The main interest of including ILs in the polymers is the fact that they are capable to modify and enhance the interactions established between the analytes and the final sorbent phase.

The majority of IL-modified polymers are synthetized by the free radical polymerization of the monomer (or monomers) and the IL [184, 193, 196 - 198, 200] In many cases, other modifiers such as porogen agents or cross-linkers are also required.

IL-based molecular imprinted polymers (IL-MIPs) have also been utilized as sorbents in µ-SPE [184, 194, 195, 199, 201]. The preparation of a MIP requires a template, which is the organic molecule that generates recognition sites in the final polymer for capturing the analytes of interest task-specifically [202]. The most common procedure to prepare an IL-MIP follows this strategy: a blank polymer is first prepared by the co-polymerization of functional monomers and cross-linkers. Then, the blank polymer reacts with the template and the IL, and, finally, the template is removed of the IL-MIP by elution. It has been observed that the presence of ILs in the MIP enhances the recognition capabilities and the selectivity of the final polymer.

Other μ-SPE works with IL-modified polymer sorbents include sol-gel materials doped with C_6Py-PF_6 [198] and IL-composite mesoporous polymers [197].

The working procedure for μ-SPE with IL-based sorbents starts by placing the sorbent in an empty SPE cartridge of polypropylene, connected in some cases to a SPE manifold with an entry to a vacuum pump. The amounts of IL-modified sorbent required in the described methods are between 40 and 500 mg. Other steps in the μ-SPE method such as drying, elution... do not differ significantly from those of conventional SPE.

The μ-SPE technique has been used in combination with HPLC-UV in the majority of the reported applications [184, 186, 188 - 191, 200]. Applications range from the determination of organics [184 - 186, 188, 189, 191, 193, 195 - 197, 200] to inorganics [187, 190, 198] in a variety of samples of quite different nature: medicinal plants, waters, foods and biological fluids.

3.3.1.2. ILs in Dispersive μ-SPE

Dispersive μ-SPE requires the dispersion of the sorbent into the matrix sample, instead of using a sorbent packed in a static SPE cartridge in which the sample goes through. It improves the contact area between the sorbent and the sample, accelerating the mass transferring of analytes between the two phases, while providing higher extraction efficiencies and lower extraction times than non-dispersive μ-SPE. The efficient dispersion of the sorbent material can be conducted with the aid of vortex, ultrasounds, or any strong stirring method.

IL-modified silica, PILs-modified silica, as well as a number of hybrid materials containing ILs have been utilized in the D-μ-SPE mode. The utilization of magnetic nanoparticles (MNPs) in combination with D-μ-SPE has shown to be particularly helpful.

Among IL-modified silica sorbents [203 - 206], it is interesting to mention the use of IL-based periodic mesoporous organosilica (PMOs) sorbents. PMOs are synthetized from bridge organosilica precursors ($(R'O)_3$-Si-R-Si- $(OR')_3$) [207].

The organic group R can be an IL, being an integral part of the mesoporous walls.

Microspheres of IL-MIPs have also been prepared for D-μ-SPE using phenylephrine as the template and AlC_2Im-Br as co-functional monomer [208].

PILs-modified silica has also been employed in D-μ-SPE, in the form of neat microspheres of cross-linked PILs [209].

The use of D-μ-SPE assisted by MNPs is lately a hot topic in analytical chemistry [210]. The nano-size of the MNPs leads to a higher superficial surface area for the SPE extraction favoring the partitioning of the analytes. Besides, their magnetic properties simplify the procedure due to the possibility of applying an external magnetic field (a strong magnet) for the separation of the sorbent material, avoiding in this way tedious centrifugation and separation steps, and minimizing the overall sample preparation time. Furthermore, MNPs can be chemically modified in a simple manner with other materials.

IL-MNPs have been used successfully in D-μ-SPE approaches, specifically with magnetite (Fe_3O_4) MNPs modified with ILs (Fe_3O_4@ILs) [204, 211 - 213], or Fe_3O_4 MNPs covered with a layer of silica before being modified with ILs (Fe_3O_4@SiO_2@IL) [214, 215]. The layer of silica improves the stability, even in acidic media, of the MNPs, avoiding their tendency to form aggregates. Furthermore, it has been described the use of the IL C_7MIm-PF_6 coated onto a Fe_3O_4-grafted graphene nanocomposite, and utilized as sorbent in the extraction of nitrobenzenes [216]. MWCNTs magnetic nanocomposites have also been used in D-μ-SPE [217, 218].

PILs-MNPs have also been utilized for D-μ-SPE applications. The PILs form *core–shell* structures in which the PIL is the *shell*, acting as the stationary phase of the sorbent. The magnetic *core* can be Fe_3O_4 MNPs coated with oleic acid (Fe_3O_4@oleic acid) [219] or Fe_3O_4 MNPs coated with silica (Fe_3O_4@SiO_2) [220, 221].

In addition, IL-based surfactants have been utilized in combination with MNPs. The electrostatic interactions between the head group of the IL-based surfactants and the oppositely charged group of the MNP surface leads the formation of mixed hemimicelles or admicelles (depending on the IL-based surfactant concentration). IL-based surfactant mixed hemimicelles have been utilized as sorbents in D-μ-SPE, using as solid supports: Fe_3O_4 MNPs [222 - 225], TiO_2 MNPs [226], or magnetic MWCNTs@SiO_2 [227].

IL-based surfactant can also be employed in D-μ-SPE using a non-magnetic support such as silica gel [228].

The amount of IL-based sorbent needed in these methods ranges between 4 and 500 mg. HPLC-UV is the selected separation technique in the majority of applications of organic compounds [186, 189, 192, 197, 201, 208, 212, 216, 219 - 221, 229 - 231], whereas FAAS [187, 198, 213, 226] is the most employed technique when inorganic species are extracted.

3.3.2. ILs in Solid-phase Microextraction

Solid-phase microextraction (SPME) is a non-exhaustive extraction and preconcentration technique introduced by Pawliszyn *et al.* in 1989 [48]. This method is based on the utilization of a stationary phase (a sorbent) linked to a solid substrate in the form of a fiber, typically of ~1 cm long and being the sorbent coating thickness between ~1 and 100 μm. The extraction and preconcentration of the analytes is generally achieved when the sorbent coating is exposed to the sample, either by direct immersion (DI-SPME) [232] or headspace (HS-SPME) [233]. Then, analytes which partitioned to the coating fiber can be easily desorbed in an analytical instrument by exposing the fiber at high temperatures if GC is used, or by dipping the fiber in low amounts of an organic solvent if HPLC or CE is used. Fig. (**4B**) shows a scheme of the overall procedure.

SPME offers many advantages over other sample preparation procedures such as simplicity, easiness of automation, portable, ability to reach high preconcentration factors and, more important, it is solvent-free [234], which is quite advisable from an environmental point of view. As disadvantages, fibers are in some way somehow fragile, they can present carry-over, and only seven types of sorbents are commercially available. Efforts in last years have therefore been shifted to the development of new sorbent materials as adequate SPME coatings [28].

ILs can be used effectively to prepare SPME fibers due to their unique physicochemical properties: high thermal stability, negligible vapor pressure and variable viscosity. In addition, the tuneability of the ILs, which refers to the ability to incorporate easily different substituents into the IL structure, could enhance specific interactions between the coating and the analytes [235]. PILs have also been explored as sorbent materials in SPME. Thereby, several types of IL- and PIL- based sorbent coatings have been described in the SPME literature.

This chapter will distinguish them according to the mode in which they are attached to the solid substrate, as: (1) sorbent coatings prepared by physical dip-coating, in which the IL (or the PIL) is retained onto the solid support due to physical interactions like adsorption, absorption, electrostatic forces..., and (2) sorbent coatings chemically bonded to the fiber support. Besides, other SPME works describe the combined use of ILs or PILs with other materials, such as polymers or carbon nanotubes, to form hybrid sorbent coatings. Table **2** [236 - 248] includes representative examples of each type of IL/PIL-based sorbent coating utilized in SPME according to this classification.

3.3.2.1. IL-Based Sorbent Coatings

3.3.2.1.1. IL-Based Sorbent Coating Physically Attached to the Fiber Support

The first application involving the utilization of an IL in SPME was carried out by Liu *et al.* in 2005 [236]. They utilized the IL C_8MIm-PF_6 in HS mode to extract several VOCs in paints, in combination with GC. The IL was retained in the fiber support during the extraction due to its high viscosity. However, after every run, the sorbent coating was lost and so it was not reusable.

Since this application, other works focused on the modification of the fiber support previous to the dip-coating of the IL. These modifications involve sequences of pretreatment steps of the support, looking for an improvement in the stability and reusability of the IL SPME fiber. Thus, Hsieh *et al.* bonded a bare silica fiber with Nafion prior to the deposition of three different ILs: C_8MIm-CF_3CO, C_4MIm-CF_3CO and PhC_3MIm-CF_3CO (being $-PhC_3$ the phenylpropyl group) [237]. The Nafion support enhanced the surface interactions between the ILs and the silica support. This stabilization provided reusable fibers

Table 2. Representative examples for applications of ILs and ILs derivatives in solid-phase microextraction (SPME).

Sorbent coating type	Stationary phase material/s	Solid support	Supporting method & polymerization comments if applicable	Analytical application	Ref.
IL physically attached to the support	C_8MIm-PF_6	fused silica fiber	dip-coating	VOCs in paints by HS-SPME-GC-FID	[236]
	C_8MIm-CF_3CO C_4MIm-CF_3CO PhC_3MIm-CF_3CO	derivatized fused silica fiber	coating with Nafion and dip-coating of the IL	PAHs in water by HS-SPME-GC-MS	[237]
	C_4MIm-PF_6 C_8MIm-PF_6	derivatized fused silica fiber	etching with ammonium hydrogen difluoride and dip-coating of the IL	PAHs in mosquito coil smoke by HS-SPME--C-FID	[238]
IL chemically bonded to the support	TPMIm-NTf_2	fused silica fiber	refluxing reaction	MTBE[a] in gasoline by HS-SPME-GC-MS	[239]
	AlMIm-PF_6–OH-TSO AlMIm-NTf_2 –OH-TSO	-	sol-gel technique	phenolic estrogens and aromatic amines in water by DI-SPME-GC-FID	[240]
PIL physically attached to the support	poly-ViC_{16}Im-NTf_2	fused silica fiber	thermal free-radical polymerization with AIBN of the IL monomer, metathesis anion exchange reaction and dip-coating	PAHs and phenols in water by DI-SPME--C-MS	[241]
	binary mixtures of poly-ViC_6Im-NTf_2, poly-ViC_{16}Im-NTf_2 , poly-ViC_6Im-Cl, or poly-ViC_{16}Im-Cl	fused silica fiber	thermal free-radical polymerization with AIBN of the IL monomer, metathesis anion exchange reaction and dip-coating	organic compounds in aqueous solutions by HS-SPME-GC-FID	[242]
PIL chemically bonded to the support	poly-ViC_8Im-PF_6	derivatized SS wire	coating with a single microstructured Ag layer by the Ag mirror reaction and free radical polymerization of the IL monomer with AIBN	organic compounds in aqueous solutions by HS or DI-SPME-GC-FID	[243]

(Table 2) contd.....

Sorbent coating type	Stationary phase material/s	Solid support	Supporting method & polymerization comments if applicable	Analytical application	Ref.
Cross-linked PIL chemically bonded to the support	(*) IL monomer: $ViC_{16}Im-NTf_2$ IL cross-linker: $(ViBzIm)_2C_{12}-2NTf_2$	fused silica fiber	thermal free-radical co-polymerization with AIBN and dip-coating of the PIL	coffee aroma by HS-SPME-GC-MS	[244]
	(*) IL monomer: $ViBzC_{16}Im-NTf_2$ IL cross-linker: $(ViBzIm)_2C_{12}-2NTf_2$	fused silica fiber	UV *in situ* free radical co-polymerization with DAROCUR 1173	PCBs[b] from ocean water & bovine milk by HS-SPME-GC-ECD /MS	[245]
	IL monomer: $ViC_9OHIm-Br$[c] IL cross-linker: $(ViIm)_2C_4-2Br$	derivatized and anodized Ti wire	anodization of Ti wire in two steps in ethylene glycol electrolyte, functionalization with a coupling agent and free-radical co-polymerization with AIBN	alcohols and volatile fatty acids in aqueous solutions by HS-SPME-GC-FID	[246]
Hybrid sorbent coating	MOF: IRMOF-3 IL: $C_4MIm-PF_6$ Polymer: PDMS	SS wire	*in situ* growth of the MOF on SS wire and protection with a IL and PDMS layer	PAHs in water by HS-SPME-GC-MS	[247]
	MWCNTs Cross-linked PIL: -IL monomer: $ViC_4Im-NTf_2$ -IL cross-linker: $(ViIm)_2C_{12}-2NTf_2$	fused silica fiber	thermal free-radical co-polymerization with AIBN in the presence of MWCNTs and dip-coating of the PIL	PAHs in water by HS-SPME-GC-FID	[248]

(*) selected as the best sorbent coating among others.

[a] methyl-*tert*-butyl-ether.

[b] polychlorinated biphenyls.

[c] being $-C_9OH$ the nonanol group.

(up to ~50 extractions). In a recent work, a mixture of Nafion and methyltrioctylammonium thiosalicylate was dipped onto a PTFE tube or a fused silica fiber, and it was employed for the determination of Hg by DI-SPME [249].

Huang *et al.* developed IL-based SPME fibers in which fused-silica capillaries were etched with ammonium hydrogen difluoride prior to the coating with $C_4MIm-PF_6$ or $C_8MIm-PF_6$ [238]. The obtained fibers showed better extraction efficiencies for polycyclic aromatic hydrocarbons (PAHs) when compared with Nafion precoated capillary fibers. In another application, for the determination of acetone in human plasma, the fused silica capillary was subjected to a sodium chloride pretreatment, followed by the dip-coating with selected ILs [250]. It has also been described the use of a fused silica fiber impregnated with a polymer, which was formed by a silicone elastomer cross-linked to the IL $EeMIm-NTf_2$ (being –Ee the ethoxyethyl group) [251]. This SPME fiber was utilized in the determination of amphetamines from human urine samples.

3.3.2.1.2. IL-Based Sorbent Coating Chemically Bonded to the Fiber Support

The obtaining of reusable IL-based SPME fibers with a higher lifetime can be accomplished by chemically bonding the IL to the fiber support. Thereby, Amini *et al.* have developed a SPME fiber in which the IL $TPMIm-NTf_2$ (being –TP the trimethoxysilyl propyl group) was chemically bonded to a fused silica fiber [239]. In this approach, the bonding between the IL and the silica was acquired by a refluxing reaction under nitrogen atmosphere. The obtained fiber was used up to 16 times while the analogous physically-coated $TPMIm-NTf_2$ fiber was not reusable.

IL-based coatings derived from sol-gel techniques have also been developed to obtain more thermal stable and resistant SPME coatings [41]. Malik *et al.* were pioneers in developing the sol-gel technology, based on the incorporation of organic components into inorganic polymers structures [252]. Thus, $AlMIm-PF_6$ and $AlMIm-NTf_2$ (being –Al the allyl group) have been chemically bonded to a derivatized fused silica fiber *via* a sol-gel free-radical cross-linking reaction, with the aid of γ-methacryloxypropyltrimethoxysilane as coupling agent [240]. In this case, the IL acts as the stationary phase of the sorbent coating, being the final stationary phase a sol-gel IL-hydroxy terminated silicone oil coating. Similar sol-gel coatings have been developed since this original work [253 - 256]. It is important to note that this kind of material prepared by sol-gel, in which the IL got finally bonded to silica, involves other inorganic components for its preparation, and so it could also be categorized as hybrid material.

3.3.2.2. PIL-Based Sorbent Coatings

PILs are synthetized from IL monomers. Two different reactions are normally involved during their synthesis: (1) a free radical polymerization of the IL monomer precursor; and (2) an anion exchange reaction which involves anion interchange. Depending on the order of these two reactions, two different synthetic routes can be distinguished [19].

Recently, a new type of PIL based on the co-polymerizaction of two different ILs has been described [40]. The resultant polymer, termed cross-linked PIL, is normally formed by a monocationic IL monomer cross-linked with a dicationic IL [40], entailing a higher active surface in the final PIL.

In comparison with ILs, PILs possess higher viscosity, thermal stability and mechanical strength [257]. Thus, PILs offer many advantages for their utilization as sorbent coatings in SPME, especially in GC applications when high temperatures are required to desorb the analytes from the fiber. In addition, the integrity of the fiber is ensured when working in DI-SPME mode.

3.3.2.2.1. PIL-Based Sorbent Coating Physically Attached to the Fiber Support

Zhao *et al.*, from Anderson's group, were the first to explore the utility of PILs as SPME sorbent coatings in 2008 [258]. Authors studied three imidazolium-based PILs, poly-ViC_6Im-NTf_2, poly-$ViC_{12}Im$-NTf_2, and poly-$ViC_{16}Im$-NTf_2, for the extraction of esters and fatty acid methyl esters from water and wines using HS-SPME-GC-FID. Since then, several works have reported the utilization of other PILs in HS-SPME, normally employing monocationic imidazolium IL monomers [242, 259 - 266]. López-Darias *et al.* also demonstrated the utility of PILs fibers in DI-SPME [241], and since then, other works have appeared in the literature [267, 268]. PILs based on dicationic imidazolium IL monomers have also been studied as SPME sorbent coatings, specifically poly-$(ViIm)_2C_4$-$2NTf_2$ and poly-$(ViIm)_2C_{12}$-$2NTf_2$ [269]. Better stabilities have also been described for PILs if the fused silica fiber is previously etched with ammonium hydrogen difluoride [270].

The most common synthesis of PILs involves a free radical polymerization of the corresponding IL monomer using 2,2′-azobis(2-methylpropionitrile) (AIBN) as radical initiator [241, 242, 258 - 269]. After the polymerization reaction, the next step is the dip-coating of the PIL onto the fiber support. Generally, the fiber supported is dipped into a solution of the PIL. Fused silica fiber is the most common support employed for this purpose, but its surface must be previously free of the polyimide polymer [241, 242, 258 - 269].

It is worthy of mentioning that Graham *et al.* (also from Anderson's group) have explored the utilization of mixtures of PILs as SPME sorbent coatings [242]. These PILs, based on imidazolium cations with NTf_2^- or Cl^- as anions, were combined in different ratios previous to the dip-coating of the silica fiber. Authors demonstrated the utility of the PIL-based sorbent coatings to extract selectively *n*-alcohols through the modification of the amount of the counter anions present in the fiber, thus pointing out the influence of anions in the performance of the final PIL for SPME [262].

3.3.2.2.2. PIL-Based Sorbent Coating Chemically Bonded to the Fiber Support

With regards to PIL-based sorbent coatings chemically bonded to the fiber support, it is important to distinguish between neat PILs and cross-linked PILs.

Neat PILs have been chemically immobilized onto derivatizated stainless steel (SS) wire [243, 271 - 273]. The derivatization of the SS can be accomplished with different procedures: coating with a Si interlayer followed by hydroxylation and mercaptopropyl-functionalization [271]; coating with Au as interlayer, followed by a mercaptopropyl-functionalization and re-coating with a silica layer [272]; coating with a single microstructured Ag layer by the silver mirror reaction [243], or coating with the same microstructured Ag layer followed by 1,8-octanedithio- -functionalization [244]. In the latter application, a new type of PIL-based sorbent coating, termed as double-confined PIL coatings, was prepared by the co-polymerization of the cation and anion of the IL ViC_8Im-*p*StS (being *p*StS the anion *p*-styrenesulfonate) on the derivatized SS.

The majority of works in the last years have been shifted to the preparation of cross-linked PIL-based sorbent coatings chemically bonded to fused silica [244, 245, 259, 261, 274]. The co-polymerization of the monocationic IL monomer and the dicationic IL cross-linker is normally conducted *via* a thermal free radical reaction assisted by AIBN [244, 246, 261, 275]. More recently, a UV free radical co-polymerization initiated with 2-hydroxy-2-methylpropiophenone (DAROCOUR 1173) have been utilized [245, 259, 261, 274]. In the latter applications involving DAROCOUR 117, all reactions needed to prepare the SPME fiber: the derivatization of the silica support, the co-polymerization, and the chemical bonding between the cross-linked PIL and the fiber, are conducted in a single step and onto the fiber support. Thus, the procedure is termed as UV *in situ* co-polymerization. In addition to fused silica fibers, other metallic fiber supports have been tested to prepare cross-linked PIL-based sorbent coatings. Thus, microstructured Ag coated on SS wire [275] and anodized Ti wire [246] have been described as proper metallic supports for chemically bonded cross-linked PILs in SPME.

3.3.2.3. Hybrid ILs/PILs Sorbent Coatings

IL/PILs-based hybrid sorbent coatings refer to SPME fibers in which the stationary phase is formed by an IL (or PIL) and another material. The combination of the unique properties of the ILs (or PILs) and the advantages derived from the utilization of other materials such as polymers of quite different nature, carbon nanotubes (CNTs), or even silica, among others, create stable and resistant SPME fibers with unique performance.

Section 3.2.1.2 described the utilization of sol-gel based sorbent coatings, in which the own IL was the main extractant component of the stationary phase. However, ILs can also serve as co-solvents (or porogenic agents) of the stationary phase during the sol-gel reaction. Thus, their presence enhances the porous surface of the coating and improves the final stability of the material. Shearrow *et al.* utilized this methodology to prepare sorbent coatings for *in-tube* SPME applications [276]. During the synthesis, two different ILs based on phosphonium and pyridinium cations ($C_6C_6C_6C_{14}P$-BF_4 and C_4(4-MPy)-BF_4, respectively) were combined with four different polymers to generate the hybrid material, and finally immobilized onto a fused silica fiber. Similar works based on this sol-gel procedure are able to generate IL-based hybrid organic-inorganic co-polymer coatings [277].

Recent works have described the combination of ILs with metal-organic frameworks (MOFs) for the preparation of SPME fibers [247, 278]. In the reported applications, the IL ($C_6MIm-BF_4$) acted in one case as the porogenic solvent to generate a hybrid MOF-polymeric monolithic SPME fiber [278]. In another application, the IL (together with polydimethysyloxane) formed a protection layer of the MOF, which was grown on a SS wire [247].

Hybrid sorbents composed by dicationic ILs and PILs bonded onto silica and derivatized silica particles have also been reported [279]. The materials had layers of 50 μm and were glued onto flexible supports of wire.

In addition, IL-polyaniline (PANI) composites have been tested as sorbent materials in SPME [280, 281]. These composites were coated onto a Pt fiber by electrochemical deposition. Under these conditions, authors were able to obtain highly controllable film thicknesses.

Cross-linked PILs have also been combined with multi-walled carbon nanotubes (MWCNTs) and supported onto a fused silica fiber [248] or onto a SS wire modified with a Ag layer [282]. The MWCNTs could be dispersed in the PIL phase, resulting in a significant increase in the contact surface area analyte-resulting stationary phase.

A cross-linked PIL has also been prepared using an IL monomer co-polymerized with methylacryloyl-substituted polyhedral oligomeric silsesquioxane (POSS) as cross-linker [248]. The developed PIL-POSS phase was chemically bonded onto a Ti wire as solid support.

3.3.2.4. Analytical Applications

There are a large number of current works involving the utilization of IL and PILs in SPME for analytical determinations, as it has been covered by recent reviews [28, 40, 41, 235].

Thus, it has been covered the determination of a number of organic compounds including PAHs [238, 241 - 243, 245, 267, 268, 271, 278], phenols [241 - 243, 259, 268, 271], *n*-alcohols [242, 246, 247, 262, 274, 275], VOCs [236, 260], BTEX [242, 243], aliphatic hydrocarbons [264], fatty acid methyl esters [258,

164], among many others. These analytes can be extracted from food, environmental, pharmaceutical or biological samples [40] in HS- mode [236 - 238, 242 - 248, 251, 258, 260, 262 - 266, 269, 270, 273 - 275] but also in DI-mode [241, 243, 248, 249, 254, 267, 268, 271 - 274, 282]. Multiple headspace SPME mode has also been described for PILs [283]. These applications combine the SPME technique mainly with GC [236 - 238, 241 - 248, 251, 254, 258 - 275, 283, 284] and HPLC [282, 285].

Although the majority of the applications utilizes *in fiber*-SPME, there are some works which report the utilization of ILs in *in tube*-SPME [276, 277, 286]. Recently, a SPME in membrane method has been developed [285]. The membrane was synthesized *via* a surface chain transfer reaction using ViC_4Im-Br and methylmethacrylate. Bianchi *et al.* recently utilized PILs-based sorbent coatings contained in homemade devices (necklaces and pins) for the *in situ* monitoring of ketamine in recreational places [269].

Many of the reported applications of ILs/PILs SPME works compare the extraction performance of the sorbent coatings with commercially available SPME fibers such as polyacrylate (PA) [241, 244, 261, 268, 266, 270], polydimethylsyloxane (PDMS) [241, 245, 248, 258, 260, 267 - 270, 272], divinylbencil-polydimethylsyloxane (DVB-PDMS) [260, 261, 269] or carboxen-polydimethylsyloxane (CAR-PDMS) [259, 260, 265], to prove the superior performance of the novel materials despite low coating thicknesses achieved in the in-lab synthetic approaches.

Finally, it is also worthy to mention that applications of ILs in SPME not only cover their use as sorbent material, but also as desorption solvents [287], with superior ability to that of common organic solvents used in the off–line desorption step of SPME.

3.3.3. ILs in Stir Bar Sorptive Extraction

Stir bar sorptive extraction (SBSE) is another non-exhaustive technique introduced by Baltussen *et al.* in 1999 [288]. SBSE is based on the same principles that SPME: the partitioning of the analytes between the sample matrix and a stationary phase. The difference between the two techniques lies in the solid support in which the stationary phase is retained. Thus, SBSE employs stir bars of ~10 – 20 mm covered with the sorbent phase, while SPME uses fibers. Besides, the stir bars are coated with a higher amount of stationary phase, generally 20 to 125 μL of PDMS (in SPME, the maximum volume of commercially PDMS coated onto the fiber is ~0.5 μL) [289].This increased amount of sorbent allows better extraction performance and enrichment factors 100 times higher than that of SPME [290]. It takes longer times for analytes to reach the equilibrium state in SBSE if compared to SPME. Nevertheless, it is also possible to work under non-equilibrium conditions if the extraction time is accurately controlled. The stir bars commercially available are formed by a magnetic rod encapsulated in a glass jacket on which the PDMS phase is placed [289]. However, a drawback of SBSE is that only PDMS, ethylene glycol (EG)-silicone, and PA, are currently the commercially available coatings, thus limiting their extraction efficiency toward polar and less polar compounds. Furthermore, complete automation is not possible with SBSE. Another disadvantage is the presence of carry–over (or memory effect) during the desorption step. Fig. (**4C**) includes a scheme of the SBSE procedure. As SPME, SBSE can be used in a DI– mode, introducing the stir bar in the sample solution, but also in HS-. In the last extraction mode, special devices to hold the stir bar are available [291]. Desorption of analytes retained on the stir bars can also be accomplished at high temperatures (in a GC injector) or with proper solvents using off-line desorption (for HPLC or CE) [289].

To the best of our knowledge, there is not any work which uses ILs, PILs, or hybrid materials as sorbent for SBSE. There is one SBSE application which employs the IL $C_8MIm\text{-}BF_4$ as modifier of the desorption solvent, methanol [290]. The proposed SBSE method was effectively applied to the determination of carvedilol from human serum samples, coupled with HPLC–UV using PDMS and poly-methyl-methacrylate/ethylenglycol–dimethacrylate as sorbents. Authors affirm that the utilization of the IL during the desorption step favors the elimination of the carry–over effect of the stir bar.

3.3.4. ILs in Stir Cake Sorptive Extraction

Stir cake sorptive extraction (SCSE) is a recent technique introduced by Huang *et al.* in 2011 [182]. The method is based on the utilization of a monolithic cake of sorbent as stationary phase. The monolithic cake is placed in a homemade glass holder with an iron wire which provides the possibility of the application of magnetic stirring during the extraction (Fig. **4D**), in a procedure that resembles SBSE.

SCSE presents advantages over the SBSE because the stationary phase does not touch the bottom of the sample vessel [182]. It eliminates the friction between the vessel and the sorbent coating. Besides, the protection of the extraction phase commonly permits the utilization of higher stirring rates. As a result, the monolithic cake can be used up to ~100–300 extractions (~1000 h) whereas the stir bars can be damaged with the vessels and they can only be used for ~20–60 extractions (~200 h) [182, 292, 293]. In addition, monoliths are microporous polymer materials with a high specific surface area. They are formed by the co–polymerization of a monomer mixture with a porogen solvent. Their utilization to form the sorbent cake provides a simple operational procedure and high extractive capacity, while being low cost [182].

Poly–vinylimidazole-divinylbenzene has been used as main monolithic cake in SCSE determinations. In order to obtain new polymeric materials, PILs have also been utilized [294, 295]. Thereby, the IL ViBzMIm–Cl (being –ViBz the vinylbenzyl group) has been co–polymerized with divinylbenzene in the presence of dimethylsulfoxide as porogenic agent to obtain a monolithic cake [294]. The resulting sorbent material was able to extract polar aromatic amines and apolar parabens. Furthermore, the obtained sorbent is able to participate in anionic-exchange processes. Another PIL-based monolithic cake has been prepared by the co–polymerization of ethylene dimethacrylate and the IL AlMIm-Cl, in the presence of dimethylformamide as porogenic solvent [295]. The sorbent material was successfully utilized in the determination of inorganic anions present in water samples.

CONCLUSION AND TRENDS

Microextraction techniques have benefited enormously from the utilization of ILs and derivatives as novel extractant solvents or sorbents. The impressive solvation abilities of ILs, their tuneability making even simple to obtain task-specific ILs, together with their negligible vapor pressures at room temperature, among many other interesting physicochemical properties, are the main reasons contributing to this success. There are a number of IL derivatives, such as IL-based surfactants and PILs, also presenting high outstanding properties. These IL derivatives have extended the applicability of neat ILs in microextraction procedures. In this sense, ILs, IL-based surfactants and PILs, as well as hybrid materials incorporating them, are expected to continue expanding their applications in both liquid-phase microextraction and solid-based microextraction techniques in Analytical Chemistry. Particularly, the employment of hybrid materials, including silica, carbon nanotubes, molecular imprinted-polymers, and magnetic-based materials, is a field that which we believe will continue to flourish in combination with microextraction methods.

CONFLICT OF INTEREST

The authors confirm that they have no conflict of interest to declare for this publication.

ACKNOWLEDGEMENTS

V.P. thanks the Spanish Ministry of Economy and Competitiveness (MINECO) for the Ramón y Cajal contract with the University of La Laguna (ULL). M.J.T.-R. acknowledges Fundación CajaCanarias for her PhD fellowship. J.H.A. and V.P. thank funding from Fundación CajaCanarias ref. project SPDs-AGUA05. V. Pino also would like to acknowledge project ref. MAT2013-43101-R from MINECO.

LIST OF ABBREVIATIONS

AAS	atomic absorption spectrometry
AIBN 2	2′-azobis(2-methylpropionitrile)
CAR-PDMS	carboxen-polydimethylsyloxane
CE	capillary electrophoresis
c-MWCNTs	carboxylated multiwalled carbon nanotubes
CNTs	carbon nanotubes
D-μ-SPE	dispersive micro-solid-phase extraction

DAROCOUR 1173	2-hydroxy-2-methylpropiophenone
DI-SDME	direct immersion single-drop microextraction
DI-SPME	direct immersion solid-phase microextraction
DLLME	dispersive liquid-liquid microextraction
DSDME	directly suspended droplet microextraction
DVB	divinylbenzene
DVB-PDMS	divinylbencil-polydimethylsyloxane
ECD	electron capture detection
EG	ethylene glycol
FID	flame-ionization detector
GC	gas chromatography
HF-LPME	hollow fiber liquid-phase microextraction
HS-SDME	headspace single-drop microextraction
HS-SPME	headspace solid-phase microextraction
ICP	inductively coupled plasma
IL-MIPs	ionic liquid-based molecular imprinted polymers
ILs	ionic liquids
IMS	ion mobility spectrometry
LLE	liquid-liquid extraction
LLME	liquid-liquid microextraction
LPME	liquid-phase microextraction
MALDI	matrix-assisted laser desorption ionization
MNPs	magnetic nanoparticles
MOFs	metal-organic frameworks
MS	mass spectrometry
MWCNTs	multi-walled carbon nanotubes
PA	polyacrylate
PAHs	polycyclic aromatic hydrocarbons
PANI	polyaniline
PCBs	polychlorinated biphenyls
PDMS	polydimethylsyloxane
PILs	polymeric ionic liquids
PLE	pressurized liquid extraction
PMOs	periodic mesoporous organosilica

POSS	polyhedral oligomeric silsesquioxane
RTILs	room temperature ionic liquids
SBSE	stir-bar sorptive extraction
SCSE	stir-cake sorptive extraction
SDME	single-drop microextraction
SFO	solidification of the floating organic
SPE	solid-phase extraction
µ-SPE	micro-solid-phase extraction
SPME	solid-phase microextraction
SS	stainless steel
UV	ultraviolet
VAEME	vortex-assisted emulsification microextraction
VOCs	volatile organic compounds

REFERENCES

[1] Welton, T. Room-temperature ionic liquids. Solvents for synthesis and catalysis. *Chem. Rev.,* **1999**, *99*(8), 2071-2084.
[http://dx.doi.org/10.1021/cr980032t] [PMID: 11849019]

[2] Rogers, R.D.; Seddon, K.R. Chemistry. Ionic liquids solvents of the future? *Science,* **2003**, *302*(5646), 792-793.
[http://dx.doi.org/10.1126/science.1090313] [PMID: 14593156]

[3] Hallett, J.P.; Welton, T. Room-temperature ionic liquids: solvents for synthesis and catalysis. 2. *Chem. Rev.,* **2011**, *111*(5), 3508-3576.
[http://dx.doi.org/10.1021/cr1003248] [PMID: 21469639]

[4] Huddleston, J.G.; Visser, A.E.; Reichert, W.M.; Willauer, H.D.; Broker, G.A.; Rogers, R.D. Characterization and comparison of hydrophilic and hydrophobic room temperature ionic liquids incorporating the imidazolium cation. *Green Chem.,* **2001**, *3*, 156-164.
[http://dx.doi.org/10.1039/b103275p]

[5] Liu, J.F.; Jonsson, J.A.; Jiang, G.B. Application of ionic liquids in analytical chemistry. *Trends Analyt. Chem.,* **2005**, *24*, 20-27.
[http://dx.doi.org/10.1016/j.trac.2004.09.005]

[6] Poole, C.F.; Poole, S.K. Extraction of organic compounds with room temperature ionic liquids. *J. Chromatogr. A,* **2010**, *1217*(16), 2268-2286.
[http://dx.doi.org/10.1016/j.chroma.2009.09.011] [PMID: 19766228]

[7] Ruiz-Ángel, M.J.; Berthod, A. Reversed phase liquid chromatography of alkyl-imidazolium ionic liquids. *J. Chromatogr. A,* **2006**, *1113*(1-2), 101-108.
[http://dx.doi.org/10.1016/j.chroma.2006.01.124] [PMID: 16487533]

[8] Ho, T.D.; Zhang, C.; Hantao, L.W.; Anderson, J.L. Ionic liquids in analytical chemistry: fundamentals, advances, and perspectives. *Anal. Chem.,* **2014**, *86*(1), 262-285. [http://dx.doi.org/10.1021/ac4035554] [PMID: 24205989]

[9] Anderson, J.L.; Ding, J.; Welton, T.; Armstrong, D.W. Characterizing ionic liquids on the basis of multiple solvation interactions. *J. Am. Chem. Soc.,* **2002**, *124*(47), 14247-14254. [http://dx.doi.org/10.1021/ja028156h] [PMID: 12440924]

[10] Davis, J.H. Task-specific ionic liquids. *Chem. Lett.,* **2004**, *33*, 1072-1077. [http://dx.doi.org/10.1246/cl.2004.1072]

[11] Swatloski, R.P.; Holbrey, J.D.; Rogers, R.D. Ionic liquids are not always green: hydrolysis of 1-buty--3-methylimidazolium hexafluorophosphate. *Green Chem.,* **2003**, *5*, 361-363. [http://dx.doi.org/10.1039/b304400a]

[12] Pham, T.P.; Cho, C.W.; Yun, Y.S. Environmental fate and toxicity of ionic liquids: a review. *Water Res.,* **2010**, *44*(2), 352-372. [http://dx.doi.org/10.1016/j.watres.2009.09.030] [PMID: 19854462]

[13] Cevasco, G.; Chiappe, C. Are ionic liquids a proper solution to current environmental challenges? *Green Chem.,* **2014**, *16*, 2375-2385. [http://dx.doi.org/10.1039/c3gc42096e]

[14] Welton, T. Ionic liquids in green chemistry. *Green Chem.,* **2011**, *13*, 225. [http://dx.doi.org/10.1039/c0gc90047h]

[15] Tan, Z.; Liu, J.; Pang, L. Advances in analytical chemistry using the unique properties of ionic liquids. *Trends Analyt. Chem.,* **2013**, *39*, 218-227. [http://dx.doi.org/10.1016/j.trac.2012.06.005]

[16] Joshi, M.D.; Anderson, J.L. Recent advances of ionic liquids in separation science and mass spectrometry. *Advances,* **2012**, *2*, 5470-5484.

[17] Sun, P.; Armstrong, D.W. Ionic liquids in analytical chemistry. *Anal. Chim. Acta,* **2010**, *661*(1), 1-16. [http://dx.doi.org/10.1016/j.aca.2009.12.007] [PMID: 20113709]

[18] Berthod, A.; Ruiz-Ángel, M.J.; Carda-Broch, S. Ionic liquids in separation techniques. *J. Chromatogr. A,* **2008**, *1184*(1-2), 6-18. [http://dx.doi.org/10.1016/j.chroma.2007.11.109] [PMID: 18155711]

[19] Mecerreyes, D. Polymeric ionic liquids: Broadening the properties and applications of polyelect-rolytes. *Prog. Polym. Sci.,* **2011**, *36*, 1629-1648. [http://dx.doi.org/10.1016/j.progpolymsci.2011.05.007]

[20] Pino, V.; Germán-Hernández, M.; Martín-Pérez, A.; Anderson, J.L. Ionic liquid-based surfactants in separation science. *Sep. Sci. Technol.,* **2012**, *47*, 264-276. [http://dx.doi.org/10.1080/01496395.2011.620589]

[21] Leong, M.I.; Fuh, M.R.; Huang, S.D. Beyond dispersive liquid-liquid microextraction. *J. Chromatogr. A,* **2014**, *1335*, 2-14. [http://dx.doi.org/10.1016/j.chroma.2014.02.021] [PMID: 24582396]

[22] Spietelun, A.; Marcinkowski, Ł.; de la Guardia, M.; Namieśnik, J. Green aspects, developments and

perspectives of liquid phase microextraction techniques. *Talanta,* **2014**, *119*, 34-45. [http://dx.doi.org/10.1016/j.talanta.2013.10.050] [PMID: 24401382]

[23] Prosen, H. Applications of liquid-phase microextraction in the sample preparation of environmental solid samples. *Molecules,* **2014**, *19*(5), 6776-6808. [http://dx.doi.org/10.3390/molecules19056776] [PMID: 24858267]

[24] Kocúrová, L.; Balogh, I.S.; Andruch, V. Solvent microextraction: A review of recent efforts at automation. *Microchem. J.,* **2013**, *110*, 599-607. [http://dx.doi.org/10.1016/j.microc.2013.07.009]

[25] Huang, Z.; Lee, H.K. Materials-based approaches to minimizing solvent usage in analytical sample preparation. *Trends Analyt. Chem.,* **2012**, *39*, 228-244. [http://dx.doi.org/10.1016/j.trac.2012.05.007]

[26] Deng, J.; Yang, Y.; Wang, X.; Luan, T. Strategies for coupling solid-phase microextraction with mass spectrometry. *Trends Analyt. Chem.,* **2014**, *55*, 55-67. [http://dx.doi.org/10.1016/j.trac.2013.12.004]

[27] He, M.; Chen, B.; Hu, B. Recent developments in stir bar sorptive extraction. *Anal. Bioanal. Chem.,* **2014**, *406*(8), 2001-2026. [http://dx.doi.org/10.1007/s00216-013-7395-y] [PMID: 24136250]

[28] Xu, J.; Zheng, J.; Tian, J.; Zhu, F.; Zeng, F.; Su, C.; Ouyang, G. New materials in solid-phase microextraction. *Trends Analyt. Chem.,* **2013**, *47*, 68-83. [http://dx.doi.org/10.1016/j.trac.2013.02.012]

[29] Vuckovic, D. High-throughput solid-phase microextraction in multi-well-plate format. *Trends Analyt. Chem.,* **2013**, *45*, 136-153. [http://dx.doi.org/10.1016/j.trac.2013.01.004]

[30] Silva, E.A.; Risticevic, S.; Pawliszyn, J. Recent trends in SPME concerning sorbent materials, configurations and *in vivo* applications. *Trends Analyt. Chem.,* **2013**, *43*, 24-36. [http://dx.doi.org/10.1016/j.trac.2012.10.006]

[31] Gałuszka, A.; Migaszewski, Z.; Namieśnik, J. The 12 principles of green analytical chemistry and the SIGNIFICANCE mnemonic of green analytical practices. *Trends Analyt. Chem.,* **2013**, *50*, 78-84. [http://dx.doi.org/10.1016/j.trac.2013.04.010]

[32] Płotka, J.; Tobiszewski, M.; Sulej, A.M.; Kupska, M.; Górecki, T.; Namieśnik, J. Green chromatography. *J. Chromatogr. A,* **2013**, *1307*, 1-20. [http://dx.doi.org/10.1016/j.chroma.2013.07.099] [PMID: 23932374]

[33] Li, Y.; Fabiano-Tixier, A.S.; Vian, M.A.; Chemat, F. Solvent-free microwave extraction of bioactive compounds provides a tool for green analytical chemistry. *Trends Analyt. Chem.,* **2013**, *47*, 1-11. [http://dx.doi.org/10.1016/j.trac.2013.02.007]

[34] Farré, M.; Pérez, S.; Gonçalves, C.; Alpendurada, M.F.; Barcelo, D. Green analytical chemistry in the determination of organic pollutants in the aquatic environment. *Trends Analyt. Chem.,* **2010**, *29*, 1347-1362. [http://dx.doi.org/10.1016/j.trac.2010.07.016]

[35] Tobiszewski, M.; Mechlińska, A.; Namieśnik, J. Green analytical chemistry theory and practice.

Chem. Soc. Rev., **2010**, *39*(8), 2869-2878.
[http://dx.doi.org/10.1039/b926439f] [PMID: 20502819]

[36] Zhang, P.; Hu, L.; Lu, R.; Zhou, W.; Gao, H. Application of ionic liquids for liquid–liquid microextraction. *Anal. Methods,* **2013**, *5*, 5376-5385.
[http://dx.doi.org/10.1039/c3ay40597d]

[37] Ruiz-Aceituno, L.; Sanz, M.L.; Ramos, L. Use of ionic liquids in analytical sample preparation of organic compounds from food and environmental samples. *Trends Analyt. Chem.,* **2013**, *43*, 121-145.
[http://dx.doi.org/10.1016/j.trac.2012.12.006]

[38] Han, D.; Tang, B.; Lee, Y.R.; Row, K.H. Application of ionic liquid in liquid phase microextraction technology. *J. Sep. Sci.,* **2012**, *35*(21), 2949-2961.
[http://dx.doi.org/10.1002/jssc.201200486] [PMID: 23002004]

[39] Trujillo-Rodríguez, M.J.; Rocío-Bautista, P.; Pino, V.; Afonso, A.M. Ionic liquids in dispersive liquid-liquid microextraction. *Trends Analyt. Chem.,* **2013**, *51*, 87-106.
[http://dx.doi.org/10.1016/j.trac.2013.06.008]

[40] Yu, H.; Ho, T.D.; Anderson, J.L. Ionic liquid and polymeric ionic liquid coatings in solid-phase microextraction. *Trends Analyt. Chem.,* **2013**, *45*, 219-232.
[http://dx.doi.org/10.1016/j.trac.2012.10.016]

[41] Ho, T.D.; Canestraro, A.J.; Anderson, J.L. Ionic liquids in solid-phase microextraction: a review. *Anal. Chim. Acta,* **2011**, *695*(1-2), 18-43.
[http://dx.doi.org/10.1016/j.aca.2011.03.034] [PMID: 21601027]

[42] Pena-Pereira, F.; Lavilla, I.; Bendicho, C. Liquid-phase microextraction techniques within the framework of green chemistry. *Trends Analyt. Chem.,* **2010**, *29*, 617-628.
[http://dx.doi.org/10.1016/j.trac.2010.02.016]

[43] Sarafraz Yazdi, A.; Eshaghi, Z. Surfactant enhanced liquid-phase microextraction of basic drugs of abuse in hair combined with high performance liquid chromatography. *J. Chromatogr. A,* **2005**, *1094*(1-2), 1-8.
[http://dx.doi.org/10.1016/j.chroma.2005.07.110] [PMID: 16257282]

[44] Jeannot, M.A.; Cantwell, F.F. Solvent microextraction into a single drop. *Anal. Chem.,* **1996**, *68*(13), 2236-2240.
[http://dx.doi.org/10.1021/ac960042z] [PMID: 21619310]

[45] Liu, H.; Dasgupta, P.K. Analytical chemistry in a drop. Solvent extraction in a microdrop. *Anal. Chem.,* **1996**, *68*(11), 1817-1821.
[http://dx.doi.org/10.1021/ac960145h] [PMID: 21619093]

[46] Liu, S.; Dasgupta, P.K. Liquid droplet. A renewable gas sampling interface. *Anal. Chem.,* **1995**, *67*, 2042-2049.
[http://dx.doi.org/10.1021/ac00109a023]

[47] Cardoso, A.A.; Dasgupta, P.K. Analytical chemistry in a liquid film/droplet. *Anal. Chem.,* **1995**, *67*, 2562-2566.
[http://dx.doi.org/10.1021/ac00111a011]

[48] Arthur, C.L.; Pawliszyn, J. Solid-phase microextraction with thermal desorption using fused silica

optical fibers. *Anal. Chem.,* **1990**, *62*, 2145-2148.
[http://dx.doi.org/10.1021/ac00218a019]

[49] Wu, J.; Xiang, B.; Xia, J. Application of ultrasound-assisted emulsification-microextraction combined with high performance liquid chromatography to the determination of propoxur in environmental and beverage samples. *Mikrochim. Acta,* **2009**, *166*, 157-162.
[http://dx.doi.org/10.1007/s00604-009-0179-6]

[50] Jain, A.; Verma, K.K. Recent advances in applications of single-drop microextraction: a review. *Anal. Chim. Acta,* **2011**, *706*(1), 37-65.
[http://dx.doi.org/10.1016/j.aca.2011.08.022] [PMID: 21995911]

[51] Cuadros-Rodríguez, L.; García-Campaña, A.M.; Jiménez-Linares, C.; Roman-Ceba, M. Estimation of performance characteristics of an analytical method using the data set of the calibration experiment. *Anal. Lett.,* **1993**, *26*, 1243-1258.
[http://dx.doi.org/10.1080/00032719308019900]

[52] Khalili Zanjani, M.R.; Yamini, Y.; Shariati, S.; Jönsson, J.A. A new liquid-phase microextraction method based on solidification of floating organic drop. *Anal. Chim. Acta,* **2007**, *585*(2), 286-293.
[http://dx.doi.org/10.1016/j.aca.2006.12.049] [PMID: 17386676]

[53] Ruiz-Palomero, C.; Soriano, M.L.; Valcárcel, M. Ternary composites of nanocellulose, carbonanotubes and ionic liquids as new extractants for direct immersion single drop microextraction. *Talanta,* **2014**, *125*, 72-77.
[http://dx.doi.org/10.1016/j.talanta.2014.02.055] [PMID: 24840417]

[54] Wen, X.; Deng, Q.; Wang, J.; Yang, S.; Zhao, X. A new coupling of ionic liquid based-single drop microextraction with tungsten coil electrothermal atomic absorption spectrometry. *Spectroc. Acta Pt. A-Molec. Biomolec. Spectr.,* **2013**, *105*, 320-325.
[http://dx.doi.org/10.1016/j.saa.2012.12.040]

[55] Zhang, A.; Zha, Y. Single-drop liquid phase microextraction accelerated by surface acoustic wave. *J. Sep. Sci.,* **2013**, *36*(6), 1085-1089.
[http://dx.doi.org/10.1002/jssc.201200973] [PMID: 23355443]

[56] Ahmad, F.; Wu, H.F. Characterization of pathogenic bacteria using ionic liquid *via* single drop microextraction combined with MALDI-TOF MS. *Analyst (Lond.),* **2011**, *136*(19), 4020-4027.
[http://dx.doi.org/10.1039/c1an15350a] [PMID: 21804984]

[57] Wen, X.; Deng, Q.; Guo, J. Ionic liquid-based single drop microextraction of ultra-trace copper in food and water samples before spectrophotometric determination. *Spectroc. Acta Pt. A-Molec. Biomolec. Spectr,* **2011**, *79*, 1941-1945.

[58] Wu, C.; Chen, D.; Deng, H.; Liu, Y.; Zhou, A. Determination of anilines in water samples using ionic liquid-based single drop microextraction coupled with high performance liquid chromatography. **2011**.
[http://dx.doi.org/10.1109/icbbe.2011.5780773]

[59] Wang, Q.; Qiu, H.; Li, J.; Han, H.; Liu, X.; Jiang, S. Novel approach to improve the detection of colchicine *via* online coupling of ionic liquid-based single-drop microextraction with capillary electrophoresis. *J. Sep. Sci.,* **2011**, *34*(5), 594-600.
[http://dx.doi.org/10.1002/jssc.201000686] [PMID: 21268261]

[60] Wang, Q.; Qiu, H.; Li, J.; Liu, X.; Jiang, S. On-line coupling of ionic liquid-based single-drop microextraction with capillary electrophoresis for sensitive detection of phenols. *J. Chromatogr. A,* **2010**, *1217*(33), 5434-5439. [http://dx.doi.org/10.1016/j.chroma.2010.06.059] [PMID: 20621300]

[61] Manzoori, J.L.; Amjadi, M.; Abulhassani, J. Ultra-trace determination of lead in water and food samples by using ionic liquid-based single drop microextraction-electrothermal atomic absorption spectrometry. *Anal. Chim. Acta,* **2009**, *644*(1-2), 48-52. [http://dx.doi.org/10.1016/j.aca.2009.04.029] [PMID: 19463561]

[62] Manzoori, J.L.; Amjadi, M.; Abulhassani, J. Ionic liquid-based single drop microextraction combined with electrothermal atomic absorption spectrometry for the determination of manganese in water samples. *Talanta,* **2009**, *77*(4), 1539-1544. [http://dx.doi.org/10.1016/j.talanta.2008.09.045] [PMID: 19084676]

[63] Xia, L.; Li, X.; Wu, Y.; Hu, B.; Chen, R. Ionic liquids based single drop microextraction combined with electrothermal vaporization inductively coupled plasma mass spectrometry for determination of Co, Hg and Pb in biological and environmental samples. *Spectroc. Acta Pt. B-Atom. Spectr.,* **2008**, *63*, 1290-1296.

[64] Zhou, Q.X.; Xiao, J.P.; Ye, C.L.; Wang, X.M. Enhancement of sensitivity for determination of phenols in environmental water samples by single-drop liquid phase microextraction using ionic liquid prior to HPLC. *Chin. Chem. Lett.,* **2006**, *17*, 1073-1076.

[65] Sarafraz-Yazdi, A.; Mofazzeli, F. Ionic liquid-based submerged single drop microextraction: a new method for the determination of aromatic amines in environmental water samples. *Chromatographia,* **2010**, *72*, 867-873. [http://dx.doi.org/10.1365/s10337-010-1731-6]

[66] Vidal, L.; Chisvert, A.; Canals, A.; Salvador, A. Ionic liquid-based single-drop microextraction followed by liquid chromatography-ultraviolet spectrophotometry detection to determine typical UV filters in surface water samples. *Talanta,* **2010**, *81*(1-2), 549-555. [http://dx.doi.org/10.1016/j.talanta.2009.12.042] [PMID: 20188961]

[67] Pena-Pereira, F.; Lavilla, I.; Bendicho, C.; Vidal, L.; Canals, A. Speciation of mercury by ionic liquid-based single-drop microextraction combined with high-performance liquid chromatography-photodiode array detection. *Talanta,* **2009**, *78*(2), 537-541. [http://dx.doi.org/10.1016/j.talanta.2008.12.003] [PMID: 19203620]

[68] Vidal, L.; Chisvert, A.; Canals, A.; Salvador, A. Sensitive determination of free benzophenone-3 in human urine samples based on an ionic liquid as extractant phase in single-drop microextraction prior to liquid chromatography analysis. *J. Chromatogr. A,* **2007**, *1174*(1-2), 95-103. [http://dx.doi.org/10.1016/j.chroma.2007.07.077] [PMID: 17720175]

[69] Guo, X.; Yin, D.; Peng, J.; Hu, X. Ionic liquid-based single-drop liquid-phase microextraction combined with high-performance liquid chromatography for the determination of sulfonamides in environmental water. *J. Sep. Sci.,* **2012**, *35*(3), 452-458. [http://dx.doi.org/10.1002/jssc.201100777] [PMID: 22258811]

[70] Liu, Q.; Liu, Y.; Chen, S.; Liu, Q. Ionic liquid for single-drop microextraction followed by high-performance liquid chromatography-ultraviolet detection to determine carbonyl compounds in

environmental waters. *J. Sep. Sci.,* **2010**, *33*(15), 2376-2382.
[http://dx.doi.org/10.1002/jssc.201000051] [PMID: 20574954]

[71] Yao, C.; Pitner, W.R.; Anderson, J.L. Ionic liquids containing the tris(pentafluoroethyl) trifluorophosphate anion: a new class of highly selective and ultra hydrophobic solvents for the extraction of polycyclic aromatic hydrocarbons using single drop microextraction. *Anal. Chem.,* **2009**, *81*(12), 5054-5063.
[http://dx.doi.org/10.1021/ac900719m] [PMID: 19419149]

[72] Amjadi, M.; Manzoori, J.L.; Abulhassani, J. Ionic liquid-based, single-drop microextraction for preconcentration of cobalt before its determination by electrothermal atomic absorption spectrometry. *J. AOAC Int.,* **2010**, *93*(3), 985-991.
[PMID: 20629404]

[73] Martinis, E.M.; Bertón, P.; Altamirano, J.C.; Hakala, U.; Wuilloud, R.G. Tetradecyl(trihexyl)phosphonium chloride ionic liquid single-drop microextraction for electrothermal atomic absorption spectrometric determination of lead in water samples. *Talanta,* **2010**, *80*(5), 2034-2040.
[http://dx.doi.org/10.1016/j.talanta.2009.11.012] [PMID: 20152449]

[74] Yang, J.; Wei, H.; Teng, X.; Zhang, H.; Shi, Y. Dynamic ultrasonic nebulisation extraction coupled with headspace ionic liquid-based single-drop microextraction for the analysis of the essential oil in *Forsythia suspensa. Phytochem. Anal.,* **2014**, *25*(2), 178-184.
[http://dx.doi.org/10.1002/pca.2490] [PMID: 24347206]

[75] Jiang, C.; Wei, S.; Li, X.; Zhao, Y.; Shao, M.; Zhang, H.; Yu, A. Ultrasonic nebulization headspace ionic liquid-based single drop microextraction of flavour compounds in fruit juices. *Talanta,* **2013**, *106*, 237-242.
[http://dx.doi.org/10.1016/j.talanta.2013.01.028] [PMID: 23598122]

[76] He, S.-W.; Shen, C.-Y.; Wei, X.-Q.; Jin, M.-C.; Cai, M.-Q. Determination of trace PCBs in water by GC-MS with ionic liquid based headspace single-drop microextraction under ultrasound. *Advanced Materials Research,* **2013**, *726-731*, 74-80.
[http://dx.doi.org/10.4028/www.scientific.net/AMR.726-731.74]

[77] He, X.; Zhang, F.; Jiang, Y. An improved ionic liquid-based headspace single-drop microextraction-liquid chromatography method for the analysis of camphor and trans-anethole in compound liquorice tablets. *J. Chromatogr. Sci.,* **2012**, *50*(6), 457-463.
[http://dx.doi.org/10.1093/chromsci/bms038] [PMID: 22496513]

[78] Cha, D.; Qiang, N. Determining o-nitrophenol in wastewater. *Filtr.,* **2012**, *49*, 38-41.
[http://dx.doi.org/10.1016/S0015-1882(12)70113-1]

[79] Rahmani, M.; Kaykhaii, M. Determination of methylcyclopentadienyl-manganese tricarbonyl in gasoline and water *via* ionic-liquid headspace single drop microextraction and electrothermal atomic absorption spectrometry. *Mikrochim. Acta,* **2011**, *174*, 413-419.
[http://dx.doi.org/10.1007/s00604-011-0648-6]

[80] Sheikhloie, H.; Saber-Tehrani, M.; Abrumand-Azar, P.; Waqif-Husain, S. Analysis of tributyltin and triphenyltin in water by ionic liquid–headspace single-drop microextraction then HPLC with fluorimetric detection. *Acta Chromatogr.,* **2009**, *21*, 577-589.
[http://dx.doi.org/10.1556/AChrom.21.2009.4.5]

[81] Zhou, Q.; Ye, C. Ionic liquid for improved single-drop microextraction of aromatic amines in water samples. *Mikrochim. Acta,* **2008**, *162*, 153-159. [http://dx.doi.org/10.1007/s00604-007-0857-1]

[82] Aguilera-Herrador, E.; Lucena, R.; Cárdenas, S.; Valcárcel, M. Direct coupling of ionic liquid based single-drop microextraction and GC/MS. *Anal. Chem.,* **2008**, *80*(3), 793-800. [http://dx.doi.org/10.1021/ac071555g] [PMID: 18154274]

[83] Carrillo-Carrión, C.; Simonet, B.M.; Valcárcel, M. (CdSe/ZnS QDs)-ionic liquid-based headspace single drop microextraction for the fluorimetric determination of trimethylamine in fish. *Analyst (Lond.),* **2012**, *137*(5), 1152-1159. [http://dx.doi.org/10.1039/c2an15914g] [PMID: 22268066]

[84] Chisvert, A.; Román, I.P.; Vidal, L.; Canals, A. Simple and commercial readily-available approach for the direct use of ionic liquid-based single-drop microextraction prior to gas chromatography determination of chlorobenzenes in real water samples as model analytical application. *J. Chromatogr. A,* **2009**, *1216*(9), 1290-1295. [http://dx.doi.org/10.1016/j.chroma.2008.12.078] [PMID: 19144344]

[85] Vidal, L.; Domini, C.E.; Grané, N.; Psillakis, E.; Canals, A. Microwave-assisted headspace single-drop microextration of chlorobenzenes from water samples. *Anal. Chim. Acta,* **2007**, *592*(1), 9-15. [http://dx.doi.org/10.1016/j.aca.2007.03.066] [PMID: 17499064]

[86] Márquez-Sillero, I.; Cárdenas, S.; Valcárcel, M. Direct determination of 2,4,6-tricholoroanisole in wines by single-drop ionic liquid microextraction coupled with multicapillary column separation and ion mobility spectrometry detection. *J. Chromatogr. A,* **2011**, *1218*(42), 7574-7580. [http://dx.doi.org/10.1016/j.chroma.2011.06.032] [PMID: 21737082]

[87] Márquez-Sillero, I.; Aguilera-Herrador, E.; Cárdenas, S.; Valcárcel, M. Determination of 2,4,6-tricholoroanisole in water and wine samples by ionic liquid-based single-drop microextraction and ion mobility spectrometry. *Anal. Chim. Acta,* **2011**, *702*(2), 199-204. [http://dx.doi.org/10.1016/j.aca.2011.06.046] [PMID: 21839198]

[88] Yang, J.; Wei, H.; Yu, C.; Shi, Y.; Zhang, H. Extraction of the volatile and semivolatile compounds in seeds of Cuminum cyminum L. using hydrodistillation followed by headspace-ionic liquid-based single-drop microextraction. *Chromatographia,* **2012**, *75*, 1435-1443. [http://dx.doi.org/10.1007/s10337-012-2335-0]

[89] Vallecillos, L.; Borrull, F.; Pocurull, E. Determination of musk fragrances in sewage sludge by pressurized liquid extraction coupled to automated ionic liquid-based headspace single-drop microextraction followed by GC-MS/MS. *J. Sep. Sci.,* **2012**, *35*(20), 2735-2742. [http://dx.doi.org/10.1002/jssc.201200326] [PMID: 23019133]

[90] Vallecillos, L.; Pocurull, E.; Borrull, F. Fully automated ionic liquid-based headspace single drop microextraction coupled to GC-MS/MS to determine musk fragrances in environmental water samples. *Talanta,* **2012**, *99*, 824-832. [http://dx.doi.org/10.1016/j.talanta.2012.07.036] [PMID: 22967629]

[91] Aguilera-Herrador, E.; Lucena, R.; Cárdenas, S.; Valcárcel, M. Ionic liquid-based single drop microextraction and room-temperature gas chromatography for on-site ion mobility spectrometric analysis. *J. Chromatogr. A,* **2009**, *1216*(29), 5580-5587.

[http://dx.doi.org/10.1016/j.chroma.2009.05.071] [PMID: 19523638]

[92] Zhao, H.; Lu, S.; Du, W.; Zeng, B. Ionic liquid-based headspace single-drop microextraction coupled to gas chromatography for the determination of chlorobenzene derivatives. *Mikrochim. Acta,* **2009**, *165*, 29-33.
[http://dx.doi.org/10.1007/s00604-008-0092-4]

[93] Aguilera-Herrador, E.; Lucena, R.; Cárdenas, S.; Valcárcel, M. Determination of trihalomethanes in waters by ionic liquid-based single drop microextraction/gas chromatographic/mass spectrometry. *J. Chromatogr. A,* **2008**, *1209*(1-2), 76-82.
[http://dx.doi.org/10.1016/j.chroma.2008.09.030] [PMID: 18817919]

[94] Zhao, F.Q.; Li, J.; Zeng, B.Z. Coupling of ionic liquid-based headspace single-drop microextraction with GC for sensitive detection of phenols. *J. Sep. Sci.,* **2008**, *31*(16-17), 3045-3049.
[http://dx.doi.org/10.1002/jssc.200800308] [PMID: 18704999]

[95] Aguilera-Herrador, E.; Lucena, R.; Cárdenas, S.; Valcárcel, M. Ionic liquid-based single-drop microextraction/gas chromatographic/mass spectrometric determination of benzene, toluene, ethylbenzene and xylene isomers in waters. *J. Chromatogr. A,* **2008**, *1201*(1), 106-111.
[http://dx.doi.org/10.1016/j.chroma.2008.06.010] [PMID: 18586254]

[96] Yao, C.; Twu, P.; Anderson, J.L. Headspace single drop microextraction using micellar ionic liquid extraction solvents. *Chromatographia,* **2010**, *72*, 393-402.
[http://dx.doi.org/10.1365/s10337-010-1675-x]

[97] Li, M.; Li, Z.; Li, G.; Chen, L.; Wang, L.; Jiang, C.; Li, L.; Zhang, Z. Symmetrical 1,3-dialkylimidazolium ionic liquid based single-drop microextraction coupled to GC-MS for the determination of esters in natural flavors. *Acta Chimi. Sin.,* **2012**, *70*, 1625-1630.
[http://dx.doi.org/10.6023/A12040155]

[98] Martinis, E.M.; Wuilloud, R.G. Cold vapor ionic liquid-assisted headspace single-drop microextraction: A novel preconcentration technique for mercury species determination in complex matrix samples. *J. Anal. At. Spectrom.,* **2010**, *25*, 1432-1439.
[http://dx.doi.org/10.1039/c004678g]

[99] Pedersen-Bjergaard, S.; Rasmussen, K.E. Liquid-liquid-liquid microextraction for sample preparation of biological fluids prior to capillary electrophoresis. *Anal. Chem.,* **1999**, *71*(14), 2650-2656.
[http://dx.doi.org/10.1021/ac990055n] [PMID: 10424162]

[100] Audunsson, G. Aqueous/aqueous extraction by means of a liquid membrane for sample cleanup and preconcentration of amines in a flow system. *Anal. Chem.,* **1986**, *58*, 2714-2723.
[http://dx.doi.org/10.1021/ac00126a030]

[101] Thordarson, E.; Pálmarsdóttir, S.; Mathiasson, L.; Jönsson, J.Å. Sample preparation using a miniaturized supported liquid membrane device connected on-line to packed capillary liquid chromatography. *Anal. Chem.,* **1996**, *68*(15), 2559-2563.
[http://dx.doi.org/10.1021/ac950929f] [PMID: 21619202]

[102] Ouyang, G.; Zhao, W.; Pawliszyn, J. Automation and optimization of liquid-phase microextraction by gas chromatography. *J. Chromatogr. A,* **2007**, *1138*(1-2), 47-54.
[http://dx.doi.org/10.1016/j.chroma.2006.10.093] [PMID: 17116305]

[103] Ghambarian, M.; Yamini, Y.; Esrafili, A. Developments in hollow fiber based liquid-phase microextraction: principles and applications. *Mikrochim. Acta,* **2012**, *177*, 271-294. [http://dx.doi.org/10.1007/s00604-012-0773-x]

[104] Kjelsen, I.J.; Gjelstad, A.; Rasmussen, K.E.; Pedersen-Bjergaard, S. Low-voltage electromembrane extraction of basic drugs from biological samples. *J. Chromatogr. A,* **2008**, *1180*(1-2), 1-9. [http://dx.doi.org/10.1016/j.chroma.2007.12.006] [PMID: 18164716]

[105] Liu, W.; Wei, Z.; Zhang, Q.; Wu, F.; Lin, Z.; Lu, Q.; Lin, F.; Chen, G.; Zhang, L. Novel multifunctional acceptor phase additive of water-miscible ionic liquid in hollow-fiber protected liquid phase microextraction. *Talanta,* **2012**, *88*, 43-49. [http://dx.doi.org/10.1016/j.talanta.2011.08.017] [PMID: 22265467]

[106] Ma, X.; Huang, M.; Li, Z.; Wu, J. Hollow fiber supported liquid-phase microextraction using ionic liquid as extractant for preconcentration of benzene, toluene, ethylbenzene and xylenes from water sample with gas chromatography-hydrogen flame ionization detection. *J. Hazard. Mater.,* **2011**, *194*, 24-29. [http://dx.doi.org/10.1016/j.jhazmat.2011.07.066] [PMID: 21871732]

[107] Basheer, C.; Alnedhary, A.A.; Madhava Rao, B.S.; Balasubramanian, R.; Lee, H.K. Ionic liquid supported three-phase liquid-liquid-liquid microextraction as a sample preparation technique for aliphatic and aromatic hydrocarbons prior to gas chromatography-mass spectrometry. *J. Chromatogr. A,* **2008**, *1210*(1), 19-24. [http://dx.doi.org/10.1016/j.chroma.2008.09.040] [PMID: 18823629]

[108] Zeng, C.; Hu, Y.; Luo, J. Ionic liquid-based hollow fiber supported liquid membrane extraction combined with thermospray flame furnace AAS for the determination of cadmium. *Mikrochim. Acta,* **2012**, *177*, 53-58. [http://dx.doi.org/10.1007/s00604-011-0748-3]

[109] Chen, H.; Han, J.; Wang, Y.; Hu, Y.; Ni, L.; Liu, Y.; Kang, W.; Liu, Y. Hollow fiber liquid-phase microextraction of cadmium(II) using an ionic liquid as the extractant. *Microchim. Acta,* **2014**, *181*, 1455-1461. [http://dx.doi.org/http://dx.doi.org/10.1007/s00604-014-1274-x]

[110] Zou, Y.; Zhang, Z.; Shao, X.; Chen, Y.; Wu, X.; Yang, L.; Zhu, J.; Zhang, D. Application of three-phase hollow fiber LPME using an ionic liquid as supported phase for preconcentration of malachite green from water samples with HPLC detection. *Bull. Korean Chem. Soc.,* **2014**, *35*, 371-376. [http://dx.doi.org/10.5012/bkcs.2014.35.2.371]

[111] Abulhassani, J.; Manzoori, J.L.; Amjadi, M. Hollow fiber based-liquid phase microextraction using ionic liquid solvent for preconcentration of lead and nickel from environmental and biological samples prior to determination by electrothermal atomic absorption spectrometry. *J. Hazard. Mater.,* **2010**, *176*(1-3), 481-486. [http://dx.doi.org/10.1016/j.jhazmat.2009.11.054] [PMID: 19962233]

[112] Liu, Y.; Wang, Y.; Hu, Y.; Ni, L.; Han, J.; Chen, T. *Chen, H.; Liu, Y. Ionic liquid-based hollow fiber liquid-phase microextraction for the determination of trace lead (II) in environmental water and tea drinks samples by graphite furnace atomic absorption spectrometry*; J. Iran Chem. Soc, **2014**.

[113] Nomngongo, P.N.; Ngila, J.C.; Msagati, T.A.; Moodley, B. Chemometric optimization of hollow

fiber-liquid phase microextraction for preconcentration of trace elements in diesel and gasoline prior to their ICP-OES determination. *Mikrochim. Acta,* **2014**, *114*, 141-147. [http://dx.doi.org/10.1016/j.microc.2013.12.013]

[114] Peng, J-F.; Liu, J-F.; Hu, X-L.; Jiang, G-B. Direct determination of chlorophenols in environmental water samples by hollow fiber supported ionic liquid membrane extraction coupled with high-performance liquid chromatography. *J. Chromatogr. A,* **2007**, *1139*(2), 165-170. [http://dx.doi.org/10.1016/j.chroma.2006.11.006] [PMID: 17113589]

[115] Tao, Y.; Liu, J-F.; Hu, X-L.; Li, H-C.; Wang, T.; Jiang, G-B. Hollow fiber supported ionic liquid membrane microextraction for determination of sulfonamides in environmental water samples by high-performance liquid chromatography. *J. Chromatogr. A,* **2009**, *1216*(35), 6259-6266. [http://dx.doi.org/10.1016/j.chroma.2009.06.025] [PMID: 19632683]

[116] Wang, S.-R.; Wang, S. Ionic liquid-based hollow fiber-supported liquid phase microextraction enhanced electrically for the determination of neutral red. *J. Food Drug Anal.*, **2014**, 22(4), 418-424. [http://dx.doi.org/10.1016/j.jfda.2014.03.006] (acceded August 2014).

[117] Zou, Y.; Zhang, Z.; Shao, X.; Chen, Y.; Wu, X.; Yang, L.; Zhu, J.; Zhang, D. Hollow-fiber-supported liquid-phase microextraction using an ionic liquid as the extractant for the pre-concentration of bisphenol A, 17-β-estradiol, estrone and diethylstilbestrol from water samples with HPLC detection. *Water Sci. Technol.,* **2014**, *69*(5), 1028-1035. [http://dx.doi.org/10.2166/wst.2013.824] [PMID: 24622552]

[118] Ge, D.; Lee, H.K. Ionic liquid based hollow fiber supported liquid phase microextraction of ultraviolet filters. *J. Chromatogr. A,* **2012**, *1229*, 1-5. [http://dx.doi.org/10.1016/j.chroma.2011.12.110] [PMID: 22307149]

[119] Nitiyanontakit, S.; Varanusupakul, P.; Miró, M. Hybrid flow analyzer for automatic hollow-fibe--assisted ionic liquid-based liquid-phase microextraction with in-line membrane regeneration. *Anal. Bioanal. Chem.,* **2013**, *405*(10), 3279-3288. [http://dx.doi.org/10.1007/s00216-013-6744-1] [PMID: 23386000]

[120] Rezaee, M.; Assadi, Y.; Milani Hosseini, M.R.; Aghaee, E.; Ahmadi, F.; Berijani, S. Determination of organic compounds in water using dispersive liquid-liquid microextraction. *J. Chromatogr. A,* **2006**, *1116*(1-2), 1-9. [http://dx.doi.org/10.1016/j.chroma.2006.03.007] [PMID: 16574135]

[121] Maya, F.; Horstkotte, B.; Estela, J.M.; Cerdà, V. Automated in-syringe dispersive liquid-liquid microextraction. *Trends Analyt. Chem.,* **2014**, *59*, 1-8. [http://dx.doi.org/10.1016/j.trac.2014.03.009]

[122] Abu-Bakar, N.B.; Makahleh, A.; Saad, B. Vortex-assisted liquid-liquid microextraction coupled with high performance liquid chromatography for the determination of furfurals and patulin in fruit juices. *Talanta,* **2014**, *120*, 47-54. [http://dx.doi.org/10.1016/j.talanta.2013.11.081] [PMID: 24468341]

[123] Liu, Y.; Zhao, E.; Zhu, W.; Gao, H.; Zhou, Z. Determination of four heterocyclic insecticides by ionic liquid dispersive liquid-liquid microextraction in water samples. *J. Chromatogr. A,* **2009**, *1216*(6), 885-891. [http://dx.doi.org/10.1016/j.chroma.2008.11.076] [PMID: 19118833]

[124] Khani, R.; Ghasemi, J.B.; Shemirani, F. Simultaneous multicomponent spectrophotometric monitoring of methyl and propyl parabens using multivariate statistical methods after their preconcentration by robust ionic liquid-based dispersive liquid–liquid microextraction. *Spectroc. Acta Pt. A-Molec. Biomolec. Spectr,* **2014**, *122*, 295-303.

[125] Fan, Y.; Liu, S.; Xie, Q. Rapid determination of phthalate esters in alcoholic beverages by conventional ionic liquid dispersive liquid-liquid microextraction coupled with high performance liquid chromatography. *Talanta,* **2014**, *119*, 291-298. [http://dx.doi.org/10.1016/j.talanta.2013.11.023] [PMID: 24401417]

[126] Xin, L.; Rao, F.; Min, L.; Li-Ping, G.; Li, Y. Ionic liquid-based dispersive liquid-liquid microextraction coupled with capillary electrophoresis to determine drugs of abuse in urine. *Chin. J. Anal. Chem.,* **2013**, *41*, 1919-1922. [http://dx.doi.org/10.1016/S1872-2040(13)60697-0]

[127] Hayama, T.; Yabuuchi, Y.; Iwamatsu, T.; Tamashima, E.; Kawami, Y.; Itoyama, M.; Yoshida, H.; Yamaguchi, M.; Nohta, H. Concerted derivatization and concentration method with dispersive liquid-liquid microextraction for liquid chromatographic analysis of 5-hydroxyindoles in human serum. *Talanta,* **2013**, *117*, 27-31. [http://dx.doi.org/10.1016/j.talanta.2013.08.035] [PMID: 24209305]

[128] Escudero, L.B.; Martinis, E.M.; Olsina, R.A.; Wuilloud, R.G. Arsenic speciation analysis in mono-varietal wines by on-line ionic liquid-based dispersive liquid-liquid microextraction. *Food Chem.,* **2013**, *138*(1), 484-490. [http://dx.doi.org/10.1016/j.foodchem.2012.10.054] [PMID: 23265515]

[129] Asghari, A.; Ghazaghi, M.; Rajabi, M.; Behzad, M.; Ghaedi, M. Ionic liquid-based dispersive liquid–liquid microextraction combined with high performance liquid chromatography–UV detection for the simultaneous pre-concentration and determination of Ni, Co, Cu and Zn in water samples. *J. Serb. Chem. Soc,* **2014**, *79*, 63-76. [http://dx.doi.org/10.2298/JSC062212081A]

[130] Rajabi, M.; Kamalabadi, M.; Jamali, M.R.; Zolgharnein, J.; Asanjarani, N. Application of response surface methodology for optimization of ionic liquid-based dispersive liquid-liquid microextraction of cadmium from water samples. *Hum. Exp. Toxicol.,* **2013**, *32*(6), 620-631. [http://dx.doi.org/10.1177/0960327112455672] [PMID: 22893353]

[131] Abadi, M.D.; Chamsaz, M.; Arbab-Zavar, M.H. Ionic liquid-based dispersive liquid-liquid microextraction combined with flame atomic absorption spectrometry for determination of silver in water samples. *Asian J. Chem.,* **2013**, *25*, 3984-3988.

[132] Cruz-Vera, M.; Lucena, R.; Cárdenas, S.; Valcárcel, M. One-step in-syringe ionic liquid-based dispersive liquid-liquid microextraction. *J. Chromatogr. A,* **2009**, *1216*(37), 6459-6465. [http://dx.doi.org/10.1016/j.chroma.2009.07.040] [PMID: 19674753]

[133] Zhou, Q.; Bai, H.; Xie, G.; Xiao, J. Trace determination of organophosphorus pesticides in environmental samples by temperature-controlled ionic liquid dispersive liquid-phase microextraction. *J. Chromatogr. A,* **2008**, *1188*(2), 148-153. [http://dx.doi.org/10.1016/j.chroma.2008.02.094] [PMID: 18346747]

[134] Baghdadi, M.; Shemirani, F. Cold-induced aggregation microextraction: a novel sample preparation

technique based on ionic liquids. *Anal. Chim. Acta,* **2008**, *613*(1), 56-63. [http://dx.doi.org/10.1016/j.aca.2008.02.057] [PMID: 18374702]

[135] del Mar Parrilla Vázquez, M.; Martínez Galera, M.; Parrilla Vázquez, P.; Uclés Moreno, A. Trace analysis of herbicides in wastewaters by a dispersive liquid-liquid microextraction approach and liquid chromatography with quadrupole linear ion trap mass spectrometry: evaluation of green parameters. *J. Sep. Sci.,* **2014**, *37*(12), 1511-1520. [http://dx.doi.org/10.1002/jssc.201400148] [PMID: 24687968]

[136] Zhang, J.; Liang, Z.; Guo, H.; Gao, P.; Lu, R.; Zhou, W.; Zhang, S.; Gao, H. Ionic liquid-based totally organic solvent-free emulsification microextraction coupled with high performance liquid chromatography for the determination of three acaricides in fruit juice. *Talanta,* **2013**, *115*, 556-562. [http://dx.doi.org/10.1016/j.talanta.2013.06.010] [PMID: 24054632]

[137] Zeeb, M.; Tayebi Jamil, P.; Berenjian, A.; Ganjali, M.R.; Talei Bavil Olyai, M.R. Quantitative analysis of piroxicam using temperature-controlled ionic liquid dispersive liquid phase microextraction followed by stopped-flow injection spectrofluorimetry. *Daru,* **2013**, *21*(1), 63. [http://dx.doi.org/10.1186/2008-2231-21-63] [PMID: 23895461]

[138] Mirzaei, M.; Amirtaimoury, N. Temperature induced aggregation ionic liquid dispersive liquid-liquid microextraction method for separation trace amount of cobalt ion. *J. Anal. Chem.,* **2014**, *69*, 503-508. [http://dx.doi.org/10.1134/S1061934814060112]

[139] Bamdad, F.; Ardalani, M.; Sangi, M.R. Trace determination of cadmium ions by flame atomic absorption spectrometry after pre-concentration using temperature-controlled ionic liquid dispersive liquid phase microextraction. *J. Braz. Chem. Soc.,* **2014**, *25*, 246-252.

[140] Zeeb, M.; Ganjali, M.R.; Norouzi, P. Preconcentration and trace determination of chromium using modified ionic liquid cold-induced aggregation dispersive liquid–liquid microextraction: application to different water and food samples. *Food Anal. Methods,* **2013**, *6*, 1398-1406. [http://dx.doi.org/10.1007/s12161-012-9557-5]

[141] Zhou, Q.; Zhang, X.; Xiao, J. Ultrasound-assisted ionic liquid dispersive liquid-phase micro-extraction: a novel approach for the sensitive determination of aromatic amines in water samples. *J. Chromatogr. A,* **2009**, *1216*(20), 4361-4365. [http://dx.doi.org/10.1016/j.chroma.2009.03.046] [PMID: 19329122]

[142] Li, S.; Cai, S.; Hu, W.; Chen, H.; Liu, H. Ionic liquid-based ultrasound-assisted dispersive liquid-liquid microextraction combined with electrothermal atomic absorption spectrometry for a sensitive determination of cadmium in water samples, *Spectroc. Acta Pt. B-Atom. Spectr.,* **2009**, *64*, 666-671.

[143] Vázquez, M.M.; Vázquez, P.P.; Galera, M.M.; Moreno, A.U. Comparison of two ionic liquid dispersive liquid-liquid microextraction approaches for the determination of benzoylurea insecticides in wastewater using liquid chromatography-quadrupole-linear ion trap-mass spectrometry: evaluation of green parameters. *J. Chromatogr. A,* **2014**, *1356*, 1-9. [http://dx.doi.org/10.1016/j.chroma.2014.06.032] [PMID: 24993054]

[144] Sun, J.N.; Chen, J.; Shi, Y.P. Multiple functional ionic liquids based dispersive liquid-liquid microextraction combined with high performance chromatography for the determination of phenolic compounds in water samples. *Talanta,* **2014**, *125*, 329-335. [http://dx.doi.org/10.1016/j.talanta.2014.03.013] [PMID: 24840452]

[145] Hatami, M.; Karimnia, E.; Farhadi, K. Determination of salmeterol in dried blood spot using an ionic liquid based dispersive liquid-liquid microextraction coupled with HPLC. *J. Pharm. Biomed. Anal.*, **2013**, *85*, 283-287. [http://dx.doi.org/10.1016/j.jpba.2013.07.025] [PMID: 23973634]

[146] Dong, S.; Hu, Q.; Yang, Z.; Liu, R.; Huang, G.; Huang, T. An ionic liquid-based ultrasound assisted dispersive liquid–liquid microextraction procedure followed by HPLC for the determination of low concentration of phytocides in soil. *Microchem. J.*, **2013**, *110*, 221-226. [http://dx.doi.org/10.1016/j.microc.2013.03.018]

[147] Parrilla Vázquez, M.M.; Parrilla Vázquez, P.; Martínez Galera, M.; Gil García, M.D.; Uclés, A. Ultrasound-assisted ionic liquid dispersive liquid-liquid microextraction coupled with liquid chromatography-quadrupole-linear ion trap-mass spectrometry for simultaneous analysis of pharmaceuticals in wastewaters. *J. Chromatogr. A,* **2013**, *1291*, 19-26. [http://dx.doi.org/10.1016/j.chroma.2013.03.066] [PMID: 23587315]

[148] Rajabi, M.; Asemipour, S.; Barfi, B.; Jamali, M.R.; Behzad, M. Ultrasound-assisted ionic liquid based dispersive liquid–liquid microextraction and flame atomic absorption spectrometry of cobalt, copper, and zinc in environmental water samples. *J. Mol. Liq.*, **2014**, *194*, 166-171. [http://dx.doi.org/10.1016/j.molliq.2014.01.026]

[149] Stanisz, E.; Werner, J.; Matusiewicz, H. Mercury species determination by task specific ionic liquid-based ultrasound-assisted dispersive liquid–liquid microextraction combined with cold vapour generation atomic absorption spectrometry. *Microchem. J.*, **2013**, *110*, 28-35. [http://dx.doi.org/10.1016/j.microc.2013.01.006]

[150] López-García, I.; Vicente-Martínez, Y.; Hernández-Córdoba, M. Nonchromatographic speciation of selenium in edible oils using dispersive liquid-liquid microextraction and electrothermal atomic absorption spectrometry. *J. Agric. Food Chem.*, **2013**, *61*(39), 9356-9361. [http://dx.doi.org/10.1021/jf4027537] [PMID: 24011330]

[151] Xu, X.; Su, R.; Zhao, X.; Liu, Z.; Li, D.; Li, X.; Zhang, H.; Wang, Z. Determination of formaldehyde in beverages using microwave-assisted derivatization and ionic liquid-based dispersive liquid-liquid microextraction followed by high-performance liquid chromatography. *Talanta,* **2011**, *85*(5), 2632-2638. [http://dx.doi.org/10.1016/j.talanta.2011.08.037] [PMID: 21962694]

[152] Xu, X.; Su, R.; Zhao, X.; Liu, Z.; Zhang, Y.; Li, D.; Li, X.; Zhang, H.; Wang, Z. Ionic liquid-based microwave-assisted dispersive liquid-liquid microextraction and derivatization of sulfonamides in river water, honey, milk, and animal plasma. *Anal. Chim. Acta,* **2011**, *707*(1-2), 92-99. [http://dx.doi.org/10.1016/j.aca.2011.09.018] [PMID: 22027124]

[153] Xu, X.; Liu, Z.; Zhao, X.; Su, R.; Zhang, Y.; Shi, J.; Zhao, Y.; Wu, L.; Ma, Q.; Zhou, X.; Zhang, H.; Wang, Z. Ionic liquid-based microwave-assisted surfactant-improved dispersive liquid-liquid microextraction and derivatization of aminoglycosides in milk samples. *J. Sep. Sci.,* **2013**, *36*(3), 585-592. [http://dx.doi.org/10.1002/jssc.201200801] [PMID: 23303586]

[154] Asensio-Ramos, M.; Hernández-Borges, J.; Borges-Miquel, T.M.; Rodríguez-Delgado, M.A. Ionic liquid-dispersive liquid-liquid microextraction for the simultaneous determination of pesticides and

metabolites in soils using high-performance liquid chromatography and fluorescence detection. *J. Chromatogr. A,* **2011**, *1218*(30), 4808-4816.
[http://dx.doi.org/10.1016/j.chroma.2010.11.030] [PMID: 21145553]

[155] Ye, L.; Liu, J.; Yang, X.; Peng, Y.; Xu, L. Orthogonal array design for the optimization of ionic liquid-based dispersive liquid-liquid microextraction of benzophenone-type UV filters. *J. Sep. Sci.,* **2011**, *34*(6), 700-706.
[http://dx.doi.org/10.1002/jssc.201000552] [PMID: 21290603]

[156] Xiao, C.; Tang, M.; Li, J.; Yin, C-R.; Xiang, G.; Xu, L. Determination of sildenafil, vardenafil and aildenafil in human plasma by dispersive liquid-liquid microextraction-back extraction based on ionic liquid and high performance liquid chromatography-ultraviolet detection. *J. Chromatogr. B Analyt. Technol. Biomed. Life Sci.,* **2013**, *931*, 111-116.
[http://dx.doi.org/10.1016/j.jchromb.2013.05.013] [PMID: 23774245]

[157] Khan, S.; Soylak, M.; Kazi, T.G. Room temperature ionic liquid-based dispersive liquid phase microextraction for the separation/preconcentration of trace Cd($^{2+}$) as 1-(2-pyridylazo)-2-naphthol (PAN) complex from environmental and biological samples and determined by FAAS. *Biol. Trace Elem. Res.,* **2013**, *156*(1-3), 49-55.
[http://dx.doi.org/10.1007/s12011-013-9853-y] [PMID: 24197606]

[158] Lai, X.; Ruan, C.; Liu, R.; Liu, C. Application of ionic liquid-based dispersive liquid-liquid microextraction for the analysis of ochratoxin A in rice wines. *Food Chem.,* **2014**, *161*, 317-322.
[http://dx.doi.org/10.1016/j.foodchem.2014.04.033] [PMID: 24837956]

[159] Rao, R.N.; Raju, S.S.; Vali, R.M. Ionic-liquid based dispersive liquid-liquid microextraction followed by high performance liquid chromatographic determination of anti-hypertensives in rat serum. *J. Chromatogr. B Analyt. Technol. Biomed. Life Sci.,* **2013**, *931*, 174-180.
[http://dx.doi.org/10.1016/j.jchromb.2013.05.027] [PMID: 23797113]

[160] Ku, Y.C.; Leong, M.I.; Wang, W.T.; Huang, S.D. Up-and-down shaker-assisted ionic liquid-based dispersive liquid-liquid microextraction of benzophenone-type ultraviolet filters. *J. Sep. Sci.,* **2013**, *36*(8), 1470-1477.
[http://dx.doi.org/10.1002/jssc.201201108] [PMID: 23495210]

[161] Wu, H.; Guo, J-B.; Du, L-M.; Tian, H.; Hao, C-X.; Wang, Z-F.; Wang, J-Y. A rapid shaking-based ionic liquid dispersive liquid phase microextraction for the simultaneous determination of six synthetic food colourants in soft drinks, sugar- and gelatin-based confectionery by high-performance liquid chromatography. *Food Chem.,* **2013**, *141*(1), 182-186.
[http://dx.doi.org/10.1016/j.foodchem.2013.03.015] [PMID: 23768345]

[162] Baghdadi, M.; Shemirani, F. *In situ* solvent formation microextraction based on ionic liquids: a novel sample preparation technique for determination of inorganic species in saline solutions. *Anal. Chim. Acta,* **2009**, *634*(2), 186-191.
[http://dx.doi.org/10.1016/j.aca.2008.12.017] [PMID: 19185118]

[163] Yao, C.; Anderson, J.L. Dispersive liquid-liquid microextraction using an *in situ* metathesis reaction to form an ionic liquid extraction phase for the preconcentration of aromatic compounds from water. *Anal. Bioanal. Chem.,* **2009**, *395*(5), 1491-1502.
[http://dx.doi.org/10.1007/s00216-009-3078-0] [PMID: 19756541]

[164] Fernández, E.; Vidal, L.; Iniesta, J.; Metters, J.P.; Banks, C.E.; Canals, A. Screen-printed electrode-based electrochemical detector coupled with *in-situ* ionic-liquid-assisted dispersive liquid-liquid microextraction for determination of 2,4,6-trinitrotoluene. *Anal. Bioanal. Chem.,* **2014**, *406*(8), 2197-2204.
[http://dx.doi.org/10.1007/s00216-013-7415-y] [PMID: 24247549]

[165] Agudelo Mesa, L.B.; Padró, J.M.; Reta, M. Analysis of non-polar heterocyclic aromatic amines in beefburguers by using microwave-assisted extraction and dispersive liquid-ionic liquid microextraction. *Food Chem.,* **2013**, *141*(3), 1694-1701.
[http://dx.doi.org/10.1016/j.foodchem.2013.04.076] [PMID: 23870880]

[166] Stanisz, E.; Zgoła-Grześkowiak, A. *In situ* metathesis ionic liquid formation dispersive liquid-liquid microextraction for copper determination in water samples by electrothermal atomic absorption spectrometry. *Talanta,* **2013**, *115*, 178-183.
[http://dx.doi.org/10.1016/j.talanta.2013.04.063] [PMID: 24054576]

[167] Yu, C.; Zhang, S.; Zhang, J.; Li, S.; Zhou, W.; Gao, H.; Lu, R. An *in situ* ionic liquid dispersive liquid-liquid microextraction method for the detection of pyrethroids by LC-UV in environmental water samples. *J. Braz. Chem. Soc.,* **2013**, *246*, 1034-1040.

[168] López-García, I.; Vicente-Martínez, Y.; Hernández-Córdoba, M. Determination of lead and cadmium using an ionic liquid and dispersive liquid-liquid microextraction followed by electrothermal atomic absorption spectrometry. *Talanta,* **2013**, *110*, 46-52.
[http://dx.doi.org/10.1016/j.talanta.2013.02.015] [PMID: 23618174]

[169] Zhao, R.S.; Wang, X.; Sun, J.; Hu, C.; Wang, X.K. Determination of triclosan and triclocarban in environmental water samples with ionic liquid/ionic liquid dispersive liquid-liquid microextraction prior to HPLC-ESI-MS/MS. *Mikrochim. Acta,* **2011**, *174*, 145-151.
[http://dx.doi.org/10.1007/s00604-011-0607-2]

[170] Zhao, R.S.; Wang, X.; Li, F.W.; Wang, S.S.; Zhang, L.L.; Cheng, C.G. Ionic liquid/ionic liquid dispersive liquid-liquid microextraction. *J. Sep. Sci.,* **2011**, *34*(7), 830-836.
[http://dx.doi.org/10.1002/jssc.201000802] [PMID: 21360666]

[171] Zhao, R.S.; Wang, X.; Zhang, L.L.; Wang, S.S.; Yuan, J.P. Ionic liquid/ionic liquid dispersive liquid-liquid microextraction, a new sample enrichment procedure for the determination of hexabromocyclododecane diastereomers in environmental water samples. *Anal. Methods,* **2011**, *3*, 831-836.
[http://dx.doi.org/10.1039/c0ay00708k]

[172] Gao, S.; Yang, X.; Yu, W.; Liu, Z.; Zhang, H. Ultrasound-assisted ionic liquid/ionic liquid-dispersive liquid-liquid microextraction for the determination of sulfonamides in infant formula milk powder using high-performance liquid chromatography. *Talanta,* **2012**, *99*, 875-882.
[http://dx.doi.org/10.1016/j.talanta.2012.07.050] [PMID: 22967637]

[173] Qin, H.; Li, B.; Liu, M.S.; Yang, Y.L. Separation and pre-concentration of glucocorticoids in water samples by ionic liquid supported vortex-assisted synergic microextraction and HPLC determination. *J. Sep. Sci.,* **2013**, *36*(8), 1463-1469.
[http://dx.doi.org/10.1002/jssc.201200989] [PMID: 23418157]

[174] Ravelo-Pérez, L.M.; Hernández-Borges, J.; Asensio-Ramos, M.; Rodríguez-Delgado, M.A. Ionic

liquid based dispersive liquid-liquid microextraction for the extraction of pesticides from bananas. *J. Chromatogr. A,* **2009**, *1216*(43), 7336-7345.
[http://dx.doi.org/10.1016/j.chroma.2009.08.012] [PMID: 19700165]

[175] Germán-Hernández, M.; Pino, V.; Anderson, J.L.; Afonso, A.M. A novel *in situ* preconcentration method with ionic liquid-based surfactants resulting in enhanced sensitivity for the extraction of polycyclic aromatic hydrocarbons from toasted cereals. *J. Chromatogr. A,* **2012**, *1227*, 29-37.
[http://dx.doi.org/10.1016/j.chroma.2011.12.097] [PMID: 22265775]

[176] Delgado, B.; Pino, V.; Anderson, J.L.; Ayala, J.H.; Afonso, A.M.; González, V. An *in-situ* extraction-preconcentration method using ionic liquid-based surfactants for the determination of organic contaminants contained in marine sediments. *Talanta,* **2012**, *99*, 972-983.
[http://dx.doi.org/10.1016/j.talanta.2012.07.073] [PMID: 22967651]

[177] Arvand, M.; Bozorgzadeh, E.; Shariati, S.; Zanjanchi, M.A. Ionic liquid-based dispersive liquid-liquid microextraction for the determination of formaldehyde in wastewaters and detergents. *Environ. Monit. Assess.,* **2012**, *184*(12), 7597-7605.
[http://dx.doi.org/10.1007/s10661-012-2521-4] [PMID: 22258742]

[178] Peng, B.; Zhang, J.; Wu, C.; Li, S.; Li, Y.; Gao, H.; Lu, R.; Zhou, W. Use of ionic liquid-based dispersive liquid-liquid microextraction and high-performance liquid chromatography to detect formaldehyde in air, water, and soil samples. *J. Liquid Chromatogr. Relat. Technol.,* **2014**, *37*, 815-828.
[http://dx.doi.org/10.1080/10826076.2012.758145]

[179] Páleníková, A.; Hrouzková, S. Microextraction in packed syringe: solvent-minimized sample preparation technique. *Monatsh. Chem.,* **2014**, *145*, 537-549.
[http://dx.doi.org/10.1007/s00706-013-1119-z]

[180] Spietelun, A.; Marcinkowski, Ł.; de la Guardia, M.; Namieśnik, J. Recent developments and future trends in solid phase microextraction techniques towards green analytical chemistry. *J. Chromatogr. A,* **2013**, *1321*, 1-13.
[http://dx.doi.org/10.1016/j.chroma.2013.10.030] [PMID: 24238710]

[181] Risticevic, S.; Niri, V.H.; Vuckovic, D.; Pawliszyn, J. Recent developments in solid-phase microextraction. *Anal. Bioanal. Chem.,* **2009**, *393*(3), 781-795.
[http://dx.doi.org/10.1007/s00216-008-2375-3] [PMID: 18836706]

[182] Huang, X.; Chen, L.; Lin, F.; Yuan, D. Novel extraction approach for liquid samples: stir cake sorptive extraction using monolith. *J. Sep. Sci.,* **2011**, *34*(16-17), 2145-2151.
[PMID: 21674796]

[183] Sigma-Aldrich. Title of material: SPE Cartridge (Tube) Configuration Guide. http://www.sigmaaldrich.com/analytical-chromatography/sample-preparation/spe/tube-configuration-guide.htm, Accessed 19 October 2016.

[184] Vidal, L.; Riekkola, M.L.; Canals, A. Ionic liquid-modified materials for solid-phase extraction and separation: a review. *Anal. Chim. Acta,* **2012**, *715*, 19-41.
[http://dx.doi.org/10.1016/j.aca.2011.11.050] [PMID: 22244164]

[185] Tian, M.; Yan, H.; Row, K.H. Solid-phase extraction of tanshinones from *Salvia Miltiorrhiza Bunge* using ionic liquid-modified silica sorbents. *J. Chromatogr. B Analyt. Technol. Biomed. Life Sci.,* **2009**,

877(8-9), 738-742.
[http://dx.doi.org/10.1016/j.jchromb.2009.02.012] [PMID: 19237325]

[186] Zhao, H.; Dai, B.; Xu, L.; Wang, X.; Qiao, X.; Xu, Z. Preparation and application of immobilised ionic liquid in solid-phase extraction for determination of trace acrylamide in food samples coupled with high-performance liquid chromatography. *J. Sci. Food Agric.,* **2014**, *94*(9), 1787-1793.
[http://dx.doi.org/10.1002/jsfa.6493] [PMID: 24282135]

[187] Cheng, J.F.; Ma, X.; Wu, Y. Silica gel chemically modified with ionic liquid as novel sorbent for solid-phase extraction and preconcentration of lead from beer and tea drink samples followed by flame atomic absorption spectrometric determination. *Food Anal. Methods,* **2014**, *7*, 1083-1089.
[http://dx.doi.org/10.1007/s12161-013-9716-3]

[188] Zhang, H.; Row, K.H. Application of ionic liquid modified silica for solid-phase extraction of polysaccharides from *Laminaria japonica. J. Carbohydr. Chem.,* **2014**, *33*, 225-237.
[http://dx.doi.org/10.1080/07328303.2014.913059]

[189] Bi, W.; Row, K.H. Application of anion-exchange imidazolium silica for the multiphase dispersive extraction of phenolic acids. *J. Sep. Sci.,* **2013**, *36*(16), 2621-2628.
[http://dx.doi.org/10.1002/jssc.201300195] [PMID: 23861179]

[190] Sadeghi, S.; Moghaddam, A.Z. Solid-phase extraction and HPLC-UV detection of Cr(III) and Cr(VI) using ionic liquid-functionalized silica as a hydrophobic sorbent. *Anal. Methods,* **2014**, *6*, 4867-4877.
[http://dx.doi.org/10.1039/c4ay00493k]

[191] Su, P.; Wang, R.Y.; Yu, Y.; Yang, Y. Microwave-assisted synthesis of ionic liquid-modified silica as a sorbent for the solid-phase extraction of phenolic compounds from water. *Anal. Methods,* **2014**, *6*, 704-709.
[http://dx.doi.org/10.1039/C3AY41771A]

[192] Vidal, L.; Robin, O.; Parshintsev, J.; Mikkola, J.P.; Riekkola, M.L. Quaternary ammonium-functionalized silica sorbents for the solid-phase extraction of aromatic amines under normal phase conditions. *J. Chromatogr. A,* **2013**, *1285*, 7-14.
[http://dx.doi.org/10.1016/j.chroma.2013.02.003] [PMID: 23473515]

[193] Fontanals, N.; Ronka, S.; Borrull, F.; Trochimczuk, A.W.; Marcé, R.M. Supported imidazolium ionic liquid phases: a new material for solid-phase extraction. *Talanta,* **2009**, *80*(1), 250-256.
[http://dx.doi.org/10.1016/j.talanta.2009.06.068] [PMID: 19782223]

[194] Gao, X.; Fan, J.; Wang, X.L.; Zhang, Y.S. Synthesis of a novel 1-ethyl-3-methylimidazolium chloride ionic liquid molecularly imprinted polymer and its properties of specific adsorption and solid phase extraction. *Acta Chimi. Sin.,* **2013**, *71*, 1411-1420.
[http://dx.doi.org/10.6023/A13040434]

[195] Yan, H.; Sun, N.; Han, Y.; Yang, C.; Wang, M.; Wu, R. Ionic liquid-mediated molecularly imprinted solid-phase extraction coupled with gas chromatography-electron capture detector for rapid screening of dicofol in vegetables. *J. Chromatogr. A,* **2013**, *1307*, 21-26.
[http://dx.doi.org/10.1016/j.chroma.2013.07.054] [PMID: 23932224]

[196] Han, D.D.; Tang, B.K.; Row, K.H. Dispersive solid phase extraction with an ionic liquid modified polymer for determination of cyanazine and atrazine in tomatoes. *Anal. Lett.,* **2013**, *46*, 2359-2371.
[http://dx.doi.org/10.1080/00032719.2013.800533]

[197] Tian, M.L.; Qiao, J.; Row, K.H. Facile preparation of an ionic liquid composite mesoporous polymer as a solid phase extraction adsorbent for the separation and purification of flavonoids from *Chamaecyparis obtusa. Anal. Lett.,* **2013**, *46*, 1331-1341.
[http://dx.doi.org/10.1080/00032719.2012.763170]

[198] Abdolmohammad-Zadeh, H.; Naseri, A.; Galeh-Assadi, M.; Shabkhizan, S. Optimization of solid-phase extraction based on a new sol-gel material using a response surface methodology for the determination of copper in water samples by flame atomic absorption spectrometry. *Int. J. Environ. Anal. Chem.,* **2013**, *93*, 279-297.
[http://dx.doi.org/10.1080/03067319.2011.609936]

[199] Gao, X.Y.; Pan, M.F.; Fang, G.Z.; Jing, W.; He, S.Y.; Wang, S. An ionic liquid modified dummy molecularly imprinted polymer as a solid-phase extraction material for the simultaneous determination of nine organochlorine pesticides in environmental and food samples. *Anal. Methods,* **2013**, *5*, 6128-6134.
[http://dx.doi.org/10.1039/c3ay41083h]

[200] Bi, W.; Tian, M.; Row, K.H. Solid-phase extraction of liquiritin and glycyrrhizin from licorice using porous alkyl-pyridinium polymer sorbent. *Phytochem. Anal.,* **2010**, *21*(5), 496-501.
[http://dx.doi.org/10.1002/pca.1227] [PMID: 20931626]

[201] Bi, W.; Tian, M.; Row, K.H. Evaluation of molecularly imprinted anion-functionalized poly(ionic liquid)s by multi-phase dispersive extraction of flavonoids from plant. *J. Chromatogr. B Analyt. Technol. Biomed. Life Sci.,* **2013**, *913-914*, 61-68.
[http://dx.doi.org/10.1016/j.jchromb.2012.11.018] [PMID: 23270940]

[202] Lombardo-Agüí, M.; García-Campaña, A.M.; Gámiz-Gracia, L.; Cruces Blanco, C. Laser induced fluorescence coupled to capillary electrophoresis for the determination of fluoroquinolones in foods of animal origin using molecularly imprinted polymers. *J. Chromatogr. A,* **2010**, *1217*(15), 2237-2242.
[http://dx.doi.org/10.1016/j.chroma.2010.02.016] [PMID: 20227081]

[203] Galán-Cano, F.; Lucena, R.; Cárdenas, S.; Valcárcel, M. Dispersive micro-solid phase extraction with ionic liquid-modified silica for the determination of organophosphate pesticides in water by ultra performance liquid chromatography. *Microchem. J.,* **2013**, *106*, 311-317.
[http://dx.doi.org/10.1016/j.microc.2012.08.016]

[204] Ghaedi, M.; Elhamifar, D.; Roosta, M.; Moshkelgosha, R. Ionic liquid based periodic mesoporous organosilica: An efficient support for removal of sunset yellow from aqueous solutions under ultrasonic conditions. *J. Ind. Eng. Chem.,* **2014**, *20*, 1703-1712.
[http://dx.doi.org/10.1016/j.jiec.2013.08.020]

[205] Li, M.; Pham, P.J.; Wang, T.; Pittman, C.U., Jr; Li, T. Solid phase extraction and enrichment of essential fatty acid methyl esters from soy-derived biodiesel by novel π-complexing sorbents. *Bioresour. Technol.,* **2009**, *100*(24), 6385-6390.
[http://dx.doi.org/10.1016/j.biortech.2009.07.054] [PMID: 19692237]

[206] Li, M.; Pham, P.J.; Wang, T.; Pittman, C.U., Jr; Li, T. Selective extraction and enrichment of polyunsaturated fatty acid methyl esters from fish oil by novel π-complexing sorbents. *Separ. Purif. Tech.,* **2009**, *66*, 1-8.
[http://dx.doi.org/10.1016/j.seppur.2008.12.009]

[207] Inagaki, S.; Guan, S.; Fukushima, Y.; Ohsuna, T.; Terasaki, O. Novel mesoporous materials with a uniform distribution of organic groups and inorganic oxide in their frameworks. *J. Am. Chem. Soc.,* **1999**, *121*, 9611-9614.
[http://dx.doi.org/10.1021/ja9916658]

[208] Yan, H.; Liu, S.; Gao, M.; Sun, N. Ionic liquids modified dummy molecularly imprinted microspheres as solid phase extraction materials for the determination of clenbuterol and clorprenaline in urine. *J. Chromatogr. A,* **2013**, *1294*, 10-16.
[http://dx.doi.org/10.1016/j.chroma.2013.04.024] [PMID: 23659982]

[209] Wang, X.; Xing, L.; Shu, Y.; Chen, X.; Wang, J. Novel polymeric ionic liquid microspheres with high exchange capacity for fast extraction of plasmid DNA. *Anal. Chim. Acta,* **2014**, *837*, 64-69.
[http://dx.doi.org/10.1016/j.aca.2014.06.002] [PMID: 25000859]

[210] Ríos, A.; Zougagh, M.; Bouri, M. Magnetic (nano)materials as an useful tool for sample preparation in analytical methods. A review. *Anal. Methods,* **2013**, *5*, 4558-4573.
[http://dx.doi.org/10.1039/c3ay40306h]

[211] Cui, C.; Hu, B.; Chen, B.B.; He, M. Ionic liquid-based magnetic solid phase extraction coupled with inductively coupled plasma-optical emission spectrometry for the determination of Cu, Cd, and Zn in biological samples. *J. Anal. At. Spectrom.,* **2013**, *28*, 1110-1117.
[http://dx.doi.org/10.1039/c3ja50108f]

[212] Parham, H.; Saeed, S. Ultrasound-assisted solid phase extraction of nitro- and chloro-(phenols) using magnetic iron oxide nanoparticles and Aliquat 336 ionic liquid. *J. Chromatogr. A,* **2014**, *1336*, 34-42.
[http://dx.doi.org/10.1016/j.chroma.2014.02.012] [PMID: 24582398]

[213] Khan, S.; Kazi, T.G.; Soylak, M. Rapid ionic liquid-based ultrasound assisted dual magnetic microextraction to preconcentrate and separate cadmium-4-(2-thiazolylazo)-resorcinol complex from environmental and biological samples. *Spectroc. Acta Pt. A-Molec. Biomolec. Spectr.,* **2014**, *123*, 194-199.
[http://dx.doi.org/10.1016/j.saa.2013.12.065]

[214] Galán-Cano, F.; Alcudia-León, Mdel.C.; Lucena, R.; Cárdenas, S.; Valcárcel, M. Ionic liquid coated magnetic nanoparticles for the gas chromatography/mass spectrometric determination of polycyclic aromatic hydrocarbons in waters. *J. Chromatogr. A,* **2013**, *1300*, 134-140.
[http://dx.doi.org/10.1016/j.chroma.2013.03.026] [PMID: 23601292]

[215] Deng, M.; Jiang, C.; Jia, L. N-methylimidazolium modified magnetic particles as adsorbents for solid phase extraction of genomic deoxyribonucleic acid from genetically modified soybeans. *Anal. Chim. Acta,* **2013**, *771*, 31-36.
[http://dx.doi.org/10.1016/j.aca.2013.02.005] [PMID: 23522109]

[216] Cao, X.; Shen, L.; Ye, X.; Zhang, F.; Chen, J.; Mo, W. Ultrasound-assisted magnetic solid-phase extraction based ionic liquid-coated Fe_3O_4@graphene for the determination of nitrobenzene compounds in environmental water samples. *Analyst (Lond.),* **2014**, *139*(8), 1938-1944.
[http://dx.doi.org/10.1039/c3an01937c] [PMID: 24575420]

[217] Polo-Luque, M.L.; Simonet, B.M.; Valcárcel, M. Ionic liquid combined with carbon nanotubes: a soft material for the preconcentration of PAHs. *Talanta,* **2013**, *104*, 169-172.
[http://dx.doi.org/10.1016/j.talanta.2012.10.025] [PMID: 23597905]

[218] Saljooqi, A.; Shamspur, T.; Mohamadi, M.; Mostafavi, A. Application of a thiourea-containing task-specific ionic liquid for the solid-phase extraction cleanup of lead ions from red lipstick, pine leaves, and water samples. *J. Sep. Sci.,* **2014**, *37*(14), 1856-1861.
[http://dx.doi.org/10.1002/jssc.201400332] [PMID: 24811255]

[219] Yan, H.; Gao, M.; Yang, C.; Qiu, M. Ionic liquid-modified magnetic polymeric microspheres as dispersive solid phase extraction adsorbent: a separation strategy applied to the screening of sulfamonomethoxine and sulfachloropyrazine from urine. *Anal. Bioanal. Chem.,* **2014**, *406*(11), 2669-2677.
[http://dx.doi.org/10.1007/s00216-014-7665-3] [PMID: 24573579]

[220] Bi, W.; Wang, M.; Yang, X.; Row, K.H. Facile synthesis of poly(ionic liquid)-bonded magnetic nanospheres as a high-performance sorbent for the pretreatment and determination of phenolic compounds in water samples. *J. Sep. Sci.,* **2014**, *37*(13), 1632-1639.
[http://dx.doi.org/10.1002/jssc.201400142] [PMID: 24737646]

[221] Zhang, R.; Su, P.; Yang, L.; Yang, Y. Microwave-assisted preparation of poly(ionic liquids)-modified magnetic nanoparticles for pesticide extraction. *J. Sep. Sci.,* **2014**, *37*(12), 1503-1510.
[http://dx.doi.org/10.1002/jssc.201400125] [PMID: 24687971]

[222] He, H.; Yuan, D.; Gao, Z.; Xiao, D.; He, H.; Dai, H.; Peng, J.; Li, N. Mixed hemimicelles solid-phase extraction based on ionic liquid-coated Fe_3O_4/SiO_2 nanoparticles for the determination of flavonoids in bio-matrix samples coupled with high performance liquid chromatography. *J. Chromatogr. A,* **2014**, *1324*, 78-85.
[http://dx.doi.org/10.1016/j.chroma.2013.11.021] [PMID: 24290172]

[223] Liu, X.; Lu, X.; Huang, Y.; Liu, C.; Zhao, S. Fe_3O_4@ionic liquid@methyl orange nanoparticles as a novel nano-adsorbent for magnetic solid-phase extraction of polycyclic aromatic hydrocarbons in environmental water samples. *Talanta,* **2014**, *119*, 341-347.
[http://dx.doi.org/10.1016/j.talanta.2013.11.039] [PMID: 24401423]

[224] Zhang, Q.; Yang, F.; Tang, F.; Zeng, K.; Wu, K.; Cai, Q.; Yao, S. Ionic liquid-coated Fe_3O_4 magnetic nanoparticles as an adsorbent of mixed hemimicelles solid-phase extraction for preconcentration of polycyclic aromatic hydrocarbons in environmental samples. *Analyst (Lond.),* **2010**, *135*(9), 2426-2433.
[http://dx.doi.org/10.1039/c0an00245c] [PMID: 20668754]

[225] Zhao, X.; Shi, Y.; Wang, T.; Cai, Y.; Jiang, G. Preparation of silica-magnetite nanoparticle mixed hemimicelle sorbents for extraction of several typical phenolic compounds from environmental water samples. *J. Chromatogr. A,* **2008**, *1188*(2), 140-147.
[http://dx.doi.org/10.1016/j.chroma.2008.02.069] [PMID: 18329033]

[226] Amjadi, M.; Samadi, A. Modified ionic liquid-coated nanometer TiO_2 as a new solid phase extraction sorbent for preconcentration of trace nickel. *Colloids Surf. A Physicochem. Eng. Asp.,* **2013**, *434*, 171-177.
[http://dx.doi.org/10.1016/j.colsurfa.2013.04.059]

[227] Xiao, D.; Yuan, D.; He, H.; Pham-Huy, C.; Dai, H.; Wang, C.; Zhang, C. Mixed hemimicelle solid-phase extraction based on magnetic carbon nanotubes and ionic liquids for the determination of flavonoids. *Carbon,* **2014**, *72*, 274-286.

[http://dx.doi.org/10.1016/j.carbon.2014.01.075]

[228] Saitoh, T.; Kondo, T.; Hiraide, M. Concentration of chlorophenols in water to dialkyated catinonic surfactant-silica gel admicelles. *J. Chromatogr. A,* **2007**, *1164*(1-2), 40-47. [http://dx.doi.org/10.1016/j.chroma.2007.07.021] [PMID: 17678937]

[229] Zhu, T.; Yoon, C.H.; Row, K.H. Solid phase extraction of three organic acids from *Salicornia herbacea* L. Using amino imidazolium ionic liquid-based cartridge. *Asian J. Chem.,* **2013**, *25*, 3731-3734.

[230] He, Z.; Liu, D.; Zhou, Z.; Wang, P. Ionic-liquid-functionalized magnetic particles as an adsorbent for the magnetic SPE of sulfonylurea herbicides in environmental water samples. *J. Sep. Sci.,* **2013**, *36*(19), 3226-3233.
[PMID: 23894051]

[231] Zhang, L.; Yu, R.; Wang, Z.; Li, N.; Zhang, H.; Yu, A. Determination of triazine herbicides in vegetables by ionic liquid foam floatation solid phase extraction high performance liquid chromatography. *J. Chromatogr. B Analyt. Technol. Biomed. Life Sci.,* **2014**, *953-954*, 132-137. [http://dx.doi.org/10.1016/j.jchromb.2014.02.011] [PMID: 24607694]

[232] Louch, D.; Motlagh, S.; Pawliszyn, J. Dynamics of organic compound extraction from water using liquid-coated fused silica fibers. *Anal. Chem.,* **1992**, *64*, 1187-1199.
[http://dx.doi.org/10.1021/ac00034a020]

[233] Zhang, Z.; Pawliszyn, J. Headspace solid-phase microextraction. *Anal. Chem.,* **1993**, *65*, 1843-1852. [http://dx.doi.org/10.1021/ac00062a008]

[234] Endres, F.; Zein El Abedin, S. Air and water stable ionic liquids in physical chemistry. *Phys. Chem. Chem. Phys.,* **2006**, *8*(18), 2101-2116.
[http://dx.doi.org/10.1039/b600519p] [PMID: 16751868]

[235] Ho, T.D.; Zhang, C.; Hantao, L.W.; Anderson, J.L. Ionic liquids in analytical chemistry: fundamentals, advances, and perspectives. *Anal. Chem.,* **2014**, *86*(1), 262-285.
[http://dx.doi.org/10.1021/ac4035554] [PMID: 24205989]

[236] Liu, J.F.; Li, N.; Jiang, G.B.; Liu, J-M.; Jönsson, J.Å.; Wen, M.J. Disposable ionic liquid coating for headspace solid-phase microextraction of benzene, toluene, ethylbenzene, and xylenes in paints followed by gas chromatography-flame ionization detection. *J. Chromatogr. A,* **2005**, *1066*(1-2), 27-32.
[http://dx.doi.org/10.1016/j.chroma.2005.01.024] [PMID: 15794551]

[237] Hsieh, Y.N.; Huang, P.C.; Sun, I-W.; Whang, T-J.; Hsu, C-Y.; Huang, H-H.; Kuei, C-H. Nafion membrane-supported ionic liquid–solid phase microextraction for analyzing ultra trace PAHs in water samples. *Anal. Chim. Acta,* **2006**, *557*, 321-328.
[http://dx.doi.org/10.1016/j.aca.2005.10.019]

[238] Huang, K.P.; Wang, G.R.; Huang, B.Y.; Liu, C.Y. Preparation and application of ionic liquid-coated fused-silica capillary fibers for solid-phase microextraction. *Anal. Chim. Acta,* **2009**, *645*(1-2), 42-47. [http://dx.doi.org/10.1016/j.aca.2009.04.037] [PMID: 19481629]

[239] Amini, R.; Rouhollahi, A.; Adibi, M.; Mehdinia, A. A novel reusable ionic liquid chemically bonded fused-silica fiber for headspace solid-phase microextraction/gas chromatography-flame ionization

detection of methyl tert-butyl ether in a gasoline sample. *J. Chromatogr. A,* **2011**, *1218*(1), 130-136. [http://dx.doi.org/10.1016/j.chroma.2010.10.114] [PMID: 21130999]

[240] Liu, M.; Zhou, X.; Chen, Y.; Liu, H.; Feng, X.; Qiu, G.; Liu, F.; Zeng, Z. Innovative chemically bonded ionic liquids-based sol-gel coatings as highly porous, stable and selective stationary phases for solid phase microextraction. *Anal. Chim. Acta,* **2010**, *683*(1), 96-106. [http://dx.doi.org/10.1016/j.aca.2010.10.004] [PMID: 21094387]

[241] López-Darias, J.; Pino, V.; Anderson, J.L.; Graham, C.M.; Afonso, A.M. Determination of water pollutants by direct-immersion solid-phase microextraction using polymeric ionic liquid coatings. *J. Chromatogr. A,* **2010**, *1217*(8), 1236-1243. [http://dx.doi.org/10.1016/j.chroma.2009.12.041] [PMID: 20044092]

[242] Graham, C.M.; Meng, Y.; Ho, T.; Anderson, J.L. Sorbent coatings for solid-phase microextraction based on mixtures of polymeric ionic liquids. *J. Sep. Sci.,* **2011**, *34*(3), 340-346. [http://dx.doi.org/10.1002/jssc.201000367] [PMID: 21268258]

[243] Feng, J.; Sun, M.; Xu, L.; Li, J.; Liu, X.; Jiang, S. Preparation of a polymeric ionic liquid-coated solid-phase microextraction fiber by surface radical chain-transfer polymerization with stainless steel wire as support. *J. Chromatogr. A,* **2011**, *1218*(43), 7758-7764. [http://dx.doi.org/10.1016/j.chroma.2011.08.076] [PMID: 21924728]

[244] Toledo, B.R.; Hantao, L.W.; Ho, T.D.; Augusto, F.; Anderson, J.L. A chemometric approach toward the detection and quantification of coffee adulteration by solid-phase microextraction using polymeric ionic liquid sorbent coatings. *J. Chromatogr. A,* **2014**, *1346*, 1-7. [http://dx.doi.org/10.1016/j.chroma.2014.04.035] [PMID: 24786655]

[245] Joshi, M.D.; Ho, T.D.; Cole, W.T.; Anderson, J.L. Determination of polychlorinated biphenyls in ocean water and bovine milk using crosslinked polymeric ionic liquid sorbent coatings by solid-phase microextraction. *Talanta,* **2014**, *118*, 172-179. [http://dx.doi.org/10.1016/j.talanta.2013.10.014] [PMID: 24274285]

[246] Jia, J.; Liang, X.; Wang, L.; Guo, Y.; Liu, X.; Jiang, S. Nanoporous array anodic titanium-supported co-polymeric ionic liquids as high performance solid-phase microextraction sorbents for hydrogen bonding compounds. *J. Chromatogr. A,* **2013**, *1320*, 1-9. [http://dx.doi.org/10.1016/j.chroma.2013.10.042] [PMID: 24210297]

[247] Zheng, J.; Li, S.; Wang, Y.; Li, L.; Su, C.; Liu, H.; Zhu, F.; Jiang, R.; Ouyang, G. *In situ* growth of IRMOF-3 combined with ionic liquids to prepare solid-phase microextraction fibers. *Anal. Chim. Acta,* **2014**, *829*, 22-27. [http://dx.doi.org/10.1016/j.aca.2014.04.039] [PMID: 24856398]

[248] Zhang, C.; Anderson, J.L. Polymeric ionic liquid bucky gels as sorbent coatings for solid-phase microextraction. *J. Chromatogr. A,* **2014**, *1344*, 15-22. [http://dx.doi.org/10.1016/j.chroma.2014.04.013] [PMID: 24768123]

[249] Stanisz, E.; Werner, J.; Matusiewicz, H. Task specific ionic liquid-coated PTFE tube for solid-phase microextraction prior to chemical and photo-induced mercury cold vapour generation. *Microchem. J.,* **2014**, *114*, 229-237. [http://dx.doi.org/10.1016/j.microc.2014.01.007]

[250] Yang, P.; Lau, C.; Liu, X.; Lu, J. Direct solid-support sample loading for fast cataluminescence

determination of acetone in human plasma. *Anal. Chem.,* **2007**, *79*(22), 8476-8485. [http://dx.doi.org/10.1021/ac0702488] [PMID: 17939643]

[251] He, Y.; Pohl, J.; Engel, R.; Rothman, L.; Thomas, M. Preparation of ionic liquid based solid-phase microextraction fiber and its application to forensic determination of methamphetamine and amphetamine in human urine. *J. Chromatogr. A,* **2009**, *1216*(24), 4824-4830. [http://dx.doi.org/10.1016/j.chroma.2009.04.028] [PMID: 19426983]

[252] Kumar, A.; Gaurav, ; Malik, A.K.; Tewary, D.K.; Singh, B. A review on development of solid phase microextraction fibers by sol-gel methods and their applications. *Anal. Chim. Acta,* **2008**, *610*(1), 1-14. [http://dx.doi.org/10.1016/j.aca.2008.01.028] [PMID: 18267134]

[253] Zhou, X.; Shang, J.; Liu, M.; Liu, H.; Hao, R.; Feng, X.; Liu, F. Preparation of novel ionic liquids-based sol-gel coatings for solid-phase microextraction. *Acta Chimi. Sin.,* **2010**, *68*, 1749-1757.

[254] Tian, M.; Cheng, R.; Ye, J.; Liu, X.; Jia, Q. Preparation and evaluation of ionic liquid-calixarene solid-phase microextraction fibres for the determination of triazines in fruit and vegetable samples. *Food Chem.,* **2014**, *145*, 28-33. [http://dx.doi.org/10.1016/j.foodchem.2013.08.029] [PMID: 24128445]

[255] Zhou, X.; Shao, X.; Shu, J.J.; Liu, M.M.; Liu, H.L.; Feng, X.H.; Liu, F. Thermally stable ionic liquid-based sol-gel coating for ultrasonic extraction-solid-phase microextraction-gas chromatography determination of phthalate esters in agricultural plastic films. *Talanta,* **2012**, *89*, 129-135. [http://dx.doi.org/10.1016/j.talanta.2011.12.001] [PMID: 22284470]

[256] Gao, Z.; Deng, Y.; Hu, X.; Yang, S.; Sun, C.; He, H. Determination of organophosphate esters in water samples using an ionic liquid-based sol-gel fiber for headspace solid-phase microextraction coupled to gas chromatography-flame photometric detector. *J. Chromatogr. A,* **2013**, *1300*, 141-150. [http://dx.doi.org/10.1016/j.chroma.2013.02.089] [PMID: 23541656]

[257] Yuan, J.; Mecerreyes, D.; Antonietti, M. Poly(ionic liquid)s: An update. *Prog. Polym. Sci.,* **2013**, *38*, 1009-1036. [http://dx.doi.org/10.1016/j.progpolymsci.2013.04.002]

[258] Zhao, F.; Meng, Y.; Anderson, J.L. Polymeric ionic liquids as selective coatings for the extraction of esters using solid-phase microextraction. *J. Chromatogr. A,* **2008**, *1208*(1-2), 1-9. [http://dx.doi.org/10.1016/j.chroma.2008.08.071] [PMID: 18805539]

[259] Augusto, F.; Carasek, E.; Silva, R.G.; Rivellino, S.R.; Batista, A.D.; Martendal, E. New sorbents for extraction and microextraction techniques. *J. Chromatogr. A,* **2010**, *1217*(16), 2533-2542. [http://dx.doi.org/10.1016/j.chroma.2009.12.033] [PMID: 20035942]

[260] González-Álvarez, J.; Blanco-Gomis, D.; Arias-Abrodo, P.; Pello-Palma, J.; Ríos-Lombardía, N.; Busto, E.; Gotor-Fernández, V.; Gutiérrez-Álvarez, M.D. Analysis of beer volatiles by polymeric imidazolium-solid phase microextraction coatings: Synthesis and characterization of polymeric imidazolium ionic liquids. *J. Chromatogr. A,* **2013**, *1305*, 35-40. [http://dx.doi.org/10.1016/j.chroma.2013.07.012] [PMID: 23891206]

[261] Ho, T.D.; Cole, W.T.; Augusto, F.; Anderson, J.L. Insight into the extraction mechanism of polymeric ionic liquid sorbent coatings in solid-phase microextraction. *J. Chromatogr. A,* **2013**, *1298*, 146-151. [http://dx.doi.org/10.1016/j.chroma.2013.05.009] [PMID: 23746645]

[262] Meng, Y.; Pino, V.; Anderson, J.L. Role of counteranions in polymeric ionic liquid-based solid-phase microextraction coatings for the selective extraction of polar compounds. *Anal. Chim. Acta,* **2011**, *687*(2), 141-149.
[http://dx.doi.org/10.1016/j.aca.2010.11.046] [PMID: 21277416]

[263] Wasielewska, M.; Zygmunt, B.; Anderson, J.L. Application of HS-SPME with poly(1-vinyl-3-hexylimidazolium chloride) polymeric ionic liquid coating using GC–MS to determine volatile fatty acids in dairy wastewater. *Chromatographia,* **2014**, *77*, 151-158.
[http://dx.doi.org/10.1007/s10337-013-2572-x]

[264] Meng, Y.; Pino, V.; Anderson, J.L. Exploiting the versatility of ionic liquids in separation science: determination of low-volatility aliphatic hydrocarbons and fatty acid methyl esters using headspace solid-phase microextraction coupled to gas chromatography. *Anal. Chem.,* **2009**, *81*(16), 7107-7112.
[http://dx.doi.org/10.1021/ac901377w] [PMID: 19637849]

[265] Zhao, Q.; Anderson, J.L. Selective extraction of CO_2 from simulated flue gas using polymeric ionic liquid sorbent coatings in solid-phase microextraction gas chromatography. *J. Chromatogr. A,* **2010**, *1217*(27), 4517-4522.
[http://dx.doi.org/10.1016/j.chroma.2010.04.080] [PMID: 20537653]

[266] López-Darias, J.; Anderson, J.L.; Pino, V.; Afonso, A.M. Developing qualitative extraction profiles of coffee aromas utilizing polymeric ionic liquid sorbent coatings in headspace solid-phase microextraction gas chromatography-mass spectrometry. *Anal. Bioanal. Chem.,* **2011**, *401*(9), 2965-2976.
[http://dx.doi.org/10.1007/s00216-011-5394-4] [PMID: 21935597]

[267] Meng, Y.; Anderson, J.L. Tuning the selectivity of polymeric ionic liquid sorbent coatings for the extraction of polycyclic aromatic hydrocarbons using solid-phase microextraction. *J. Chromatogr. A,* **2010**, *1217*(40), 6143-6152.
[http://dx.doi.org/10.1016/j.chroma.2010.08.007] [PMID: 20800235]

[268] López-Darias, J.; Pino, V.; Meng, Y.; Anderson, J.L.; Afonso, A.M. Utilization of a benzyl functionalized polymeric ionic liquid for the sensitive determination of polycyclic aromatic hydrocarbons; parabens and alkylphenols in waters using solid-phase microextraction coupled to gas chromatography-flame ionization detection. *J. Chromatogr. A,* **2010**, *1217*(46), 7189-7197.
[http://dx.doi.org/10.1016/j.chroma.2010.09.016] [PMID: 20933234]

[269] Bianchi, F.; Bisceglie, F.; Dugheri, S.; Arcangeli, G.; Cupelli, V.; del Borrello, E.; Sidisky, L.; Careri, M. Ionic liquid-based solid phase microextraction necklaces for the environmental monitoring of ketamine. *J. Chromatogr. A,* **2014**, *1331*, 1-9.
[http://dx.doi.org/10.1016/j.chroma.2014.01.012] [PMID: 24472840]

[270] Ho, T.D.; Joshi, M.D.; Silver, M.A.; Anderson, J.L. Selective extraction of genotoxic impurities and structurally alerting compounds using polymeric ionic liquid sorbent coatings in solid-phase microextraction: Alkyl halides and aromatics. *J. Chromatogr. A,* **2012**, *1240*, 29-44.
[http://dx.doi.org/10.1016/j.chroma.2012.03.080] [PMID: 22537606]

[271] Xu, L.; Jia, J.; Feng, J.; Liu, J.; Jiang, S. Polymeric ionic liquid modified stainless steel wire as a novel fiber for solid-phase microextraction. *J. Sep. Sci.,* **2013**, *36*(2), 369-375.
[http://dx.doi.org/10.1002/jssc.201200644] [PMID: 23335459]

[272] Pang, L.; Liu, J-F. Development of a solid-phase microextraction fiber by chemical binding of polymeric ionic liquid on a silica coated stainless steel wire. *J. Chromatogr. A,* **2012**, *1230*, 8-14. [http://dx.doi.org/10.1016/j.chroma.2012.01.052] [PMID: 22340892]

[273] Feng, J.; Sun, M.; Xu, L.; Wang, S.; Liu, X.; Jiang, S. Novel double-confined polymeric ionic liquids as sorbents for solid-phase microextraction with enhanced stability and durability in high-ioni--strength solution. *J. Chromatogr. A,* **2012**, *1268*, 16-21. [http://dx.doi.org/10.1016/j.chroma.2012.10.037] [PMID: 23127811]

[274] Ho, T.D.; Yu, H.; Cole, W.T.; Anderson, J.L. Ultraviolet photoinitiated on-fiber copolymerization of ionic liquid sorbent coatings for headspace and direct immersion solid-phase microextraction. *Anal. Chem.,* **2012**, *84*(21), 9520-9528. [PMID: 22991947]

[275] Feng, J.; Sun, M.; Wang, X.; Liu, X.; Jiang, S. Ionic liquids-based crosslinked copolymer sorbents for headspace solid-phase microextraction of polar alcohols. *J. Chromatogr. A,* **2012**, *1245*, 32-38. [http://dx.doi.org/10.1016/j.chroma.2012.05.026] [PMID: 22673810]

[276] Shearrow, A.M.; Harris, G.A.; Fang, L.; Sekhar, P.K.; Nguyen, L.T.; Turner, E.B.; Bhansali, S.; Malik, A. Ionic liquid-mediated sol-gel coatings for capillary microextraction. *J. Chromatogr. A,* **2009**, *1216*(29), 5449-5458. [http://dx.doi.org/10.1016/j.chroma.2009.04.093] [PMID: 19515375]

[277] Shearrow, A.M.; Bhansali, S.; Malik, A. Ionic liquid-mediated bis[(3-methyldimethoxysilyl)propyl] polypropylene oxide-based polar sol-gel coatings for capillary microextraction. *J. Chromatogr. A,* **2009**, *1216*(36), 6349-6355. [http://dx.doi.org/10.1016/j.chroma.2009.07.028] [PMID: 19643422]

[278] Lin, C-L.; Lirio, S.; Chen, Y-T.; Lin, C-H.; Huang, H-Y. A novel hybrid metal-organic framework-polymeric monolith for solid-phase microextraction. *Chemistry,* **2014**, *20*(12), 3317-3321. [http://dx.doi.org/10.1002/chem.201304458] [PMID: 24615770]

[279] Wanigasekara, E.; Perera, S.; Crank, J.A.; Sidisky, L.; Shirey, R.; Berthod, A.; Armstrong, D.W. Bonded ionic liquid polymeric material for solid-phase microextraction GC analysis. *Anal. Bioanal. Chem.,* **2010**, *396*(1), 511-524. [http://dx.doi.org/10.1007/s00216-009-3254-2] [PMID: 19924406]

[280] Zhao, F.; Wang, M.; Ma, Y.; Zeng, B. Electrochemical preparation of polyaniline-ionic liquid based solid phase microextraction fiber and its application in the determination of benzene derivatives. *J. Chromatogr. A,* **2011**, *1218*(3), 387-391. [http://dx.doi.org/10.1016/j.chroma.2010.12.017] [PMID: 21185028]

[281] Gao, Z.; Li, W.; Liu, B.; Liang, F.; He, H.; Yang, S.; Sun, C. Nano-structured polyaniline-ionic liquid composite film coated steel wire for headspace solid-phase microextraction of organochlorine pesticides in water. *J. Chromatogr. A,* **2011**, *1218*(37), 6285-6291. [http://dx.doi.org/10.1016/j.chroma.2011.07.041] [PMID: 21821255]

[282] Chen, C.; Liang, X.; Wang, J.; Zou, Y.; Hu, H.; Cai, Q.; Yao, S. Development of a polymeric ionic liquid coating for direct-immersion solid-phase microextraction using polyhedral oligomeric silsesquioxane as cross-linker. *J. Chromatogr. A,* **2014**, *1348*, 80-86. [http://dx.doi.org/10.1016/j.chroma.2014.04.098] [PMID: 24857037]

[283] Feng, J.; Sun, M.; Li, L.; Wang, X.; Duan, H.; Luo, C. Multiwalled carbon nanotubes-doped polymeric ionic liquids coating for multiple headspace solid-phase microextraction. *Talanta,* **2014**, *123*, 18-24. [http://dx.doi.org/10.1016/j.talanta.2014.01.030] [PMID: 24725859]

[284] Zhao, Q.; Twu, P.; Anderson, J.L. A rapid analytical method for monitoring the enantiomeric purity of chiral molecules synthesized in ionic liquid solvents. *Chirality,* **2012**, *24*(3), 201-208. [http://dx.doi.org/10.1002/chir.21983] [PMID: 22271571]

[285] Cai, M-Q.; Wei, X.Q.; Du, C-H.; Ma, X-M.; Jin, M-C. Novel amphiphilic polymeric ionic liquid-solid phase micro-extraction membrane for the preconcentration of aniline as degradation product of azo dye Orange G under sonication by liquid chromatography-tandem mass spectrometry. *J. Chromatogr. A,* **2014**, *1349*, 24-29. [http://dx.doi.org/10.1016/j.chroma.2014.05.008] [PMID: 24857038]

[286] Wang, T-T.; Chen, Y-H.; Ma, J-F.; Hu, M-J.; Li, Y.; Fang, J-H.; Gao, H-Q. A novel ionic liquid-modified organic-polymer monolith as the sorbent for in-tube solid-phase microextraction of acidic food additives. *Anal. Bioanal. Chem.,* **2014**, *406*(20), 4955-4963. [http://dx.doi.org/10.1007/s00216-014-7923-4] [PMID: 24939131]

[287] Martín-Calero, A.; Ayala, J.H.; González, V.; Afonso, A.M. Ionic liquids as desorption solvents and memory effect suppressors in heterocyclic aromatic amines determination by SPME-HPLC fluorescence. *Anal. Bioanal. Chem.,* **2009**, *394*(4), 937-946. [http://dx.doi.org/10.1007/s00216-008-2568-9] [PMID: 19104777]

[288] Baltussen, E.; Sandra, P.; David, F.; Cramers, C. Stir bar sorptive extraction (SBSE), a novel extraction technique for aqueous samples: Theory and principles. *J. Microcolumn Sep.,* **1999**, *1110*, 737-747. [http://dx.doi.org/10.1002/(SICI)1520-667X(1999)11:10<737::AID-MCS7>3.0.CO;2-4]

[289] Lancas, F.M.; Queiroz, M.E.; Grossi, P.; Olivares, I.R. Recent developments and applications of stir bar sorptive extraction. *J. Sep. Sci.,* **2009**, *32*(5-6), 813-824. [http://dx.doi.org/10.1002/jssc.200800669] [PMID: 19278005]

[290] Talebpour, Z.; Taraji, M.; Adib, N. Stir bar sorptive extraction and high performance liquid chromatographic determination of carvedilol in human serum using two different polymeric phases and an ionic liquid as desorption solvent. *J. Chromatogr. A,* **2012**, *1236*, 1-6. [http://dx.doi.org/10.1016/j.chroma.2012.02.063] [PMID: 22458964]

[291] Zarghi, A.; Foroutan, S.M.; Shafaati, A.; Khoddam, A. Quantification of carvedilol in human plasma by liquid chromatography using fluorescence detection: application in pharmacokinetic studies. *J. Pharm. Biomed. Anal.,* **2007**, *44*(1), 250-253. [http://dx.doi.org/10.1016/j.jpba.2007.01.026] [PMID: 17391893]

[292] Huang, X.; Yuan, D. Preparation of stir bars for sorptive extraction based on monolithic material. *J. Chromatogr. A,* **2007**, *1154*(1-2), 152-157. [http://dx.doi.org/10.1016/j.chroma.2007.03.132] [PMID: 17477931]

[293] Huang, X.; Qiu, N.; Yuan, D.; Lin, Q. Preparation of a mixed stir bar for sorptive extraction based on monolithic material for the extraction of quinolones from wastewater. *J. Chromatogr. A,* **2010**, *1217*(16), 2667-2673. [http://dx.doi.org/10.1016/j.chroma.2009.09.072] [PMID: 19853255]

[294] Huang, X.; Wang, Y.; Hong, Q.; Liu, Y.; Yuan, D. Preparation a new sorbent based on polymeric ionic liquid for stir cake sorptive extraction of organic compounds and inorganic anions. *J. Chromatogr. A,* **2013**, *1314*, 7-14.
[http://dx.doi.org/10.1016/j.chroma.2013.09.011] [PMID: 24075458]

[295] Huang, X.; Chen, L.; Yuan, D.; Bi, S. Preparation of a new polymeric ionic liquid-based monolith for stir cake sorptive extraction and its application in the extraction of inorganic anions. *J. Chromatogr. A,* **2012**, *1248*, 67-73.
[http://dx.doi.org/10.1016/j.chroma.2012.06.004] [PMID: 22727559]

CHAPTER 4

Application of Nanomaterials in Solid and Liquid Microextraction

Yolanda Moliner-Martínez, Pascual Serra-Mora, Jorge Verdú-Andrés, Carmen Molins-Legua, Rosa Herráez-Hernández and **Pilar Campíns-Falcó***

Department of Analytical Chemistry, Faculty of Chemistry, University of Valencia, Dr. Moliner 50, 46100-Burjassot, Valencia, Spain

Abstract: Nanomaterials (NMs) have attracted great attention in sample preparation. In particular, because of their high surface-to-volume ratios, NMs facilitate the implementation of microextraction techniques. Additional advantages derived from NMs are the possibility of increasing the selectivity through the functionalization of their surfaces, and the improvement of mechanical and thermal stability of the extraction devices. This chapter summarizes the main uses of NMs in solid and liquid microextraction techniques, and representative examples of applications are presented.

The fabrication of coatings for fiber solid-phase microextraction (SPME) is the main objective in many of the scientific research developed in the area. For this purpose, a variety of NMs have been used such as carbon-based NMs, especially carbon nanotubes (CNTs) and graphene, metallic and silica-based NMs, and more recently, metal-organic frameworks (MOFs). NMs have also been used to prepare new sorbents for other microextraction techniques such as in-tube solid-phase microextraction(IT-SPME) and stir bar sorptive extraction (SBSE). The employment of magnetic nanoparticles (MNPs) has recently been introduced in microextraction, which leads to a new technique termed magnetic IT-SPME. Although the number of applications is still low, NMs are also receiving increasing attention in the main techniques of liquid phase microextraction namely, singe drop microextraction (SDME), hollow-fiber liquid-phase microextraction (HF-LPME) and dispersive liquid-liquid microextraction (DLLME).

* **Corresponding author Pilar Campíns-Falcó:** Department of Analytical Chemistry, Faculty of Chemistry, University of Valencia, Dr. Moliner 50, 46100-Burjassot, Valencia Spain,; Tel: 34 96 354 3002; Fax: 34 96 354 4436; E-mail: pilar.campins@uv.es

M. Valcárcel, S. Cárdenas, R. Lucena (Eds.)

Keywords: Carbon nanotubes (CNTs), Dispersive liquid-liquid microextraction (DLLME), Fiber coatings, Fiber solid phase microextraction (fiber SPME), Graphene, Head space solid-phase microextraction (HS-SPME), Hollow-fiber liquid-liquid microextraction (HF-LPME), In-tube solid phase microextraction (IT-SPME), Liquid-phase microextraction (LPME), Magnetic in-tube solid phase microextraction, Magnetic nanoparticles (MNPs), Metal organic frameworks (MOFs), Metallic nanoparticles, Multi-walled carbon nanotubes (MWCNTs), Nanomaterials (NMs), Nanoparticles (NPs), Silica nanoparticles, Single-drop microextraction (SDME), Single-walled carbon nanotubes (SWCNTs), Solid phase microextraction (SPME), Stir-bar sorptive extraction (SBSE).

4.1. INTRODUCTION

The reduction of the efforts and materials used for sample preparation is an important demand of analytical laboratories. Consequently, methodologies traditionally used to isolate and pre-concentrate the analytes such as LLE,SPE, Soxhlet extraction and others, are being gradually replaced by more simple and environmental friendly techniques. Some of those techniques are miniature versions of regular methods such as micro SPE, while others are based on the employment of a small amount of extractive phase compared to the amount of sample. The later approach is the basis of microextraction techniques. Nowadays, the most popular microextraction techniques are SPME and SBSE, both based on the employment of a sorbent as extractive medium, as well as LPME in which the extractive phase is a solvent or mixture of solvents. There are numerous variations of each of these microextraction techniques which differ in the nature of the sorbent/solvent, the extent of the extraction process, and the geometry of the extraction device.

In the last years, many of the research contributions in the area of microextraction have been focused on the development of new extractive phases that provide better selectivity and extraction efficiency. Examples of such extractive phases are new polymers including MIPs, ILs and NMs [1 - 4].

Nanosized materials have become a hot issue of research in many areas of science and technology because of their attractive physical and chemical properties, wide variety of morphologies and relative low cost. In analytical chemistry NMs have been extensively used for different purposes such as the development of sensors, the improvement of the separation performance in chromatographic and electrophoretic methods, and in sample preparation [5 - 7]. The most interesting feature of NMs for sample preparation is related to their high surface-to-volume ratios. This is particularly useful in the context of the reduction of the extractive phase dimensions. Consequently, an increasing effort is being focused on the application of NMs in microextraction. Additional advantages of NMs based sorbents are the enhancement of the selectivity through the functionalization of the NMs surfaces, and their better stability compared to other extractive media.

The term NMs is used for nanosized structures in which at least one dimension is in the nanometer size range, typically between 1 and 100 nm. Nanosized structures may be in the form of particles, tubes or pores [6]. There are several criteria to classify NMs based on their origin, chemical nature and homogeneity. In the context of their use as sorbents, NMs are typically classified as carbonaceous materials, metal/metal oxide nanoparticles (NPs) including magnetic NPs, silica NPs and MOFs [7]. Composites are also widely used in extraction because they combine the advantages of different phases in the same sorbent. So far carbon-based NMs and metallic NPs are the most explored phases in microextraction techniques. More recently, other NMs such MOFs have been tested for sample preparation but most applications still deal with SPE and μ-SPE procedures [8].

In the present chapter the main applications of NMs in microextraction will be discussed. Most attention will be focused on SPME because, as stated above, NMs are generally used as sorbents. In the last years, however, different studies have demonstrated the potential of NMs to improve the performance of LPME procedures by adding them to the sample or to the extractive phase. Consequently, the role of NMs in the main modalities of LPME will be also discussed. In Fig. (**1**) the main formats of solid and liquid microextraction techniques are depicted that have benefited from NMs.

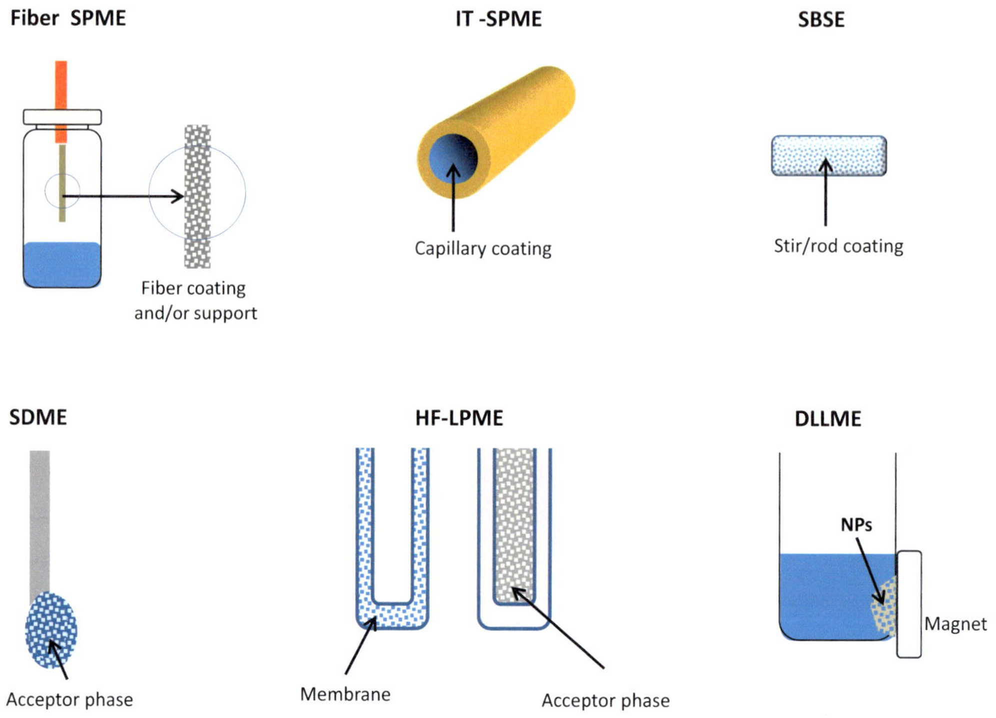

Fig. (1). NMs in the main solid and liquid microextraction techniques.

4.2. NMs IN SPME

4.2.1. SPME with Fibers

Since its introduction in the early 1990s [9], SPME has been widely used for sample preparation due to its inherent advantages such simplicity of operation, miniaturization, the reduction or even elimination of solvents and direct coupling to chromatographic and electrophoretic systems. In its classical configuration SPME is based on the distribution of the analytes between the sample (or the headspace above it) and a fused silica fiber coated with a polymeric extractive phase. After extraction the analytes are desorbed from the fibers by exposing them to high temperatures or by immersing them into a proper solvent. Thermal desorption is the option commonly selected in GC, whereas solvent desorption is used in LC. Nowadays fiber SPME is a well-established technique, and commercial fibers are available with different kinds of coatings that have been successfully used in many areas of application such as environmental, food, pharmaceutical and forensic analysis. In spite of its clear advantages, fiber SPME has a number of drawbacks such as mechanical fragility, the limited amount of extractive phase that can be placed in the fiber, the reduced operating thermal range and the lack of selectivity of available coatings, as only few coatings are commercialized (PDMS, DVB, PA, CW and CAR). These coatings are adequate for extraction of analytes of low-medium polarity but the extraction efficiency attainable for polar compounds is often low. It should be remarked that the amount of extractive phase of commercial fibers is very low (*e.g.*, 0.5 μL for a 100 μm film thickness coating) [10]. As a result, the amount of polar analytes extracted into the fiber may be insufficient to generate a suitable response in current analytical instruments.

To overcome such limitations many research efforts have been focused in the development of new extraction devices as well as in the preparation of new sorbents. The exploration of extraction devices that enable the employment of larger amounts of extractive phase is behind the development of other forms of sorbent-based microextraction such as IT-SPME. As regards the preparation of more efficient sorbents, most attention has been paid to the synthesis of polymeric coatings, as well as to the employment of NMs-based coatings [1, 11, 12]. NMs offer additional advantages such as improved mechanical, thermal and chemical stability of the fibers. This is illustrated in Fig. (**2**), which shows the main advantages provided by NMs in fibre SPME as well as in other microextraction techniques.

There are different approaches to prepare fiber coatings with NMs [13]. The method is generally selected according to the NMs and to the coating support (silica fiber, metallic wire). As regards the sorbent type, several classes of NMs have been used for preparations of fiber coatings, being carbon-based materials the most commonly used, but metallic, silica and MOFs-based coatings been also employed.

4.2.1.1. Carbon-Based NMs

Carbon-based NMs have been extensively used as sorbents for SPE and SPME because they can interact with organic analytes through hydrogen bonding, π-π stacking interactions, electrostatic and van der Waals forces, and hydrophobic interactions [6]. Carbon-based NMs comprise a variety of allotropic forms being CNTs, grapheme and fullerene the most widely used ones in SPME, directly or after chemical modification. Other allotropes such as nanofibres and nanocones/disks are also increasingly used [10].

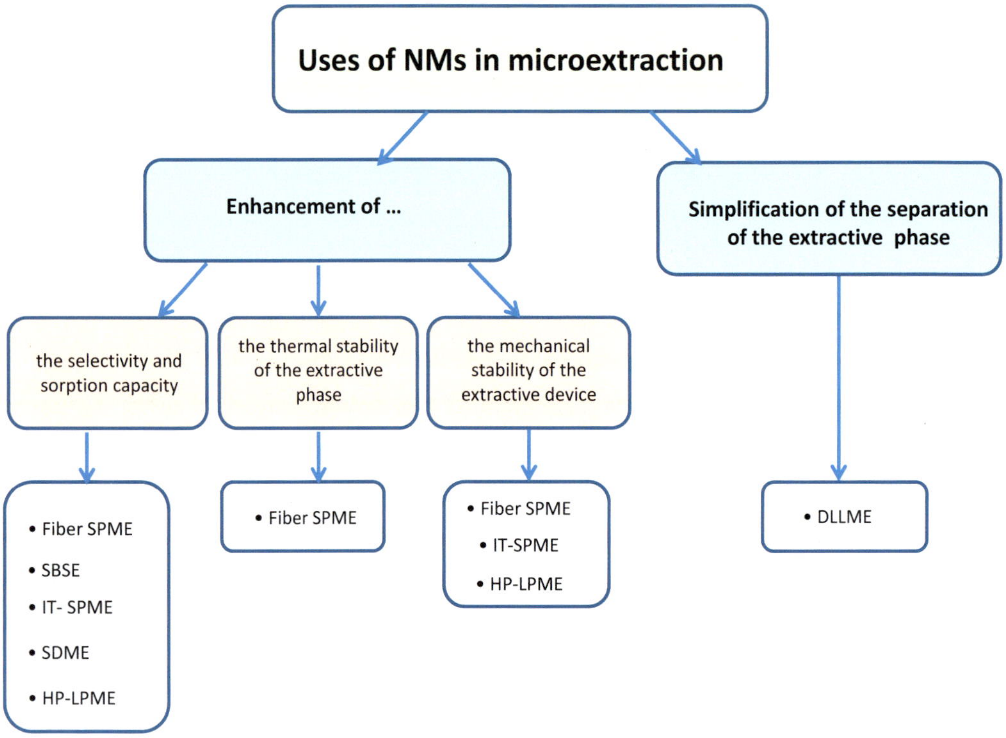

Fig. (2). Summary of the uses of NMs in microextraction techniques.

CNTs are graphite lamella sheets rolled either on a single tube (SWCNTs) or concentric SWCNTs held together by van der Waals forces (MWCNTs). The lengths of the CNTs can range from a few hundred nanometers to several microns, whereas diameters range from 1-10 nm and to 5 nm-a few hundred nm for SWCNTs and MWCNTs, respectively. The surface areas of CNTs are in the 150-1500 m^2 g^{-1} range. In general, the adsorption capacity of MWCNTs is superior than that of SWCNTs. In addition, acidic oxidation of CNTs leads to the introduction of carboxylic groups in the surface which enhances the retention efficiency for charged or polar compounds at proper pHs, and facilitates their chemical modification. Carboxylic acid functionalization is typically accomplished by treatment with a mixture of concentrated sulphuric and nitric acids.Owing to their high adsorptive capacity CNTs have been extensively used as sorbents for SPE and μ-SPE, as well as in SPME [1, 4, 14, 15].

The first application of CNTs as SPME sorbent was described by Wang *el at*, who prepared MWCNTs coatings by dipping silica fibers into a suspension of MWCNTs in dimethylformamide [16]. The fibers were used for the extraction of PBDEs in water and milk samples by immersing them directly into the samples. Then, the fibers were inserted into a GC injector at 295 ^{0}C for desorption of the analytes. Considerable enhancement factors in the responses compared to typical coatings such as PDMS and activated carbon were achieved. An important limitation of fused silica fibers is their mechanical fragility. In order to overcome this problem many procedures have been developed in which silica fibers are replaced by metallic wires (stainless steel, titanium, platinum). For example, SWCNTs were immobilised into stainless steel wires using a suspension of the CNTs in a binder made of terpineol, ethylcellulose and dibuthylphthalate [17]. After deposition, the organic binder was removed by thermal treatment of the fiber at 400 ^{0}C. The fiber was used for the extraction of some volatile ethers from urine samples by HS-SPME. Many other binders have been proposed to immobilise CNTs onto both, silica and metallic fibers. In this sense, Nafion/CNTs coatings are very attractive because Nafion acts as a binder and also as a sorbent, as reported in [18]. This coating showed excellent extraction efficiency for the target analytes (polar aromatic compounds) compared to commercial PA fibers.

The sol-gel approach has been proposed to prepare fibre coatings because it provides strong chemical bonding between the substrate and the coating. A limitation of the sol-gel approach is the low solubility of CNTs in organic solvents; thus, functionalized CNTs are generally preferred. On the other hand, the sol-gel approach is compatible with the employment of metallic supports. As an example, PEG/MWCNTs coatings have been developed which improve the extraction performance with respect PEG coated fibers for different kinds of compounds [19]. MWCNTs were first oxidized to give carboxylic acid functionalized MWCNTs, and then the carboxylic groups were converted into acyl chloride groups with $SOCl_2$, which were then made to react with PEG. The resulting MWCNTs-PEG derivative was immobilised onto a fused silica fiber (previously activated with sodium hydroxide) *via* a sol-gel process. This fibre was successfully used for the extraction of BTEX from water samples.

Another approach is the preparation of coatings by covalent bonding, as described by Liu *et al.* [20]. The authors proposed the chemical modification of a silica fiber in order to introduce amino groups, which were then made to react with oxidized MWCNTs. The silica fibers were first activated with a solution of NaOH and then treated with APTES, whereas MWCNTs were first oxidized with a mixture of H_2SO_4/HNO_3 acids to create –COOH on the surface. The coating was successfully used to extract phenols from water by direct immersion of the fibers into the samples. The stability of the fiber in acid, basic, organic solvents and high temperatures was remarkable.

Metallic supports are suitable for the preparation of coatings through the electrodeposition of charged NPs. A careful selection of the applied voltage enables a better control of the coating thickness; therefore, improved fiber-to-fiber reproducibilities are generally achieved. This methodology was used to prepare SWCNTs coatings on Pt wires [21]. The fiber was used in a SPME-LC chamber interface providing higher stability to the mobile phase than commercial fibers, according to the authors. In addition, the coating provided extraction efficiencies comparable or even higher than PA fibers for the extraction of some endocrine disruptors from water samples.

Polymeric/CNTs-based composites have been used as SPME coatings, combining the properties of CNTs and the guest components. Conductive polymers such as PPy or PANI are the most attractive polymers for such purpose. This kind of fibers are especially well suited for electrochemically enhanced SPME. The selectivity can be further modified trough the employment of the proper MIP. An example of this kind of sorbents is the molecular imprinted PPy/MWCNTs coating developed by Liu *et al.* [22]. The extractive phase was prepared by depositing MWCNTs on a Pt wire by the electrodeposition approach. Then, MIP was prepared on the surface of the MWCNTs/Pt wire by electropolymerization of pyrrole monomer in the presence of the template molecules (ofloxacin). The fibers were used for the electrochemically enhanced SPME of fluoroquinolone antibiotics from aqueous matrices.

The advantages of CNTs in fiber SPME have been extensively exploited, and many other examples of applications in different areas can be found in the literature [4 - 7, 14, 15].

Another allotropic form of carbon increasingly used in sample (micro)extraction is graphene. Graphene is a single-atom-thick two-dimensional lattice of carbon nano-sized material. It has a theoretical specific area of 2630 m^2 g^{-1}, and both sides are available for adsorption, thus, offering higher adsorption capacity than CNTs. Besides its large effective surface area, graphene has remarkable thermal and mechanical stability. Owing to its large delocalized π-electron system, graphene can form π-π stacking interactions with aromatic compounds. Graphene can be also oxidized to graphene oxidein order to introduce a variety of functional groups. An important advantage in the preparation of graphene oxide coatings is that this material possesses better water dispersibility than other carbon-based NMs. On the other hand, the purity of NMs is a key factor in the performance of the extraction. In this sense, graphene is advantageous over other carbon-based NMs because it can be obtained without using metals as catalyst [23].

Graphene can be immobilized as a fiber coating simply by immersing repeatedly the fiber into a graphene suspension (in methanol) until the desirable layer thickness is reached. This approach was used in [24] to prepare 6-8 μm coatings on stainless steel wires, which were then applied to the extraction of pyrethroid pesticides by GC with direct immersion of the fibers into the samples. The high extraction efficiencies achieved suggested a strong π-stacking interaction between graphene and the analysis. It is also possible to immobilize graphene on stainless steel fibers using silicone glue, as reported in [25]. For this purpose the fiber was successively dipped into the adhesive solution (silicone in toluene) and a graphene nanosheets powder, which was microwave synthetized by reduction of graphene oxide. This coating was tested for the HS-SPME of some organochlorine pesticides, which were then subjected to GC. According to the authors, these fibers were superior in terms of extraction efficiency and mechanical and thermal stability than commercial PDMS fibers.

To enhance the stability of the coating the sol-gel approach has been recommended in order to immobilize graphene coating on a stainless steel substrate (a plunger-in-needle microsyringe) [26]. The microextraction device was tested for PBDEs in waters by direct immersion of the fibers into the samples and GC providing better efficiencies than commercial PDMS and PA fibers, as well as excellent durability. Covalent methods have been also proposed to prepare graphene based coatings. For example, graphene oxide has been immobilized on amino-funcionalized silica fibers using a procedure similar to that described for MWCNTs [20].The coating was used for extraction of PAHs from water [27].

Finally, graphene/PPy and graphene oxide/PPy composite coatings have been prepared by electrochemical polymerization of PPy and graphene on stainless steel wires. The best extraction efficiencies for phenols were obtained with the graphene/PPy coating, being also better than those obtained with PPy coated fibers, as well as with commercial fibers [28].

Although less used, different assays have been reported on the application of other allotropic forms of carbon NMs in SPME. This is the case of fullerenes, spherical NPs in which carbon atoms are bonded in arrangements of five-six-membered rings. Fullerene C_{60}, is the most representative member of the group, and it has been successfully used as sorbent for the extraction of compounds with electron donor groups [29]. The coating was prepared by dipping a PDMS coated fiber into a solution containing a photosensitive reagent and fullerene in heptane. Then the fiber was exposed to UV radiation to complete polymerization and cured at 300 ^{0}C. The resulting fibers were used for the extraction of BTEX by HS-SPME prior to their analysis by GC with thermal desorption taking advantage of the high thermal stability of fullerene. The same authors developed a hydroxyfullere based coating using a sol-gel approach for the extraction of PCBs, PAHs and aromatic amines [30]. In both cases, the thermal stability and life span of the fibers were superior to those provided by commercial PDMS fibers.

Carbon nanocones are hollow conical shape structures originated by vapor deposition of carbon atoms on a graphite substrate. The disclination of such cones is determined by the number of pentagons defining the conical apex, which ranges from zero (nanodisks) to five. The presence of the pentagonal rings results in an enhanced electronic density at the cone apex. Nanocones/disks have been used in the preparation of a novel coating for the extraction of different volatile compounds by HS-SPME. The sorbent was immobilized on a stainless steel needle with an organic binder [31]. The lower aggregation tendency of nanocones/disks compared to other carbon NMs simplified the preparation of the coating.

Coatings with carbon nanofibres (solid carbon fibers of diameters below 100 nm) deposited on stainless steel supports have also been described [32, 33]. The preparation of the coating involves electrospinning of a proper polymer onto the support, followed by the pyrolysis of the resulting phase. The resulting fibers were used for the extraction of BTEX and different phenols in HS-SPME mode, as well as for the extraction of non-volatiles β-blockers in aqueous samples. These coatings showed no swelling when immersed in a variety of liquids, as well as excellent extraction capability. A decrease in the detection limits achieved with the nanofiber based coating of 2-3 orders of magnitude was found in relation with the values obtained with commercial PA fibers [33].

4.2.1.2. Metallic NMs

Different metals and metal oxide offer interesting advantages in sample preparation when used as NPs. Of particular interest are MNPs. These particles have a magnetic core (*i. e.* Fe_3O_4) with a proper coating, typically of silica. Silica-coated magnetic materials present the advantages of easy surface modification to improve the selectivity of the extraction. MNPs are used in dispersive SPE-based approaches. In dispersive SPE extraction the sorbent is not immobilized in a column, cartridge, disk, tip.. but directly added to the bulk sample. The resulting dispersion is stirred for a defined period of time so the analytes are selectively retained in the sorbent. Then, the sorbent is separated from the sample solution by centrifugation or filtration. Finally, the analytes are released with an organic solvent or by thermal desorption. Alternatively, the sorbent can be added to the bulk sample in order to extract undesired matrix compounds. MNPs are well-suited for dispersive extraction not only for their excellent extraction capabilities but also because after extraction the particles with the retained analytes can be easily separated from the sample solution with a magnet. Therefore, the extraction procedure is significantly simplified.

As regards SPME, some metals such as gold and silver can be easily obtained as NPs providing high surface-to-volume ratios and stability, and specific interactions for some compounds. This is the case of AuNPs, which have the ability to adsorb spontaneously organic molecules containing thiol or amino groups. The potential utility of AuNPs in the preparation of extractive phases for SPME has been illustrated in [34]. The extractive phase was prepared by repeated dipping a stainless steel wire into a solution of $HAuCl_4$, and then into a solution of 1,8-octanedithiol to form a self-assembled monolayer. This coating showed remarkable selectivity towards compounds with high hydrophobicity and large π-donor structures, as demonstrated for some PAHs. The coating showed excellent selectivity and thermal stability, and durability towards acid and alkali conditions. Nevertheless, according to the authors the layer-by-layer self-assembly process was relatively cumbersome, so an alternative technique was proposed for the preparation of silver coated fibres based on the sedimentation of this metal onto the support under alkaline conditions and in the presence of ammonia and glucose [35]. Due to the very porous structure of the coating, the fibres provided excellent extraction efficiencies for PAHs and phthalates.

Metal oxide coatings also very attractive due to their higher thermal, mechanical and chemical stability, and larger area to volume ratios compared with other NPs. Some applications have been described which use ZnO nano rod coatings. ZnO nanorods can be prepared by a hydrothermal method, simply by immersing activated silica fibers into a solution containing a mixture of zinc nitrate and hexamethylenetetramine at 90 °C [36]. The resulting fibre showed excellent efficiency for the simultaneous HS-SPME of polar and non-polar compounds, and suitable stability up to 300 °C. A similar coating but on steel wires was used for HS-SPME of aldehydes [37]. Nanostructured TiO_2 coating of about 1.2 μm thick was prepared by oxidation of titanium wires with H_2O_2 at ambient temperature in [38]. The resulting fibres,which were used for the HS-SPME of DDT and its degradation products, exhibited long life span (over 150 times). Highly oriented TiO_2 nanotubes coated fibres have been obtained by an electrochemical method in [39]. Such nanotubes were obtained using a Ti wire as a working electrode and an electrolyte containing ethylene glycol and NH_4F. The fiber provided good recoveries of PAHs, anilines, phenols and alkanes prior to GC analysis. Nanostructured surface Al_2O_3 obtained with aluminium wires by electrochemical process method was used for HS-SPME of volatile compounds [40].

As stated earlier, the employment of composites in microextraction may be advantageous. For example, a ZnO/graphene composite has been proposed for the extraction of volatile-sulphur volatiles compounds [41]. The ZnO/graphene composite was attached to a silica fiber by a sol-gel procedure. Unlike pure ZnO coated fibres, the ZnO/graphene coating showed selectivity towards sulphur compounds. In addition, the presence of graphene increased the surface area. As a result, this fiber coating provided excellent enrichment factors for the target analytes.

4.2.1.3. MOFs

Self-assembled structures of metal oxide cations and organic electron donor linkers are the basis of MOFs. These nanostructured materials are also increasingly used in sample preparation because they present surfaces areas in the 1000-10400 $m^2 g^{-1}$ range. They also offer excellent thermal stability. Examples of the utility of MOFs-based coatings for fiber SPME have been reported by Yan *et al.* for the extraction of benzene homologues from air samples [42]. The coating was fabricated by hydrothermal grow of MOF-199 (coordination of Cu(II) and 1,3,5-benzene-tricarboxilic acid) films on the etched surface of a stainless steel fiber. The extraction efficiencies obtained were higher than those achieved with commercial PDMS/DVB fibers. However, the extraction performance was affected by moisture, which is a problem commonly encountered in MOFs-based extractions. In another study, a fiber was coated with a ZIP-8MOF (coordination of Zn(II) and 2-methylimidazole); this coating was used for the extraction of alkanes from fuel and biological samples by the HS-SPME approach [43]. The fibbers were used in combination with a ZIP-8 coated capillary column for the separation of the target analytes, providing not only improved extraction recoveries compared to PDMS/DVB coatings, but also excellent resolution.

4.2.1.4. Silica NPs and Nanoporous Silica Materials

Silica NPs are very attractive sorbents because they can be easily functionalized through the silanol groups on their surface. These NPs are also frequently used to prepare nanocomposites with additional advantages. For example, silica-coated Fe_3O_4 NPs combine enhanced selectivity *via* a surface reaction and magnetic properties. Thus, this type of NPs has been extensively used in SPE and magnetic-SPE. However, their use in SPME is still insufficiently explored.

Nanoporous silica functionalized with 3-[bis(2-hydroxyethyl)amino] propyltriethoxysilane was proposed as coating for the HS-SPME of BTEX [44]. The sorbent was immobilised onto cupper wires with epoxy glue. The preparation of a PANI-silica nanocomposite coating was described in [45]. The coating was prepared by electropolymerization on the surface of a stainless steel plunger which was immersed into a solution of aniline containing silica NPs. The coating exhibited non-smooth and porous structure in comparison with PANI film coatings, which in turn resulted in higher surface area. The fibers were applied to the trace level extraction of some PAHs from water.

4.2.2. IT-SPME

IT-SPME was developed as an alternative to fiber SPME in order to overcome the problems associated with the mechanical weakness of fibers, the small dimensions of the coatings and the difficulty of combining the technique with LC. In this form of SPME the extractive phase is typically immobilized as the coating or the packing a capillary tube. The extractive capillaries are generally placed between the needle and the loop of a programmable sample injector, and the analytes are extracted by repeated draw/eject cycles of the sample solution through the capillary until equilibrium. Then, the target compounds are desorbed and transferred to the analytical column with the proper solvent. This configuration has been used in many applications within the clinical and biomedical fields. Alternatively, the capillary column can be used as the loop of an injector valve (in-valve IT-SPME), and the analytes are extracted during sample loading, and transferred to the separative column with the mobile-phase by changing the valve position. In-valve IT-SPME is well suited for the extraction of organic pollutants in water analysis because large sample volumes can be passed through the capillary. Thus, although the absolute recoveries may be low, the required sensitivity is achieved by flushing though the capillary a volume of sample as large as necessary until the amount of analyte extracted is sufficient to generate the analytical response. Indeed, the selectivity and efficiency of the extraction is determined by parameters related to the capillary column such as the type of coating and film thickness, and length and internal diameter of the capillary column [46].

Initially, GC column segments were used as extractive capillaries, taking advantage of the fact a variety of stationary phases are available for compounds of different polarities. In the past years, many efforts have been focused on the synthesis of new sorbents to improve the extraction efficiency and selectivity. The preparation of coatings with polymers, ILs and NMs, as well as packing the capillaries with monolithic phases or immunosorbents are the main options tested.

Nanostructured coatings appear as an elegant option to improve the performance of SPME, especially to increase the extraction efficiency. However, unlike fiber-SPME, the potential benefits of NMs in IT-SPME have hardly been explored.

First attempt of using CNTs in IT-SPME was described in [47]. The authors immobilised –COOH functionalised MWCNTs onto the external surface of a fused silica fiber (375 μm of external diameter) using an epoxy resin glue. Then, the fibre was inserted into a PEEK tube (1 mm internal diameter) which was used as the injection loop in an in-valve configuration. The device was applied to extract some substituted aniline compounds from water samples. The high extraction capacity of the device resulted in limits of detection much lower than those achieved by other microextraction techniques, including fiber SPME.

More recently, the modification of PDMS commercial GC columns with SWCNTs has been reported as an attempt to improve extraction rates for the extraction [48]. In this approach, a PDMS capillary column, previously activated with NaOH, was functionalised with carboxylated SWCNTs *via* glutaraldehyde [49, 50]. For this purpose APTES was passed through the capillary in order to introduce amino groups on the surface of PDMS. In a second step, glutaraldehyde prepared in borate buffer was flushed through the capillary. Finally, a mixture of the carboxylated SWCNTs and 1,3-dicyclohexylcarbodiimide in dimethylformamide was passed through the capillary column, thus the carboxylated SWCNTs were immobilized on the surface of PDMS. Fig. (**3**) shows the SEM micrographs of the inner surface of the resulting coating.

This approach leads to a significant improvement of the extraction rates of some PAHs compared with those achieved with untreated coating. Table **1** summarised the preconcentration factors and the reproducibility data (expressed as %RSD) obtained for naphthalene, benzo[b]fluorantene and dibenzo[a,h]antracene with the described extractive capillaries. The results showed that preconcentration factors from 4 to 6 were reached with the carboxylated SWNTs-PDMS capillary column. Additionally, satisfactory reproducibility values were also achieved by using this capillary column.

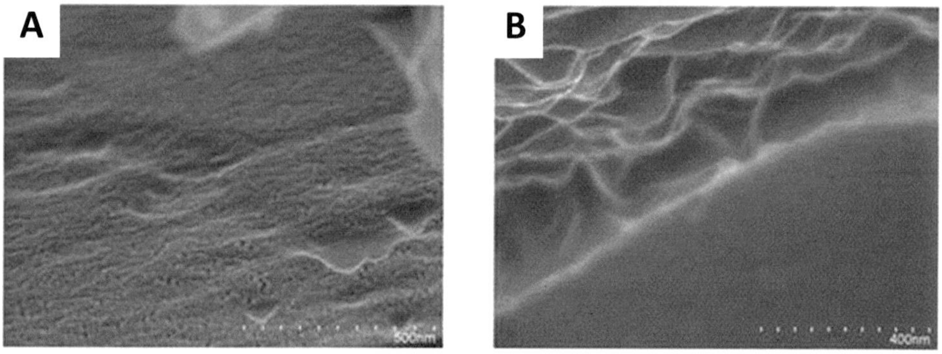

Fig. (3). SEM micrographs at two magnifications of the inner surface of the carboxylated SWNTs-PDMS capillary column.

Table 1. Preconcentration factor and reproducibility (%RSD) for naphthalene, benzo[b]fluorentene and dibenzo[a,h]antracene (100 μgL^{-1}) achieved with the proposed oxidized SWNTs-PDMS capillary columns.

	Preconcentration factor	%RSD
Naphtalene	4.0	5.7
Benzo[b]fluorantene	4.9	8.5
Dibenzo[a,h]antracene	6.3	6.8

A coating of TiO_2 NPs was reported for the extraction of phosphopeptides from protein digests [51]. The capillary was prepared by the liquid deposition method.

A fused silica capillary previously activated with NaOH was filled with a mixture of $(NH_4)_2TiF_6$ and H_3BO_3. The mixture was lead to react for 16 hours at 35 °C. Next, the capillary was washed with NaOH solution and water, and subjected to a thermal treatment at 300 °C for 2 h. The capillary was coupled to a nanoLC-MS/MS equipment with satisfactory results.A similar approach was used to form a SiO_2 coating, which was subsequently modified to introduce octadecyl substituents [52]. Activated silica capillaries were filled with a saturated solution of silica powder containing $(NH_4)_2TiF_6$, and H_3BO_3. The deposition procedure was repeated three times; then, the film was aged at 300 °C. The result was a coating with hydrophobic properties that was successfully applied to the extraction of endocrine disruptors and PAHs. In a subsequent study, a coating prepared by grafting PNIPPAm on SiO_2 NPs deposited-capillary was described [53]. First, the SiO_2-NPs coated capillary was prepared by the liquid deposition method. Next, the SiO_2-NPs were modified with 3-(triethoxysilyl) propyl methacrylate,and then functionalised with PNIPAAm through a radical initiating polymerization method. Compared to capillaries directly coated with PNIPAAm, the presence of the silica NPs in the inner-wall of the capillary provided higher specific surface area and then, more surface coverage of PNIPAAm. As a result, the extraction capability of the proposed coating for the extraction of estrogens from milk samples was clearly superior.

4.2.2.1. IT-SPME with Magnetic NPs

Magnetic NPs are very attractive for sample preparation because, as stated earlier, they can be easily isolated from the sample matrix with a magnet. Typically, the particles have a magnetic core (*i. e.* Fe_3O_4) and supported nanoparticles selected according to the application; they can also be integrated in nanocomposites. The literature shows a wide variety of examples of application of this kind of sorbents, which are used as a dispersive approach.

Magnetic IT-SPME was recently introduced in [54] as an attempt to increase absolute extraction efficiencies of ordinary IT-SPME. The coating was a magnetic hybrid material formed by Fe_3O_4 nanoparticles supported on silica immobilized in the surface of a bared fused silica capillary. To prepare the coating a solution containing the magnetic NPs, PEG, urea and tetraethylorthosilicate were treated to form a gel, which was then injected into a silica capillary previously activated. After thermal treatment the capillary was washed, so the final result was a highly porous coating in which the magnetic NPs were embedded. The capillary column was placed inside a magnetic coil that allowed the application of a magnetic field, and used in an in-valve configuration.When a magnetic field is applied, the NPs are magnetized creating around them regions with different magnetic field gradients. The diamagnetic analytes tend to be trapped in the regions in which this field gradient is minimal, thus increasing the adsorption of the analytes inside the interconnecting network created by the adsorbent SiO_2 phase. After sample loading, the magnetization of the NPs is inversed by changing the polarity of the applied magnetic field, so the analytes are released and transferred to the analytical column with the mobile phase. The utility of this methodology has been demonstrated for different kind of emerging pollutants such as some pharmaceuticals of major consumption, organophosphorous pesticides and triazines [54 - 56]. The main advantage of this technique is that the extraction efficiencies are substantially improved with respect to those achieved conventional IT-SPME systems (with non-magnetic coatings). For example, absolute extraction recoveries for the tested pharmaceuticals were increased from 10-30% to 70-100% by the magnetic IT-SPME approach, thus overcoming one of the main drawbacks of in-valve IT-SPME.

4.2.3. NMs INSBSE

SBSE is based on the extraction of the analytes with a magnetic stir bar coated with an extractive phase, typically PDMS, which is placed in the liquid sample. After extraction the bar is removed and the analytes are desorbed with a low volume of the proper solvent. Today, SBSE is well-established technology for sample extraction, and a wide variety of applications within different fields can be found in the literature [10]. PDMS, PA and ethylene glycol-silicone are available coatings for SBSE. The main advantage over fibre SPME is that the amount extractive phase is substantially increased. As a result, near quantitative recoveries for non-polar analytes can be reached. The lack of selectivity and poor efficiency for polar analytes are the major drawbacks of this technique. To overcome these limitations different configurations and extraction schemes have been tested, as well as new sorbents.

The employment of NMs in the preparation of coatings for this technique was reported in [57]. In this study a graphene/PEG-dimethacrylate composite coating was prepared by *in situ* polymerization for stir rod sorptive extraction, a modified format of SBSE proposed to avoid the contact of the stir bars with the sample vessel. The presence of graphene in the polymeric network substantially increased the specific surface. The total specific surface of graphene polymer was 149 m^2g^{-1}, whereas the specific surface for the unmodified polymer was 34 m^2 g^{-1}. This composite coating showed excellent extraction capability for PAHs with respect to that of the neat polymer [57].

More recently, different MOFs have been tested as coatings for SBSE [58]. For this purpose glass bars were activated NaOH and subjected to a sol-gel procedure using a mixture of hydroxyl terminated PDMS, methyltrimethoxysilane, poly(methylhydroxysiloxane) and the MOF in dichlorimethane/trifluoroacetic acid. Three MOFs were tested, MOF-199, MOF-5and IRMOF-3.The extraction efficiencies were compared with those achieved by coating the bars with PDMS by a similar sol-gel procedure. The coatings were tested for the extraction of some estrogens. It was observed that the presence of a MOF in the coating significantly increased the extraction efficiency for the target compounds compared to PDMS. Best results were achieved with IRMOF-3/PDMS coating, which was attributed to the presence of the amino functional groups in the ligand of the MOF. In addition, IRMOF-3 was less sensitive to the water, resulting in better stability coating. The IRMOF-3/PDMS coated stir bars could be reutilised for at least 35 times with no apparent loss in performance. The developed coating was applied to the determination of estrogens in environmental water samples.

4.2.4. LPME

As for SPME, the development of LPME is the result of demands such as simplicity, miniaturization, the employment of minimal amounts of solvents and direct coupling to analytical instruments. LPME has also benefited from the development of new phases, mainly from ILs [3]. The use of NMs is also gaining attention, but their application is still scarce. LPME encompasses different modes of microextraction with solvents, basically SDME, HF-LPME and DLLME. Next, the main application of NMs in these techniques will be discussed.

4.2.5. NMs in SDME

This is the simplest form of LPME. In this approach a drop of an organic solvent suspended from the needle of a microsyringe is immersed in an aqueous sample for a time. The extraction can also be carried out from the head space above the sample. After extraction, the drop is retracted into the syringe, and the extractive phase is introduced into the analytical system, typically the injector of a chromatograph.Whereas in fiber SPME a reduced assortment of coatings is available, a variety of solvents can be used as extractive phase in LPME. Many of the organic solvents used in classical LLE have been also used in LPME. However, the instability of the drop and the evaporation of the solvent during the extraction are problems typically encountered in SDME, and thus, alternative configurations and extractive phase have been tested.

As regards the employment of NMs, the attention has been mainly focused in the addition of metal NPs to the extractive organic solvent. As an example, the addition of AuNPs to the extractive solvent (toluene) has been proposed in [59] for the improved extraction of some peptides by SDME. The extraction is facilitated by the interaction between positively charged AuNPs and peptides at pH values above their isoelectric point. After extraction, the extractive drop was spotted onto a plate containing cyanohydroxy cinnamic acid and directly analyzed by atmospheric pressure MALDI-MS. The same authors described a similar acceptor phase but with AgNPs coated with hydrophobic molecules (dodecanethiol and octadecanethiol) for the extraction of hydrophobic peptides and proteins from urine and plasma [60]. The main advantages of this approach are that the extraction efficiency is increased and the extracts can be processed by MALDI-MS with minimum treatment.

4.2.6. NMs in HF-LPME

HF-LPME was proposed to avoid the drop instability in SDME. In its basic format the extraction device consists of a semipermeable polymeric tube which is attached to the needle of a syringe; the tube is immersed into the sample or exposed to its head space,so the analytes are first extracted into the solved used to fill the membrane pores, and then, into an acceptor solution placed inside the lumen of the fiber. After extraction the extractive solution is drawn inside a syringe, and finally injected into the analytical instrument. The versatility of the technique is derived from the variety of membranes and solvent combinations available. Several attempts have been made to improve its performance through the incorporation of new materials such CNTs.

The first attempt to take advantage of CNTs in HF-LPME was reported by Es'haghi *et al.* [61]. The authors used a solution of carboxylic acid functionalized MWCNTs in 1-octanol to fill the pores of a polypropylene hollow fiber. The analyte (caffeic acid) diffused directly from the sample to the MWCNTs-organic solvent trapped onto the membrane pores, and then to an aqueous buffer placed inside the membrane. The presence of SWCNTs in the fiber pores solution increased the extraction with respect the ordinary HP-LPME under similar operating conditions. According to the authors, the preparation of the hollow-fibers was very simple, so they were not reutilized. Because SWCNTs act as a solid sorbent in fiber SPME this approach was called hollow-fibre solid/liquid phase microextraction or hollow fiber-SPME. The nanostructured sorbent can be also immobilised inside a hollow-fiber as a fiber coating [62]. In another study, a SWCNTs reinforced polymeric phase was used in combination with polypropylene hollow fibers [63]. The extraction device was applied to the extraction of aflatoxins in the extracts obtained from cereal samples, with excellent preconcentration factors (≈3700-4000). The selectivity was further improved by using CNTs combined with a MIP, as demonstrated for the extraction of chlorogenic acid from hydroalcoholic extracts obtained from plants [64].

4.2.7. NMs in DLLME

In its typical scheme, DLLME uses a mixture of a high-density extraction solvent and a polar disperser solvent both added to the sample in order to extract the analytes. The extraction solvent must be insoluble in water and soluble in the disperser solvent; its density must be higher than that of water. The disperser solvent has to be soluble in both water and extraction solvent. The extracting solvent and the disperser solvents are rapidly injected into the aqueous sample typically placed in a conic test tube. Under such conditions a cloudy dispersion is formed with the extraction solvent dispersed as fine droplets, which enables the extraction of organic analytes within a few seconds. Finally, the mixture is centrifuged, so the extractive phase is settled at the bottom of the conical test tube and separated for further analysis.

A variety of modifications of this basic scheme have been proposed with specific purposes. Recently, a new approach based in the combination of dispersive micro-solid-phase extraction and DLLME has been described which uses modified magnetic NPs [65]. This approach, also termed DLLME-dispersive μ-SPE, avoids the employment of conical-bottom test tubes and centrifugation operations of DLLME, and also eliminates the restriction that the extracting solvent must be denser than water. This is because magnetic NPs are used to retrieve the extractive solvent (1-octanol) in the DLLME step.The extraction solvent interacts with the hydrophobic magnetic NPs (Fe_3O_4 NPs modified with 3-chloro-prop-l-triethoxysilane). Then, the 1-octanol phase in which the analyte molecules are dissolved, is separated from the sample matrix with with a magnet. Finally, the analytes are back-extracted in acetonitrile and processed by GC. The approach was tested for the extraction of some PAHs from river water. A similar scheme has been applied to the determination of 4-n-nonylphenol in water using n-octy--triethoxisilane surface-modified Fe_3O_4 magnetic NPs, and again 1-octanol as extraction solvent [66].

CONCLUSION

NMs are versatile tools that can be used to achieve one or various goals in a microextraction process. This is illustrated in Fig. (**2**), which summarizes the objectives of the incorporation of NMs in solid and liquid microextraction techniques. The improvement of the extraction efficiency and the selectivity are the main contributions of NMs-based sorbents, and when used both in SPME and LPME. In some instances the presence of NMs also provides improved thermal and mechanical stability of the extractive device, and/or facilitates the coupling with analytical instruments. However, the use of NMs in LPME is still scarce despite their potential applicability.

The purity of NMs and the strict control of the preparation of the sorbents are key factors to ensure fiber-to-fiber (or capillary-to-capillary, bar-to-bar) reproducibility. In this respect, substantial progress can be expected in forthcoming years because of the continuous advances in the methodologies used to synthetize and to immobilize NMs in microextraction devices. In Table **2** are summarized the main approaches used to prepare NMs-based sorbents for microextraction, as well as their relevant features and applications.

Table 2. Main methods used to prepare NMs in coatings for SPME and examples of application.

Methods	**Principle**	**Main features**	**NMs/**	**References/ Extraction technique***
Dipping/physical binding	The fiber or extraction support is dipped into a solution/suspension of the NMs, and then the solvent is evaporated; a binder can be also added to the NMs solution/ suspension	Simplicity Poor control of the coating thickness Poor stability	CNTs Graphene Fullerene Nanocone/disks Silica NPs	16-18[a], 47[b] 24[a],25[a] 29[a] 31[a] 44[a]
Sol-gel	The support and a colloidal suspension are brought into contact until a gel phase is formed due to the hydrolysis of a precursor and a subsequent polycondensation. The substrate has to be previously activated to introduce active groups (typically, -OH) that act as anchoring points for the sorbent phase.	Good stability Versatility, as a variety of precursors and NMs can be combined	CNTs Graphene ZnO/graphene composite Fullerenea Silica based magnetic Fe_3O_4 PDM/MOFs	19[a] 26[a] 41[a] 30[a] 54-56[b] 58[c]

(Table 2) contd.....

Methods	Principle	Main features	NMs/	References/ Extraction technique*
Covalent bonding	The NMs are attached to the extraction support by an specific chemical reaction The surface of both the support and the NMs need to be modified in order to make them to react with each other	Strong binding Specific treatment of the extraction support is necessary, thus the procedure can be complex	CNTs Graphene AuNPs	20[a], 48[b] 27[a] 34[a], 35[a]
Electrodeposition/electrochemical polymerization	The coating is formed on a metallic support acting as an electrode by direct coagulation of the NMs or by reduction of a conductive polymer.	Good control of the coating thickness though the control of the applied voltage and time	CNTs PPy/CNTs PPy/Graphene PANI/Silica composites Metal oxide	21[a] 22[a] 28[a] 45[a] 39[a], 40[a]
Electrospinning	Nanofibers generated through an electrically charged jet of polymer solution or melt are used to create a coating which are then converted into carbon coating by pyrolysis.	Only for carbon coatings	Carbon nanofibres	32[a], 33[a]
Liquid phase deposition	An oxide coating is formed from an aqueous solution of a metal-fluoro complex that is hydrolyzed by adding water, boric acid or aluminum	Limited range of NMs	Metal oxide	51-53[b]
Hydrothermal grown	The coating is formed through a reaction in aqueous media at elevated temperature	Simplicity Good stability although it has been tested only for a few MOFs	Metal oxide MOFs	37[a] 42[a]

(*) a - fiber SPME; b - IT-SPME; c - SBSE

Because of the diversity of NMs, some of them relatively unexplored (such as MOFs) the number of proposals for microextraction will increase in the future. Also, new formats of microextraction, such as magnetic IT-SPME, will extend the applicability of NMs-based microextraction.

CONFLICT OF INTEREST

The authors confirm that they have no conflict of interest to declare for this publication.

ACKNOWLEDGEMENTS

The authors are grateful to the Spanish Ministerio de Economia y Competitividad/FEDER (CTQ2014-53916-P) and to the Generalitat Valenciana (PROMETEO program 2016/109) for financial support. P. S.-M. expresses his gratitude to the MINECO-FEDER for his FPI pre-doctoral grant.

ACRONYMS

AgNPs	silver nanoparticles
APTES	3-aminopropyltriethoxysilane
AuNPs	gold nanoparticles
BTEX	benzene, toluene, ethylbenzene, xylene
CAR	carboxen
CNTs	carbon nanotubes
CW	carbowax
DLLME	dispersive liquid-liquid microextraction
D-μ-SPE	dispersive micro-solid-phase extraction
DVB	divinylbencene
GC	gas chromatography
HF-LPME	hollow-fiber liquid-phase microextraction
HS-SPME	head space solid phase microextraction
ILs	ionic liquids
IT-SPME	in-tube solid-phase microextraction
LC	liquid chromatography
LLE	liquid-liquid extraction
LPME	liquid-phase microextraction

MALDI-MS	matrix-assisted desorption/ionization mass spectroscopy
MIPs	molecular imprinted polymers
MIPPy	molecular imprinted polypyrrole
MNPs	magnetic nanoparticles
MOFs	metal-organic frameworks
MS	mass spectrometry
MWCNTs	multi-walled carbon nanotubes
µ-SPE	micro solid-phase extraction
NMs	nanomaterials
NPs	nanoparticles
PA	polyacrylate
PAHs	polycyclic aromatic hydrocarbons
PANI	polyaniline
PBDEs	polybrominated diphenyl esters
PCBs	polychlorinated biphenyls
PEG	polyethylene glycol
PDMS	polydimethylsiloxane
PNIPAAm	poly(N-isopropylacrylamide)
PPy	polypyrrole
SBSE	stir bar sorptive extraction
SDME	single drop microextraction
SPME	solid-phase microextraction
SPE	solid-phase extraction
SWCNTs	single-walled carbon nanotubes

REFERENCES

[1] Augusto, F.; Carasek, E.; Silva, R.G.; Rivellino, S.R.; Batista, A.D.; Martendal, E. New sorbents for extraction and microextraction techniques. *J. Chromatogr. A,* **2010**, *1217*(16), 2533-2542. [http://dx.doi.org/10.1016/j.chroma.2009.12.033] [PMID: 20035942]

[2] Huang, Z.; Lee, H.K. Materials-based approaches to minimizing solvent usage in analytical sample preparation. *Trends Analyt. Chem.,* **2012**, *39*, 228-244. [http://dx.doi.org/10.1016/j.trac.2012.05.007]

[3] Pena-Pereira, F.; Lavilla, I.; Bendicho, C. Liquid-phase microextraction techniques within the framework of green chemistry. *Trends Analyt. Chem.,* **2010**, *29*, 617-628. [http://dx.doi.org/10.1016/j.trac.2010.02.016]

[4] Xu, J.; Zheng, J.; Tian, J.; Zhu, F.; Zeng, F.; Su, C.; Ouyang, G. New materials in solid-phase

microextraction. *Trends Analyt. Chem.,* **2013**, *47*, 68-83.
[http://dx.doi.org/10.1016/j.trac.2013.02.012]

[5] Lucena, R.; Simonet, B.M.; Cárdenas, S.; Valcárcel, M. Potential of nanoparticles in sample preparation. *J. Chromatogr. A,* **2011**, *1218*(4), 620-637.
[http://dx.doi.org/10.1016/j.chroma.2010.10.069] [PMID: 21071033]

[6] Pyrzynska, K. Use of nanomaterials in sample preparation. *Trends Analyt. Chem.,* **2013**, *43*, 100-108.
[http://dx.doi.org/10.1016/j.trac.2012.09.022]

[7] Tian, J.; Xu, J.; Zhu, F.; Lu, T.; Su, C.; Ouyang, G. Application of nanomaterials in sample preparation. *J. Chromatogr. A,* **2013**, *1300*, 2-16.
[http://dx.doi.org/10.1016/j.chroma.2013.04.010] [PMID: 23611621]

[8] Khajeh, M.; Laurent, S.; Dastafkan, K. Nanoadsorbents: classification, preparation, and applications (with emphasis on aqueous media). *Chem. Rev.,* **2013**, *113*(10), 7728-7768.
[http://dx.doi.org/10.1021/cr400086v] [PMID: 23869773]

[9] Arthur, C.L.; Pawliszyn, J. Solid phase microextraction with thermal desorption using fused silica optical fibers. *Anal. Chem.,* **1990**, *62*, 2145-2148.
[http://dx.doi.org/10.1021/ac00218a019]

[10] Nogueira, J.M. Novel sorption-based methodologies for static microextraction analysis: A review on SBSE and related techniques. *Anal. Chim. Acta,* **2012**, *757*, 1-10.
[http://dx.doi.org/10.1016/j.aca.2012.10.033] [PMID: 23206390]

[11] Mehdinia, A.; Aziz-Zanjani, M.O. Advances for sensitive, rapid and selective extraction in different configurations of solid-phase microextraction. *Trends Analyt. Chem.,* **2013**, *51*, 13-22.
[http://dx.doi.org/10.1016/j.trac.2013.05.013]

[12] Mehdinia, A.; Aziz-Zanjani, M.O. Recent advances in nanomaterials utilized in fiber coatings for solid-phase microextraction. *Trends Analyt. Chem.,* **2013**, *42*, 205-215.
[http://dx.doi.org/10.1016/j.trac.2012.09.013]

[13] Aziz-Zanjani, M. O.; Mehdinia, A. A review on procedures for the preparation of coatings for solid phase microextraction. *Microchim. Acta.,* **2014**.
[http://dx.doi.org/10.1007/s00604-014-1265-y]

[14] Ravelo-Pérez, L.M.; Herrera-Herrera, A.V.; Hernández-Borges, J.; Rodríguez-Delgado, M.A. Carbon nanotubes: Solid-phase extraction. *J. Chromatogr. A,* **2010**, *1217*(16), 2618-2641.
[http://dx.doi.org/10.1016/j.chroma.2009.10.083] [PMID: 19914625]

[15] Zhang, B-T.; Zheng, X.; Li, H-F.; Lin, J-M. Application of carbon-based nanomaterials in sample preparation: a review. *Anal. Chim. Acta,* **2013**, *784*, 1-17.
[http://dx.doi.org/10.1016/j.aca.2013.03.054] [PMID: 23746402]

[16] Wang, J-X.; Jiang, D-Q.; Gu, Z-Y.; Yan, X-P. Multiwalled carbon nanotubes coated fibers for solid-phase microextraction of polybrominated diphenyl ethers in water and milk samples before gas chromatography with electron-capture detection. *J. Chromatogr. A,* **2006**, *1137*(1), 8-14.
[http://dx.doi.org/10.1016/j.chroma.2006.10.003] [PMID: 17055524]

[17] Rastkari, N.; Ahmadkhaniha, R.; Yunesian, M. Single-walled carbon nanotubes as an effective adsorbent in solid-phase microextraction of low level methyl *tert*-butyl ether, ethyl *tert*-butyl ether and

methyl *tert*-amyl ether from human urine. *J. Chromatogr. B Analyt. Technol. Biomed. Life Sci.,* **2009**, *877*(14-15), 1568-1574.
[http://dx.doi.org/10.1016/j.jchromb.2009.04.002] [PMID: 19395320]

[18] Liu, X.; Wang, X.; Tan, F.; Zhao, H.; Quan, X.; Chen, J.; Li, L. An electrochemically enhanced solid-phase microextraction approach based on molecularly imprinted polypyrrole/multi-walled carbon nanotubes composite coating for selective extraction of fluoroquinolones in aqueous samples. *Anal. Chim. Acta,* **2012**, *727*, 26-33.
[http://dx.doi.org/10.1016/j.aca.2012.03.054] [PMID: 22541819]

[19] Sarafraz-Yazdi, A.; Amiri, A.; Rounaghi, G.; Hosseini, H.E. A novel solid-phase microextraction using coated fiber based sol-gel technique using poly(ethylene glycol) grafted multi-walled carbon nanotubes for determination of benzene, toluene, ethylbenzene and o-xylene in water samples with gas chromatography-flame ionization detector. *J. Chromatogr. A,* **2011**, *1218*(34), 5757-5764.
[http://dx.doi.org/10.1016/j.chroma.2011.06.099] [PMID: 21782185]

[20] Ma, X.; Li, Q.; Yuan, D. Determination of endocrine-disrupting compounds in water by carbon nanotubes solid-phase microextraction fiber coupled online with high performance liquid chromatography. *Talanta,* **2011**, *85*(4), 2212-2217.
[http://dx.doi.org/10.1016/j.talanta.2011.07.066] [PMID: 21872080]

[21] Liu, H.; Li, J.; Liu, X.; Jiang, S. A novel multiwalled carbon nanotubes bonded fused-silica fiber for solid phase microextraction-gas chromatographic analysis of phenols in water samples. *Talanta,* **2009**, *78*(3), 929-935.
[http://dx.doi.org/10.1016/j.talanta.2008.12.061] [PMID: 19269452]

[22] Chen, W.; Zeng, J.; Chen, J.; Huang, X.; Jiang, Y.; Wang, Y.; Chen, X. High extraction efficiency for polar aromatic compounds in natural water samples using multiwalled carbon nanotubes/Nafion solid-phase microextraction coating. *J. Chromatogr. A,* **2009**, *1216*(52), 9143-9148.
[http://dx.doi.org/10.1016/j.chroma.2009.07.025] [PMID: 19660762]

[23] Liu, Q.; Shi, J.; Jiang, G. Application of graphene in analytical sample preparation. *Trends Analyt. Chem.,* **2012**, *37*, 1-11.
[http://dx.doi.org/10.1016/j.trac.2012.03.011]

[24] Chen, J.; Zou, J.; Zeng, J.; Song, X.; Ji, J.; Wang, Y.; Ha, J.; Chen, X. Preparation and evaluation of graphene-coated solid-phase microextraction fiber. *Anal. Chim. Acta,* **2010**, *678*(1), 44-49.
[http://dx.doi.org/10.1016/j.aca.2010.08.008] [PMID: 20869503]

[25] Ponnusamy, V.K.; Jen, J-F. A novel graphene nanosheets coated stainless steel fiber for microwave assisted headspace solid phase microextraction of organochlorine pesticides in aqueous samples followed by gas chromatography with electron capture detection. *J. Chromatogr. A,* **2011**, *1218*(39), 6861-6868.
[http://dx.doi.org/10.1016/j.chroma.2011.08.019] [PMID: 21872868]

[26] Zhang, H.; Lee, H-K. Plunger-in-needle solid-phase microextraction with graphene-based sol-gel coating as sorbent for determination of polybrominated diphenyl ethers. *J. Chromatogr. A,* **2011**, *1218*(28), 4509-4516.
[http://dx.doi.org/10.1016/j.chroma.2011.05.016] [PMID: 21636089]

[27] Xu, L.; Feng, J.; Li, J.; Liu, X.; Jiang, S. Graphene oxide bonded fused-silica fiber for solid-phase

microextraction-gas chromatography of polycyclic aromatic hydrocarbons in water. *J. Sep. Sci.,* **2012**, *35*(1), 93-100.
[http://dx.doi.org/10.1002/jssc.201100612] [PMID: 22125228]

[28] Zou, J.; Song, X.; Ji, J.; Xu, W.; Chen, J.; Jiang, Y.; Wang, Y.; Chen, X. Polypyrrole/graphene composite-coated fiber for the solid-phase microextraction of phenols. *J. Sep. Sci.,* **2011**, *34*(19), 2765-2772.
[http://dx.doi.org/10.1002/jssc.201100303] [PMID: 21823234]

[29] Xiao, C.; Han, S.; Wang, Z.; Xing, J.; Wu, C. Application of the polysilicone fullerene coating for solid-phase microextraction in the determination of semi-volatile compounds. *J. Chromatogr. A,* **2001**, *927*(1-2), 121-130.
[http://dx.doi.org/10.1016/S0021-9673(01)01046-9] [PMID: 11572381]

[30] Yu, J.; Dong, L.; Wu, C.; Wu, L.; Xing, J. Hydroxyfullerene as a novel coating for solid-phase microextraction fiber with sol-gel technology. *J. Chromatogr. A,* **2002**, *978*(1-2), 37-48.
[http://dx.doi.org/10.1016/S0021-9673(02)01347-X] [PMID: 12458943]

[31] Jiménez-Soto, J.M.; Cárdenas, S.; Valcárcel, M. Carbon nanocones/disks as new coating for solid-phase microextraction. *J. Chromatogr. A,* **2010**, *1217*(20), 3341-3347.
[http://dx.doi.org/10.1016/j.chroma.2010.03.016] [PMID: 20363473]

[32] Zewe, J.W.; Steach, J.K.; Olesik, S.V. Electrospun fibers for solid-phase microextraction. *Anal. Chem.,* **2010**, *82*(12), 5341-5348.
[http://dx.doi.org/10.1021/ac100891t] [PMID: 20503975]

[33] Newsome, T.E.; Zewe, J.W.; Olesik, S.V. Electrospun nanofibrous solid-phase microextraction coatings for preconcentration of pharmaceuticals prior to liquid chromatographic separations. *J. Chromatogr. A,* **2012**, *1262*, 1-7.
[http://dx.doi.org/10.1016/j.chroma.2012.08.046] [PMID: 23000181]

[34] Feng, J.; Sun, M.; Liu, H.; Li, J.; Liu, X.; Jiang, S. Au nanoparticles as a novel coating for solid-phase microextraction. *J. Chromatogr. A,* **2010**, *1217*(52), 8079-8086.
[http://dx.doi.org/10.1016/j.chroma.2010.10.089] [PMID: 21081234]

[35] Feng, J.; Sun, M.; Li, J.; Liu, X.; Jiang, S. A novel silver-coated solid-phase microextraction metal fiber based on electroless plating technique. *Anal. Chim. Acta,* **2011**, *701*(2), 174-180.
[http://dx.doi.org/10.1016/j.aca.2011.05.040] [PMID: 21801885]

[36] Alizadeh, R.; Najafi, N.M.; Kharrazi, S. A new solid phase micro extraction for simultaneous head space extraction of ultra traces of polar and non-polar compounds. *Anal. Chim. Acta,* **2011**, *689*(1), 117-121.
[http://dx.doi.org/10.1016/j.aca.2011.01.013] [PMID: 21338766]

[37] Ji, J.; Liu, H.; Chen, J.; Zeng, J.; Huang, J.; Gao, L.; Wang, Y.; Chen, X. ZnO nanorod coating for solid phase microextraction and its applications for the analysis of aldehydes in instant noodle samples. *J. Chromatogr. A,* **2012**, *1246*, 22-27.
[http://dx.doi.org/10.1016/j.chroma.2012.01.080] [PMID: 22342186]

[38] Cao, D.D.; Lü, J.X.; Liu, J.F.; Jiang, G.B. *In situ* fabrication of nanostructured titania coating on the surface of titanium wire: A new approach for preparation of solid-phase microextraction fiber. *Anal. Chim. Acta,* **2008**, *611*(1), 56-61.

[http://dx.doi.org/10.1016/j.aca.2008.01.067] [PMID: 18298967]

[39] Liu, H.; Wang, D.; Ji, L.; Li, J.; Liu, S.; Liu, X.; Jiang, S. A novel TiO_2 nanotube array/Ti wire incorporated solid-phase microextraction fiber with high strength, efficiency and selectivity. *J. Chromatogr. A,* **2010**, *1217*(12), 1898-1903. [http://dx.doi.org/10.1016/j.chroma.2010.01.080] [PMID: 20153861]

[40] Zhang, Z.; Wang, Q.; Li, G. Fabrication of novel nanoporous array anodic alumina solid-phase microextraction fiber coating and its potential application for headspace sampling of biological volatile organic compounds. *Anal. Chim. Acta,* **2012**, *727*, 13-19. [http://dx.doi.org/10.1016/j.aca.2012.03.033] [PMID: 22541817]

[41] Zhang, S.; Du, Z.; Li, G. Graphene-supported zinc oxide solid-phase microextraction coating with enhanced selectivity and sensitivity for the determination of sulfur volatiles in Allium species. *J. Chromatogr. A,* **2012**, *1260*, 1-8. [http://dx.doi.org/10.1016/j.chroma.2012.08.045] [PMID: 22985527]

[42] Chang, N.; Gu, Z-Y.; Wang, H-F.; Yan, X-P. Metal-organic-framework-based tandem molecular sieves as a dual platform for selective microextraction and high-resolution gas chromatographic separation of *n*-alkanes in complex matrixes. *Anal. Chem.,* **2011**, *83*(18), 7094-7101. [http://dx.doi.org/10.1021/ac2014004] [PMID: 21800908]

[43] Cui, X-Y.; Gu, Z-Y.; Jiang, D-Q.; Li, Y.; Wang, H-F.; Yan, X-P. *In situ* hydrothermal growth of metal-organic framework 199 films on stainless steel fibers for solid-phase microextraction of gaseous benzene homologues. *Anal. Chem.,* **2009**, *81*(23), 9771-9777. [http://dx.doi.org/10.1021/ac901663x] [PMID: 19894702]

[44] Hashemi, P.; Shamizadeh, M.; Badiei, A.; Poor, P.Z.; Ghiasvand, A.R.; Yarahmadi, A. Amino ethyl-functionalized nanoporous silica as a novel fiber coating for solid-phase microextraction. *Anal. Chim. Acta,* **2009**, *646*(1-2), 1-5. [http://dx.doi.org/10.1016/j.aca.2009.04.023] [PMID: 19523549]

[45] Bagheri, H.; Roostaie, A. Aniline-silica nanocomposite as a novel solid phase microextraction fiber coating. *J. Chromatogr. A,* **2012**, *1238*, 22-29. [http://dx.doi.org/10.1016/j.chroma.2012.03.027] [PMID: 22498354]

[46] Kataoka, H.; Ishizaki, A.; Nonaka, Y.; Saito, K. Developments and applications of capillary microextraction techniques: a review. *Anal. Chim. Acta,* **2009**, *655*(1-2), 8-29. [http://dx.doi.org/10.1016/j.aca.2009.09.032] [PMID: 19925911]

[47] Liu, X-Y.; Ji, Y-S.; Zhang, H.X.; Liu, M.C. Highly sensitive analysis of substituted aniline compounds in water samples by using oxidized multiwalled carbon nanotubes as an in-tube solid-phase microextraction medium. *J. Chromatogr. A,* **2008**, *1212*(1-2), 10-15. [http://dx.doi.org/10.1016/j.chroma.2008.10.034] [PMID: 18952215]

[48] Slentz, B.E.; Penner, N.A.; Lugowska, E.; Regnier, F. Nanoliter capillary electrochromatography columns based on collocated monolithic support structures molded in poly(dimethyl siloxane). *Electrophoresis,* **2001**, *22*(17), 3736-3743. [http://dx.doi.org/10.1002/1522-2683(200109)22:17<3736::AID-ELPS3736>3.0.CO;2-Y] [PMID: 11699912]

[49] Sombra, L.; Moliner-Martínez, Y.; Cárdenas, S.; Valcárcel, M. Carboxylic multi-walled carbon

nanotubes as immobilized stationary phase in capillary electrochromatography. *Electrophoresis,* **2008**, *29*(18), 3850-3857.
[http://dx.doi.org/10.1002/elps.200800275] [PMID: 18850654]

[50] González-Fuenzalida, R.A.; Moliner-Martinez, Y.; Verdú-Andrés, J.; Molins-Legua, C.; Herráez-Hernández, R.; Campíns-Falcó, P. *Nanoparticle microextraction in Sample extraction techniques for biological samples: recent advances and novel applications.,* in press

[51] Lin, B.; Li, T.; Zhao, Y.; Huang, F-K.; Guo, L.; Feng, Y-Q. Preparation of a TiO_2 nanoparticle-deposited capillary column by liquid phase deposition and its application in phosphopeptide analysis. *J. Chromatogr. A,* **2008**, *1192*(1), 95-102.
[http://dx.doi.org/10.1016/j.chroma.2008.03.043] [PMID: 18384799]

[52] Li, T.; Xu, J.; Wu, J-H.; Feng, Y-Q. Liquid-phase deposition of silica nanoparticles into a capillary for in-tube solid-phase microextraction coupled with high-performance liquid chromatography. *J. Chromatogr. A,* **2009**, *1216*(15), 2989-2995.
[http://dx.doi.org/10.1016/j.chroma.2009.01.076] [PMID: 19211102]

[53] Yu, Q-W.; Ma, Q.; Feng, Y.Q. Temperature-response polymer coating for in-tube solid-phase microextraction coupled to high-performance liquid chromatography. *Talanta,* **2011**, *84*(4), 1019-1025.
[http://dx.doi.org/10.1016/j.talanta.2011.03.003] [PMID: 21530774]

[54] Moliner-Martínez, Y.; Prima-García, H.; Ribera, A.; Coronado, E.; Campíns-Falcó, P. Magnetic in-tube solid phase microextraction. *Anal. Chem.,* **2012**, *84*(16), 7233-7240.
[http://dx.doi.org/10.1021/ac301660k] [PMID: 22861152]

[55] Moliner-Martínez, Y.; Vitta, Y.; Prima-García, H.; González-Fuenzalida, R.A.; Ribera, A.; Campíns-Falcó, P.; Coronado, E. Silica supported Fe(3)O(4) magnetic nanoparticles for magnetic solid-phase extraction and magnetic in-tube solid-phase microextraction: application to organophosphorous compounds. *Anal. Bioanal. Chem.,* **2014**, *406*(8), 2211-2215.
[http://dx.doi.org/10.1007/s00216-013-7379-y] [PMID: 24105458]

[56] González-Fuenzalida, R.A.; Moliner-Martínez, Y.; Prima-Garcia, H.; Ribera, A.; Campins-Falcó, P.; Zaragozá, R.J. Evaluation of superparamagnetic silica nanoparticles for extraction of triazines in magnetic in-tube solid phase microextraction coupled to capillary liquid chromatography. *Nanomaterials (Basel),* **2014**, *4*, 242-255.
[http://dx.doi.org/10.3390/nano4020242]

[57] Luo, Y-B.; Cheng, J.; Ma, Q.; Feng, Y.Q.; Li, J-H. Graphene polymer composite: extraction of polycyclic aromatic hydrocarbons from water samples by stir rod sorptive extraction. *Anal. Methods,* **2011**, *3*, 92-98.
[http://dx.doi.org/10.1039/C0AY00624F]

[58] Hu, C.; He, M.; Chen, B.; Zhong, C.; Hu, B. Polydimethylsiloxane/metal-organic frameworks coated stir bar sorptive extraction coupled to high performance liquid chromatography-ultraviolet detector for the determination of estrogens in environmental water samples. *J. Chromatogr. A,* **2013**, *1310*, 21-30.
[http://dx.doi.org/10.1016/j.chroma.2013.08.047] [PMID: 23992883]

[59] Sudhir, P-R.; Wu, H.F.; Zhou, Z.C. Identification of peptides using gold nanoparticle-assisted single-drop microextraction coupled with AP-MALDI mass spectrometry. *Anal. Chem.,* **2005**, *77*(22), 7380-

7385.
[http://dx.doi.org/10.1021/ac051162m] [PMID: 16285689]

[60] Shrivas, K.; Wu, H-F. Modified silver nanoparticle as a hydrophobic affinity probe for analysis of peptides and proteins in biological samples by using liquid-liquid microextraction coupled to AP-MALDI-ion trap and MALDI-TOF mass spectrometry. *Anal. Chem.,* **2008**, *80*(7), 2583-2589.
[http://dx.doi.org/10.1021/ac702309w] [PMID: 18324794]

[61] Eshaghi, Z.; Golsefidi, M.A.; Saify, A.; Tanha, A.A.; Rezaeifar, Z.; Alian-Nezhadi, Z. Carbon nanotube reinforced hollow fiber solid/liquid phase microextraction: a novel extraction technique for the measurement of caffeic acid in *Echinacea purpurea* herbal extracts combined with high-performance liquid chromatography. *J. Chromatogr. A,* **2010**, *1217*(17), 2768-2775.
[http://dx.doi.org/10.1016/j.chroma.2010.02.054] [PMID: 20227700]

[62] Sarafraz-Yazdi, A.; Amiri, A.; Rounaghi, G.; Eshtiagh-Hosseini, H. Determination of non-steroidal anti-inflammatory drugs in urine by hollow-fiber liquid membrane-protected solid-phase microextraction based on sol-gel fiber coating. *J. Chromatogr. B Analyt. Technol. Biomed. Life Sci.,* **2012**, *908*, 67-75.
[http://dx.doi.org/10.1016/j.jchromb.2012.09.040] [PMID: 23122403]

[63] Eshaghi, Z.; Sorayaei, H.; Samadi, F.; Masrournia, M.; Bakherad, Z. Fabrication of a novel nanocomposite based on sol-gel process for hollow fiber-solid phase microextraction of aflatoxins: B1 and B2, in cereals combined with high performance liquid chromatography-diode array detection. *J. Chromatogr. B Analyt. Technol. Biomed. Life Sci.,* **2011**, *879*(28), 3034-3040.
[http://dx.doi.org/10.1016/j.jchromb.2011.08.042] [PMID: 21925977]

[64] Golsefidi, M.A.; Eshaghi, Z.; Sarafraz-Yazdi, A. Design, synthesis and evaluation of a molecularly imprinted polymer for hollow fiber-solid phase microextraction of chlorogenic acid in medicinal plants. *J. Chromatogr. A,* **2012**, *1229*, 24-29.
[http://dx.doi.org/10.1016/j.chroma.2012.01.019] [PMID: 22307151]

[65] Shi, Z-G.; Lee, H.K. Dispersive liquid-liquid microextraction coupled with dispersive micro-soli--phase extraction for the fast determination of polycyclic aromatic hydrocarbons in environmental water samples. *Anal. Chem.,* **2010**, *82*(4), 1540-1545.
[http://dx.doi.org/10.1021/ac9023632] [PMID: 20092360]

[66] Tay, K-S.; Rahman, N.A.; Abas, M.R. Magnetic nanoparticle assisted dispersive liquid-liquid microextraction for the determination of 4-n-nonylphenol in water. *Anal. Methods,* **2013**, *5*, 2933-2938.
[http://dx.doi.org/10.1039/c3ay00001j]

CHAPTER 5

Principles and Developments of Solid-Phase Microextraction

Adam Kloskowski[a,*], **Łukasz Marcinkowski**[a], **Francisco Pena-Pereira**[b,c] and **Jacek Namieśnik**[b]

[a] *Gdansk University of Technology, Faculty of Chemistry, Department of Physical Chemistry, Narutowicza Str.11/12, Gdansk 80-233, Poland*

[b] *Gdansk University of Technology, Faculty of Chemistry, Department of Analytical Chemistry, Narutowicza Str.11/12, Gdansk 80-233, Poland*

[c] *University of Vigo, Faculty of Chemistry; Analytical and Food Chemistry Department; Campus As Lagoas-Marcosende s/n, 36310 Vigo, Spain*

Abstract: Sample preparation has been commonly considered a critical step of the analytical process. In this sense, remarkable efforts have been made to develop efficient sample preparation techniques which could overcome the limitations of conventional approaches. Since its inception in the early 1990's, solid-phase microextraction (SPME) has become a widespread miniaturized sample preparation technique for extraction and preconcentration of target analytes from a large variety of matrices. Interestingly, sampling, extraction, enrichment and sample introduction can be integrated into a single step in SPME. This book chapter focuses on the basic principles and current state of the art of SPME. Specifically, both thermodynamic and kinetic aspects of the SPME technique are discussed in detail. In addition, those experimental variables that show a paramount role in the extraction process, and should therefore be optimized and controlled for optimal performance, are considered. Valuable contributions that enabled the development of this solventless technique and current challenges are identified. Other related SPME devices, such as internally cooled SPME, in-tube SPME and membrane SPME, are also described.

[*] **Corresponding author Adam Kloskowski:** Gdansk University of Technology, Faculty of Chemistry, Department of Physical Chemistry, Narutowicza Str.11/12, Gdansk 80-233, Poland; Tel: 48 (58) 3472959; Fax: 48 (58)3472694; E-mail: adam.kloskowski@pg.gda.pl

M. Valcárcel, S. Cárdenas, R. Lucena (Eds.)

Keywords: Automation, Direct immersion, Divinylbenzene, Fiber coatings, Gas chromatography, Headspace, High-performance liquid chromatography, In-tube SPME, Internally cooled SPME, Kinetics, Membrane-SPME, Molecularly imprinted polymers, Poly(dimethylsiloxane), Polymeric ionic liquids, Preconcentration, Sample preparation, Sample preparation, Sol-gel sorbents, Solid-phase microextraction, Thermodynamics.

5.1. INTRODUCTION

Sample preparation has been recognized as a critical step of the analytical process, being even considered as the bottleneck of the overall process. Enrichment of target compounds, transfer of the analytes into a solvent compatible with the analytical instrumentation, minimization of potential interferences, and efficient sample clean-up, are among the main aims of sample preparation techniques. In this regard, liquid-liquid extraction and solid-phase extraction have been conventionally employed prior to the determination of relevant compounds in a variety of samples. In spite of their suitable performance, these classical techniques do not fulfill several of the challenges in analytical chemistry, including miniaturization, portability, and environmental sustainability. Furthermore, the necessity to determine relevant compounds at very low concentrations in matrices of different complexity, especially when dealing with reduced sample volumes, made the improvement of sample preparation techniques being of paramount importance. The inception of solid-phase microextraction (SPME) involved a huge advance in this sense. SPME was firstly introduced by Arthur and Pawliszyn in 1990 as an alternative to conventional sample preparation methods [1]. SPME is a non-exhaustive sample preparation technique based on the exposure of a small amount of extracting phase to the sample (or to the headspace above the sample) for extraction and preconcentration of target compounds. SPME enables the integration of sampling and sample preparation. Furthermore, its small size makes SPME being highly convenient for on-site analysis and monitoring, as well as *in vivo* analysis [2]. A variety of coating fibers for extracting analytes of different polarity and volatility are nowadays commercially available. Nevertheless, the development of novel SPME fibers with improved mechanical, chemical and thermal stability is a current trend in analytical chemistry, as discussed in sections below. Selected milestones in the development of the SPME technique are shown in Fig. (**1**).

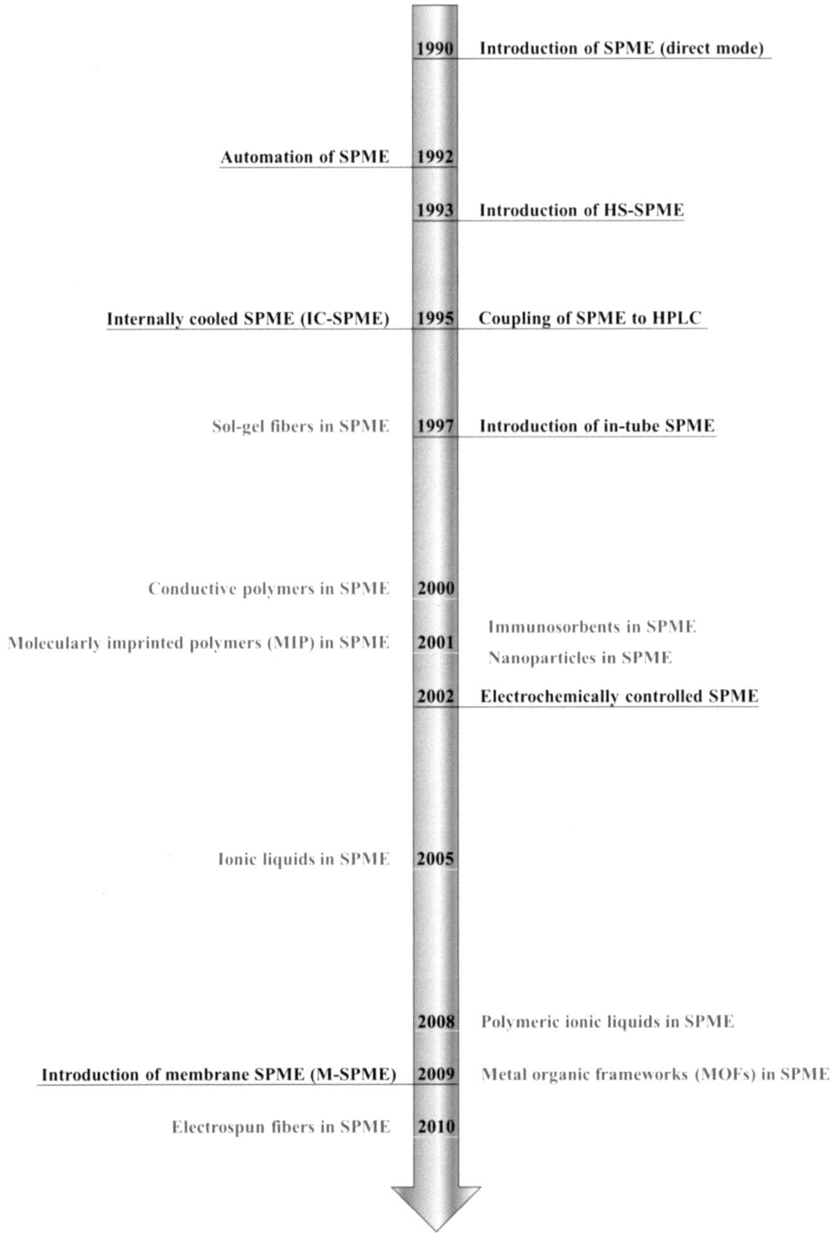

Fig. (1). Selected milestones in the development of the SPME technique.

Automation of the technique has also been addressed with the aim to improve both the sample throughput and reproducibility. The first attempt to automate SPME involved only slight modification of a gas chromatographic autosampler [3]. Supelco introduced the first commercially available automated SPME system one year later. From these initial developments, impressive advances have been achieved, and strict control of a variety of experimental factors affecting the microextraction process can nowadays be automatically performed. For instance, the use of a dual arm autosampler enabled the full automation of a microextraction procedure involving in-vial derivatization [4]. The robotic arms performed every single step of the procedure, namely, addition and mixing of buffer and derivatizing agent solutions, preparation of standard solutions by appropriate dilution from the stock solution, movement of the sample vial to the sample heater/mixer, extraction for a prescribed time, and desorption [4]. Automation of the SPME technique has been mainly directed to its combination with gas chromatography. Nevertheless, automated SPME has been also coupled with liquid chromatography (LC). In this regard, in tube-SPME was firstly automated by adapting commercial LC autosamplers [5]. The automation of SPME and coupling with LC was recently demonstrated by means of parallel extraction and desorption on a multiwall plate, providing an excellent sample throughput [6]. Progress towards the automation of SPME has been covered in the literature [7 - 9].

The environmental awareness experienced in scientific disciplines, especially boosted by the introduction of the emerging field of green chemistry [10], turned the development of environmentally sustainable analytical methodologies into a specific need [11 - 13]. In this sense, SPME has contributed significantly to the introduction of greener analytical procedures. In fact, the solvent consumption per analysis can be drastically reduced or even removed by using SPME with liquid or thermal desorption, respectively. SPME with thermal desorption can be therefore considered a paradigmatic example of green analytical chemistry.

Since its inception, more than 10,000 articles have been contributed to the development of the SPME technology, clearly showing the impact and importance of this sample preparation technique in a variety of scientific and technological fields [2]. It is worth pointing out that SPME has been identified as one of the six "great ideas of the decade" in Analytical Chemistry [14]. Several books [15 - 17] and review articles [2, 18 - 29] regarding different aspects of SPME and related techniques can be found in the literature.

In this chapter, the fundamentals of SPME are outlined. Furthermore, experimental parameters that have an impact on the performance of the SPME technique, as well as their optimization, are discussed. Finally, other SPME devices, namely, internally cooled SPME, in-tube SPME and membrane SPME, are described.

5.1.1. Theoretical Basis of SPME Technique

The main reason for the popularity of SPME is its simplicity, as proven by the large number of studies dedicated to the possible applications of this technique. While browsing through the relevant literature, one can find information about the use of SPME to isolate both organic and inorganic compounds as well as the compounds which are characterized by different polarity and volatility. Actually, it would be difficult to identify a group of analytes for which the attempts of isolating them by using the SPME fiber had not been undertaken. A similar situation takes place in the case of different sample types. SPME has been applied in the pretreatment of samples in different states of aggregation (liquid, solid and gas) as well as in a wide variety of matrix compositions. The phase in which the sample matrix occurs does not limit this technique either because SPME was successfully used for isolating various analytes from gaseous, liquid and solid (headspace method) samples. On the other hand, many reports published in the scientific literature deal with rather poorly chosen applications of SPME. This mainly encompasses situations where the use of SPME allows the analyte isolation, however, the procedure is not advantageous compared to the already existing techniques. From the point of view of analytical practice, the main strength of SPME is the simplicity of operation which results from the geometry of the system. On the other hand, the small dimensions of a microfiber force the user to follow a specific procedure when using this tool. As a consequence, the correct and optimal use of SPME technique requires an in-depth understanding of fundamental thermodynamic processes and kinetic criteria governing the mass transport during the isolation of analytes.

5.1.2. Sampling Mode

The type of procedure for isolating the analytes, which is connected to the system composition and the phase to which fiber is exposed, becomes the starting point for our further considerations. The major possibilities are presented in Fig. (**2**). In the direct immersion mode (Fig. **2A**), the fiber is directly exposed to a sample. Contrary to the remaining modes, the extraction phase in the direct immersion mode remains in physical contact with the sample matrix. As a consequence, the transfer of analytes occurs directly between these two phases in the system. In the headspace and membrane protection modes (Fig. **2B** and **2C**, respectively), the third phase, involved, plays a significant role in the process of mass transport. In the case of SPME technique, the isolation in headspace mode usually shortens the isolation time of analytes, which in the gaseous phase have diffusion coefficients ca. 4 times higher than in the liquid phase due to high convection (typically, 10^{-5} m^2/s in air, and 10^{-9} m^2/s in water) [30]. At the same time, it should be mentioned that the flux of analytes significantly depends on the contact area between the sample phase and headspace (liquid or solid) as well as the minimized distance between the fiber and sample surface. Under these considerations, the introduction of semipermeable membrane into the system does not give any advantage with regard to the kinetics of the analyte isolation process. Moreover, a further decrease in the values of diffusion coefficients of analytes in the membrane has been observed relative to, for example, their values in water; this phenomenon results from the lower mobility of molecules constituting the membrane, which is usually made from a polymer. The main aim and advantage of using a membrane is the creation of a mechanical barrier between the extractant and the sample. In this way, the fiber lifetime can be significantly extended, while for some applications, an additional influence on the selectivity of isolation process can be achieved (depending on the membrane material). The separation between the sorption phase and sample is also advantageous in the situation when the sample contains interfering substances, *e.g.* compounds with high molecular weights such as, proteins and humic acids. The application of membrane as presented in Fig. (**2D**) has a slightly different character. Its use enables the achievement of the aforementioned aims and, in addition, allows the use of sorption phases which are liquid during the extraction or desorption at higher temperatures. Contrary to the extraction modes shown in configurations B and C on Fig. (**2**), the analytes present in the membrane phase are also introduced into the further steps of analytical procedure, while in the other previously described modes, the intermediate phase (headspace or membrane) *de facto* dilutes the analytes contained in the original sample.

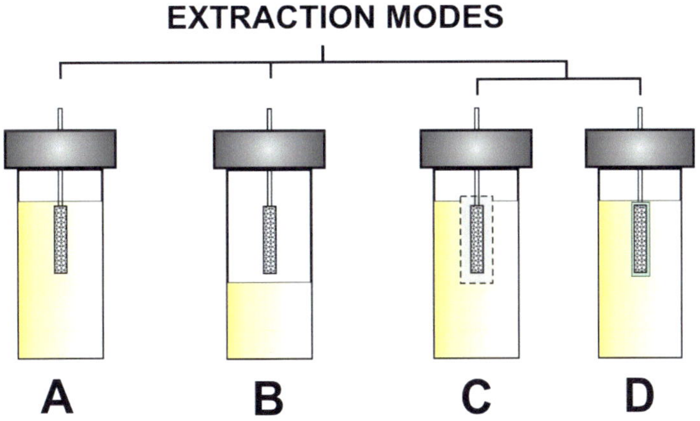

Fig. (2). Sampling modes in SPME: (**A**) direct immersion mode; (**B**) headspace mode; (**C** and **D**) membrane protection modes.

5.1.3. Sampling Thermodynamics

The extractants which are most commonly used in SPME technique are solids and pseudoliquids or mixtures of both. In the case of solids, the analyte isolation occurs as a result of their binding to the adsorbent's surface, while in the case of pseudoliquids and mixtures, the analytes dissolve in the sorption phase. The properties of mixed systems result, in a natural way, from the ratio of the two types of extractant present in the system. Although the different mechanism of analyte retention has significant practical consequences, the isolation process can be, to a certain degree, described by using similar relationships. Independent of the type of extractant phase, the fundamental phenomenon that occurs in the extraction system is the partition of analytes among all the phases in the system [31]. A quantitative description of this process for a two-phase system can be expressed as a partition coefficient in the equilibrium state, $K_{e/s}$:

$$K_{e\ s} = \frac{c_{\mathrm{e,eq}}}{c_{\mathrm{s,eq}}} \qquad \textbf{(1)}$$

The values of analyte concentration in the extractant phase ($C_{e,eq}$) and in the sample phase ($C_{s,eq}$), which have been used in the above equation, should be formally and correctly substituted with the corresponding values of activities. Nevertheless, the applied simplification is commonly used in practice because of the encountered low concentrations of analytes. Analogically, if the analyte concentration is calculated per surface area of extractant ($S_{e,eq}$), it is possible to define the partition coefficient for systems in which adsorption plays a dominant role in the process of analyte isolation, as follows:

$$K_{e\ s} = \frac{S_{\mathrm{e,eq}}}{C_{\mathrm{s,eq}}} \qquad \textbf{(2)}$$

At this point, it is worth mentioning that simplification based on the substitution of activities with the concentration values makes the correct interpretation of the above equations more difficult. In practice, it is exactly the activity values which allow - based on values of activity coefficients - the identification of factors that have a decisive influence on the values of partition coefficients. The values of activity coefficients not only describe molecular interactions in the system but they also indicate, *via* the appropriate physicochemical relationships, which physical (temperature, the volume of different phases in the system) and chemical parameters (fiber coating, pH, salt concentration) are thermodynamically significant and decisive with regard to the extraction efficiency [31]. In general, the values of partition coefficients determine the maximum values of enrichment coefficients which can be attained when the extraction is conducted under equilibrium conditions. Therefore, partition coefficients are the fundamental parameter at the optimization stage of the analyte isolation.

Because of the key role played by partition coefficients, a number of methods for determining their values have been developed [32]. One of the frequently used methods is based on the application of data originating from chromatographic separation (gas-liquid chromatography (GLC)) under isothermal conditions [33]. The mechanism of analyte retention in GLC is similar to that in the process of analyte partition between the sorbent and the sample, where the latter two components correspond to the stationary phase and carrier gas in the GLC system. Based on the determined value of net retention volume (V_n) and the volume of stationary phase (V_{fs}), we can easily calculate the value of partition coefficient for stationary phase-carrier gas system, $K_{e/g}$:

$$K_{e\ g} = \frac{V_n}{V_{fs}} \qquad \textbf{(3)}$$

The values of V_n are derived from the retention times of analytes (t_a) and an unretained compound (t_0), the pressure of carrier gas measured at the inlet (P_{in}) and outlet (P_{out}) of the chromatographic column, and the volumetric flow rate of carrier gas (F_g):

$$V_n = J_3^2 F_g\ t_a - t_0 \qquad \textbf{(4)}$$

where J_3^2 is the compressibility factor expressed as:

$$J_3^2 = \frac{3}{2}\frac{P_{in}\ P_{out}\ ^{2}-1}{P_{in}\ P_{out}\ ^{3}-1} \qquad \textbf{(5)}$$

In practice, the application of nitrogen as a carrier gas allows for using $K_{e/g}$ to approximate the value of partition coefficient in the extractant-air system. This also forms a basis for determining the values of partition coefficients in the extractant-water system, $K_{e/w}$ from the following equation:

$$\log K_{e\ w} = \log K_{e\ g} + \log K_{g\ w} \qquad \textbf{(6)}$$

where the partition coefficient for the gas-water system, $K_{g/w}$ is equal to the Henry's constant, which can be easily found in relevant literature.

The major limitation of chromatographic technique, in this context, is the necessity of using the column packed with the stationary phase which is identical to the fiber coating material. On the other hand, GLC allows a relatively simple determination of the values of partition coefficients for each analyte.

When polydimethylsiloxane (PDMS) is used as extractant, a simpler procedure can be employed. The procedure is based on the values of linear temperature programmed indices (LTPRI), which are available for many analytes in the published literature [34]. Then the values of partition coefficients, $K_{e/g}$ are derived from the following equation:

$$\log K_{e\ g}=\text{aLTPRI}+\text{b} \quad \textbf{(7)}$$

where a and b are empirical constants.

Taking into account Eq. 6, we obtain a relationship which allows us to estimate the partition coefficients in the extractant-water system:

$$\log K_{e\ w}=\text{aLTPRI}+\text{b}' \quad \textbf{(8)}$$

$$\text{where b}'= \text{b}+\log K_{g\ w}$$

At the end, it is worth mentioning that the determination of partition coefficients is also possible by using the SPME technique only [35]. In such case, the value of $K_{e/s}$ is derived from Eq. 1, however, the content of the investigated analyte in the sample and the volumes of phases in the system have to be known, obviously. The advantage of this approach is that the values of partition coefficients are determined under real conditions (*i.e.* including any possible modification in the composition of the sample matrix during the extraction process, *e.g.* pH adjustment or the addition of salt), while the chromatographic methods give only approximated values of partition coefficients.

5.1.4. Quantitative Analysis in SPME Technique

Besides the isolation of analytes from the interfering substances, the main target is to raise the concentration of analytes to the level which allows their detection using chosen technique for making final determinations. Therefore, the fundamental parameter for evaluating the applicability of sample preparation technique, including SPME, is the amount of analytes which can be isolated from a sample during the extraction. In practice, two parameters for evaluating the extraction efficiency are most frequently used, *i.e.* recovery (R) and enrichment (E) factors:

$$\text{R}=\frac{n_e}{n_{\text{os}}} \quad \textbf{(9)}$$

$$E=\frac{C_e}{C_{os}} \tag{10}$$

where n_e and C_e are, respectively, the number of moles and the concentration of analytes in an extractant, and n_{os} and C_{os} are, by analogy, the number of moles and the concentration of analytes in the original sample.

Due to the different number of phases in the system, the extraction in the aforementioned modes of analyte isolation depends on various parameters, as schematically shown below:

$$C_s \xleftrightarrow{K_{e/s};V_s;V_e} C_e \tag{11}$$

$$C_s \xleftrightarrow{K_{e/s};V_s;V_{ie}} C_{ie} \xleftrightarrow{K_{e/n};V_{ie};V_e} C_e \tag{12}$$

For a two-phase system, the values of partition coefficient and the ratio of extractant volume (V_e) to the sample volume (V_s) play a decisive role with regard to the efficiency of extraction.

Considering that the amount of analytes at each moment of the extraction process is constant, and that it equals the amount of analytes introduced into the system with the original sample, the following mass-balance equations can be formulated under equilibrium conditions:

$$C_{os}V_s=C_{e.eq}V_e+C_{s,eq}V_s \tag{13}$$

Therefore, after taking into account Eq. (1), we obtain the relationship which allows us to calculate maximum (equilibrium state) number of moles of analytes retained in the fiber coating, $n_{e,eq}$:

$$n_{e.eq}=\frac{C_{os}K_{e\ s}V_sV_e}{K_{e\ s}V_e+V_s} \tag{14}$$

By substituting the above equation into Eq. (9), we obtain a relationship which connects the recovery with the system's parameters, as follows:

$$R=\frac{K_{e\,s}V_e}{K_{e\,s}V_e+V_s} \quad \textbf{(15)}$$

Finally, by using Eq. 10, we can calculate the other parameter, *i.e.* the enrichment coefficient:

$$E=\frac{V_p}{V_e}R \quad \textbf{(16)}$$

The conclusions drawn from the above equations are significant with regard to quantitative analysis. The simplest, but rare in SPME technique, is the case when the analytes recovery is higher than 95%. In such situation, it is assumed that the extraction is exhaustive, which allows the direct calculation of analyte concentration in the original sample based on the determined number of moles of this analyte and the known sample volume. As mentioned before, due to the small volume of extractant phase used in SPME, it is a rare situation related to the analytes characterized by very large values of partition coefficients. When considering Eq. (15), it becomes apparent that it is possible to decrease the sample volume to a certain degree. However, one should keep in mind that it is also necessary to obtain the amount of analyte which is sufficient for a given technique of final determination, as indicated in Eq. (14). The opposite scenario occurs when the amount of retained analyte does not significantly lower the content of this analyte in the sample, which frequently takes place in the case of large sample volume (or low values of $K_{e/s}$ however that is basically disadvantageous). When the sample volume is significantly larger than the product $K_{e/s}V_s$, Eq. (14) assumes a simpler form and we end up with a linear relationship between the amount of retained analyte and the analyte concentration in the sample, with the proportionality ratio equal to $K_{e/s}V_e$. As a consequence, the value of the aforementioned product and the amount of retained analyte is sufficient for calculating the analyte concentration in the sample. The additional practical advantage here is that the sample volume does not have to be known. This feature has been used in certain applications, *e.g.* in the environmental pollution analysis where the fiber is exposed to the atmospheric air.

In the case of a three-phase system, we deal with headspace extraction or the application of membrane. Under this scenario, the analytes are partitioned among all the phases of the system (Eq. 12) at the ratio which depends on the sample volume (V_s), intermediate extractant volume (V_{ie}; headspace or membrane) and the volume of extractant (V_e) as well as the values of partition coefficients for the analyte being partitioned between the intermediate extractant and the sample ($K_{ie/s}$), and between the fiber coating and intermediate extractant ($K_{e/ie}$). As in a two-phase system, the mass balance can be described by the following equation:

$$C_{os}V_s = C_{e.eq}V_e + C_{ie,eq}V_{ie} + C_{s,eq}V_s \tag{17}$$

where $C_{ie,eq}$ is the analyte concentration in the intermediate extractant at the equilibrium state.

After taking into account the values of partition coefficients $K_{ie/s}$ and $K_{e/ie}$ (defined analogically as in Eq. (1)), we obtain the relationship which expresses the number of moles of analyte retained on the fiber as a function of the extraction parameters, as follows:

$$n_{e.eq} = \frac{C_{os}K_{e\ s}V_sV_e}{K_{e\ s}V_e + K_{ie\ s}V_{ie} + V_s} \tag{18}$$

where $K_{e/s}$ has been calculated in accordance with Eq. (6) as the product $K_{ie/s}K_{e/ie}$.

Although the direct application of the above equation requires the values of many parameters and is not used in practice very often, we can reach significant practical conclusions based on its analysis. First of all, it is noteworthy that the volume of intermediate extractant is only present in the denominator. This means that making determinations in modes B and C (Fig. **2**) always results in lowering the number of moles of retained analyte. In practice, in the case of headspace extraction mode, the volume of headspace phase should be kept as low as possible.

When the membrane is an integral part of the coating, the analytes are not lost (Fig. **2D**). Another reached conclusion is based on the fact that in the equilibrium state it does not matter whether the fiber is exposed directly to the sample or placed in the sample headspace. In both cases, the amount of analytes retained on the fiber will be the same. On the other hand, in the case of isolating the analytes with non-negligible volatility (high values of $K_{ie/s}$) directly from the sample, the headspace volume should be approaching zero.

5.1.5. Kinetics of Extraction in SPME Technique

Due to the fact that the extraction process with the use of SPME technique can be easily automated, the possibility of conducting the analyte isolation under non-equilibrium conditions has become more attractive. This is due to the increased control over the basic parameter in the kinetically controlled extraction, namely, the time. The optimization of extraction time has a profound meaning in practice, for example, in the situation where it is necessary to perform a large number of repeatable measurements. Because of a very long fiber exposure time to the sample, extraction under equilibrium conditions is performed in practice only when necessary, that is, when the amount of retained analytes (calculated from Eq. (14) for example) is close to the detection limits of the detector used. Moreover, it is worth mentioning that under non-equilibrium conditions some of the above mentioned parameters such as, partition coefficient, the volumes of phases or analytes concentration also play an important role because they affect the flux of analytes in the system.

Usually, the extraction of analytes is graphically illustrated with a sorption profile, which demonstrates the amount of retained analytes as a function of time, as shown in Fig. (**3**). In this figure, two distinct areas are present, *i.e.* the first area characterized by kinetic regime and the second one, described by the equilibrium conditions. In practice, the first area includes the time period during which the extraction efficiency attains 95% of its maximum ($t_{0.95}$), as derived from Eq. (14). Consequently, the equilibrium area contains all other time periods longer than $t_{0.95}$. This arbitrarily set cut-off point mainly results from the fact that the equilibrium state is theoretically reached in infinity. In practice, very often the additional subarea is distinguished inside the kinetic area (Fig. **3A**). The following assumption can be made for this subarea: the amount of analytes retained on the fiber increases linearly with increasing time. The aforementioned assumption concerns only very short extraction times because, from the theoretical point of view, the sorption curve does not have linear areas [36]. The significance of the presented area division manifests itself mainly in the mode of performing quantitative determinations. Within the area A in Fig. (**3**), it is possible to perform kinetically controlled calibration because the amount of retained analytes depends linearly on time and the analyte concentration in the sample, if the process parameters are constant [17]. On the other hand, in the area marked C, the quantitative determinations can be performed based on the know values of partition coefficients; if the extraction is exhaustive then the analyte concentration in the sample can be calculated directly by using the known sample volume.

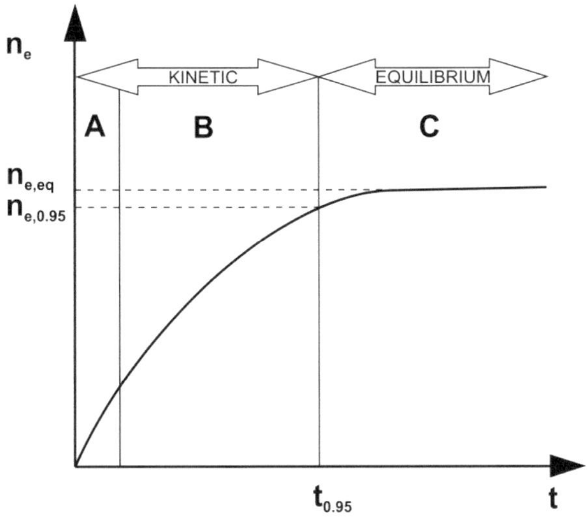

Fig. (3). Sorption profile as a function of time.

Defining the system geometry is the starting point for the quantitative analysis of extraction kinetics. A schematic representation of the fiber exposed to the sample is shown in Fig. (**4**).

According to the Fick's first law, the value of mass flux F crossing a diffusion layer in a one-dimensional space is described by the following equation:

$$F = -D_a \nabla C_a \quad \textbf{(19)}$$

where D_a is a diffusion coefficient, and ∇C_a is the concentration gradient of analyte in the diffusion layer.

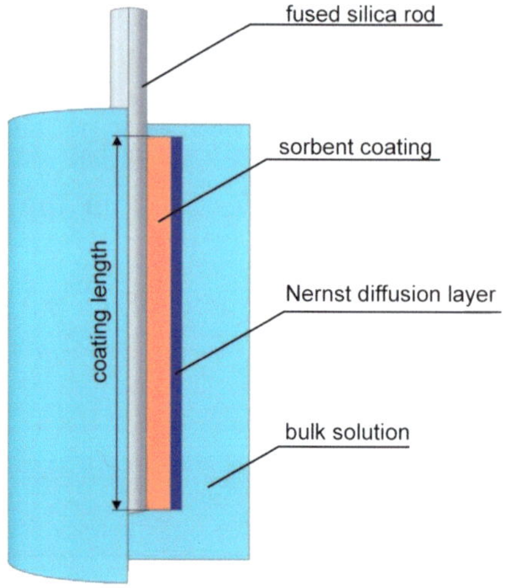

Fig. (4). Schematic representation of the SPME fiber exposed to the sample solution.

In the case of SPME technique, the above equation can only be analyzed qualitatively due to the system geometry and temporal variability of concentration. Nevertheless, the analysis of this equation leads to significant conclusions. The concentration distribution in the system where the fiber is exposed to the media characterized by high diffusion resistance (*e.g.* water) has been shown schematically in Fig. (**5**).

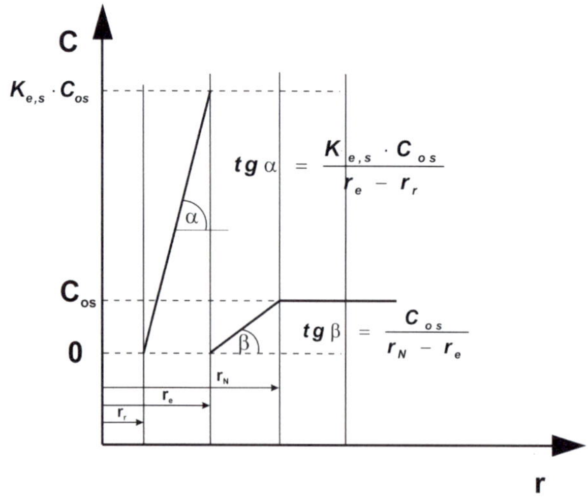

Fig. (5). Analyte concentration distribution where the SPME fiber is exposed to the sample.

It is noticeable that at the starting point (t=0) the system behaves in the most advantageous way, meaning, the highest mass flux is achieved for a given set of mixing and temperature parameters, which results from the maximum concentration gradient equal to the slope of line describing the changes in analyte concentration in the static diffusion layer (assuming $C(r_e)=C_{os}$). In the case of fiber coating, the analyte concentration on its external surface results from the partition of the analyte and can be approximated by the value of product $K_{e/s}C_{os}$, while the analyte concentration for the radius $r=r_r$ equals zero. The situation in the static diffusion layer is analogous, *i.e.* under the assumption of instantaneous sorption of analytes onto the coating material, it can be hypothesized that the analyte concentration in the sample at the coating-sample interface ($r=r_e$) is infinitesimal, while its value at the distance equal to the external surface of diffusion layer is equal to the analyte concentration in the original sample C_{os}. From the user's point of view, the only way to increase the analyte extraction rate under these circumstances is by decreasing the thickness of diffusion layer using the effective mixing of sample. Considering quantitative aspects of the extraction, it should be kept in mind that the values of diffusion coefficients in water are approximately one order of magnitude higher compared to those in the extractant layer (*e.g.* $\cong 10^{-6}$ cm^2/s for PDMS). Under ideal conditions as defined by the concentration values (Fig. **5**), it can be assumed that for t=0 the concentration difference in the Nernst diffusion layer is equal to the initial C_{os}, while the difference in analyte concentrations in the coating equals the product $K_{e/s}C_{os}$.

Based on these premises, the approximated thickness of diffusion layer, which sets the limits of mechanical mixing in case it is indispensable, can be determined. By using the Fick's equation, the following ratio can be formulated:

$$\frac{F_N}{F_e} = \frac{D_a^N \frac{C_{os}}{x_N}}{D_a^e \frac{K_{e\ s} C_{os}}{x_e}} = 1 \qquad \textbf{(20)}$$

where F_N and F_e denote the mass flux of analytes in the static Nernst layer and the fiber coating, respectively; x_N and x_e denote the respective thickness of Nernst diffusion layer and of the coating; and D_a^N and D_a^e are the values of analyte diffusion coefficients in the Nernst layer and the coating material, respectively [37].

After a simple transformation and by assuming that $D_a^N = 10D_a^e$, we obtain the following equation:

$$x_N = \frac{10}{K_{e\ s}} x_e \tag{21}$$

Bearing in mind that the coating thickness of commercially available fibers does not exceed 100μm, it should be noted that for the analytes with low partition coefficients (<100), the limit of mixing efficiency can be reached. Then the diffusion rate in the coating becomes a limiting factor.

At this point, it has to be noted that the Fick's law describes the processes taking place in the Nernst diffusion layer better than those occurring in the fiber coating, which results from the significantly lower ratio of the layer thickness to the radius. In the case of coating, more reliable results are obtained when the system geometry is taken into account. To this end, a differential diffusion equation for a cylindrical system has to be solved, as follows [30]:

$$\frac{\partial C_a}{\partial r^2} + \frac{1}{r}\frac{\partial C_a}{\partial r} = \frac{1}{D_a^e}\frac{\partial C_a}{\partial t} \tag{22}$$

As a result, we obtain the extraction process which is closer to the real one, as presented in Fig. (**6**).

In comparison to ideal conditions, the main significant difference is that the initial analyte concentration on the external surface of coating is lower than that calculated from the relevant partition coefficient. Obviously, this is caused by the finite rate of analyte transport through the Nernst layer. Moreover, the analyte concentration inside the coating increases faster than on the coating surface. As a consequence, it has to be concluded that the earlier observations about the effect of the efficiency of mechanical mixing are most applicable at the initial stage of extraction. The extended exposure time of the fiber causes that the transport of analytes in the fiber coating becomes the limiting factor with respect to extraction rate. Therefore, when aiming for the equilibrium state in the system, it is worth considering the possibility of applying alternative mixing procedures, *e.g.* sonication, which also increases the transport of analytes in the coating. When considering the changes in analyte transport rate in the Nernst layer, one should pay attention to the fact that they mainly result from the decreased rate of analyte absorption by the fiber. This, in turn, reduces the concentration gradient, and originates from the aforementioned decrease in analyte flux into the deeper area of coating. In some cases, a decrease in the concentration gradient due to the decreased analyte concentration in the sample can also be observed. Such scenario occurs when the analytes characterized by high partition coefficients are extracted from small-volume samples [30].

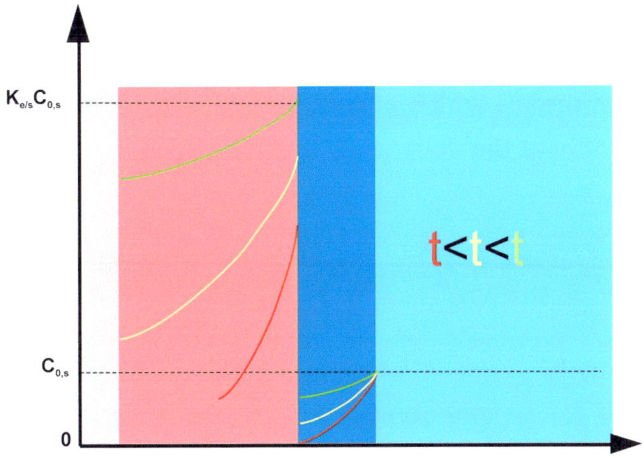

Fig. (6). Scheme of analyte concentration profiles *vs.* time in cylindrical extraction system.

5.1.6. Passive Sampling with a SPME Device

The application of SPME sampler is a separate issue, as presented in Fig. (7).

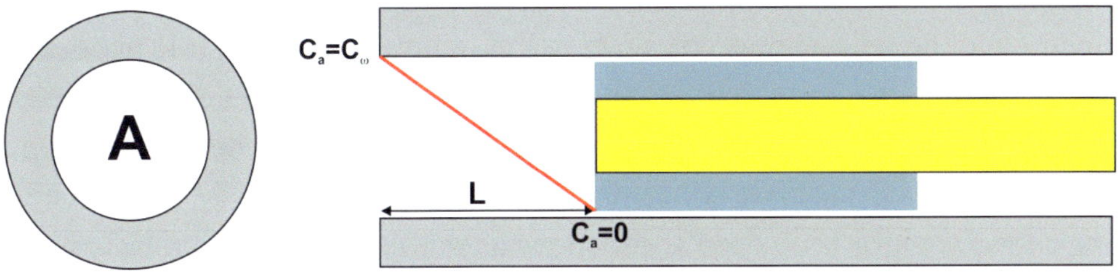

Fig. (7). Schematic of passive sampling with the SPME fiber retracted inside the needle of the assembly.

In this illustration, the element of the fiber in which the extraction takes place remains hidden inside the needle of the SPME device and therefore is not directly exposed to the sample. In this way, a static diffusion layer is obtained, with a fixed length L and the known cross-section area A. The analyte concentration profile in the diffusion layer, shown in the figure, is a direct outcome of the application of previously used Fick's law (Eq. 19). Due to the static character of diffusion layer and its length, diffusion through this layer influences the analyte isolation rate. This, in turn, causes that the isolation process becomes independent of the sample movement outside the SPME needle. When the Fick's law is expressed in relation to the constant cross-section area, as follows:

$$F=\frac{1}{A}\frac{dn}{dt} \qquad \textbf{(23)}$$

we obtain the relationship which allows the calculation of changing number of moles of the analyte n at the distance L from the needle's inlet in dependence on time [38]:

$$dn=AD_a\frac{C\ t}{L}dt \qquad \textbf{(24)}$$

where C(t) is the analyte concentration at the inlet of SPME needle as a function of time. Although for strictly formal reasons the numerator in Eq. 24 should consist of the difference between the analyte concentration at the needle's inlet and the analyte concentration at distance L, the above transcription has a significant practical meaning. This results from the fact that the major aim of the described SPME application is to obtain the time-weighted average (TWA) of analytes concentration. For this purpose, one fundamental condition has to be fulfilled, *i.e.* the value of analyte flux over time can only depend on the system geometry and the analyte concentration in the sample. Therefore it is necessary to secure conditions under which the analyte concentration in the direct vicinity of extractant is zero. This means that the difference in analyte concentration in the static layer equals the analyte concentration in the sample [39, 40]. In practice, the above considerations bear significant consequences. The extraction time is a key parameter; it should be chosen in such a way as not to exceed 5% of fiber loading relative to the analyte concentration under equilibrium conditions. It has been assumed that in this way the instantaneous analyte sorption onto extractant can be achieved which enables the fulfillment of the condition about the null analyte concentration near the fiber. On the other hand, the determination of TWA concentration is dedicated to long-time measurements. Therefore, the choice of other system parameters, which influence either the amount of retained analytes (cross-section area, the distance between coating and the needle's inlet) or the sorption capacity of extractant (extractant type and its volume), is also of significance [35]. At this point, the minimum amount of analytes which are required for conducting final determinations using selected techniques has to be kept in mind. In this context, choosing the extractant with higher affinity for the analyzed analytes is a very attractive solution (there are limited possibilities of increasing the extractant volume in SPME technique).

5.1.7. The Mechanism of Analyte Isolation

The last theoretical aspect of SPME application, which has a fundamental practical meaning, is the extractant type. Conducting the extraction with the use of solid extractants (therefore involving adsorption process) results in basic differences compared to analytical methodology for isolating analytes based on partition mechanism. The analysis of the properties of parameter $S_{e,eq}$ in Eq. (2) can be the starting point for the discussion. This parameter can be interpreted as the area occupied by a given analyte under equilibrium conditions, however, it is not equivalent to the available adsorbent area (or the number of available active sites). This is due to the fact that during the extraction, the number of available active sites diminishes because they are being occupied by the already adsorbed particles of analyte; in a more generalized approach, the number of active sites depends on all the components of sample matrix [41].

As a consequence, the application of surface coverage θ, defined in Eq. (25), is more appropriate for the purposes of quantitative analysis.

$$\theta = \frac{C^{a}_{\text{ads}}}{C_{\text{ads}}} \tag{25}$$

where C^{a}_{ads} denotes the concentration of adsorbed analyte per unit of extractant volume, and C_{ads} is the total density of active sites per unit of extractant volume. The above transcription is correct when the distribution of active sites in the extractant is homogenous. If the Langmuir isotherm model is used to describe adsorption as:

$$\theta = \frac{K_{ads\ s} C_{s.eq}}{1+K_{ads\ s} C_{s,eq}} \tag{26}$$

where $K_{ads/s}$ is the equilibrium constant, then we simply obtain a relationship between the analyte concentration in an adsorbent and the partition coefficient $K_{ads/s}$ under equilibrium conditions:

$$C_{ads}^{a} = C_{ads} \frac{K_{e\ s} C_{s.eq}}{1+K_{e\ s} C_{s,eq}} \tag{27}$$

Commonly, the lacking value of equilibrium analyte concentration in sample $C_{s,eq}$ poses a hindrance to applying the above equation. Nevertheless, the analysis of Eq. (27) allows for drawing significant conclusions. First of all, it is noticeable that the concentration of retained analyte is non-linearly dependent on the analyte concentration in the sample. Considering the calibration procedure, this finding is a vital difference in comparison to the extraction based on the absorption mechanism. The linearity of Eq. (27) can only be achieved when the product $K_{e/s}C_{s,eq}$ is significantly smaller than 1. However such scenario is completely irrational because it implies the infinitesimal affinity of analytes for the extractant (or very low concentration of analytes). As a consequence, the adsorption-based extraction is characterized by a very limited linearity range compared to partition-based extraction, which has been presented schematically in Fig. (**8**).

Another important aspect of the use of adsorbents is the interpretation of their surface properties as described by C_{ads}. In fact, C_{ads} does not describe the total number of active sites but only its fraction that is available to the analyte for which Eq. (27) has been formulated.

Although the total number of active sites in an adsorbent is constant, the number of sites that are available to a given analyte depends on the extractant affinity for other sample components. In other words, we deal with the competing processes.

Therefore in the presence of other components interacting well with the extractant or occurring at high concentrations the number of active sites available to a given analyte is only a fraction of their total number. As a result, both the maximum amount of analytes retained under equilibrium conditions and the linear range of extraction process decrease. The influence of matrix composition on the aforementioned parameters is shown schematically in Fig. (**9**).

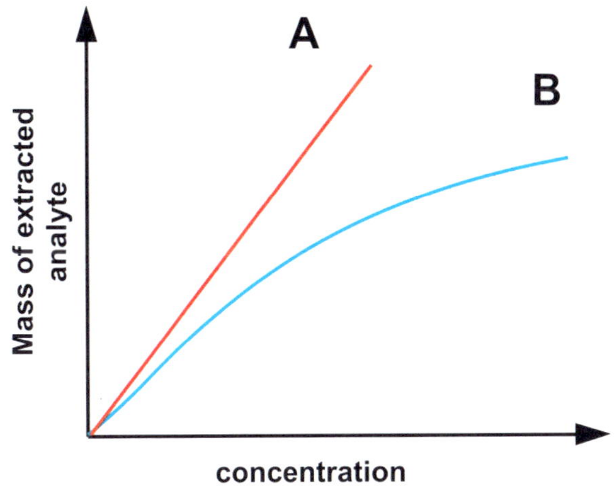

Fig. (8). Partition-based (**A**) and adsorption-based (**B**) extraction profiles.

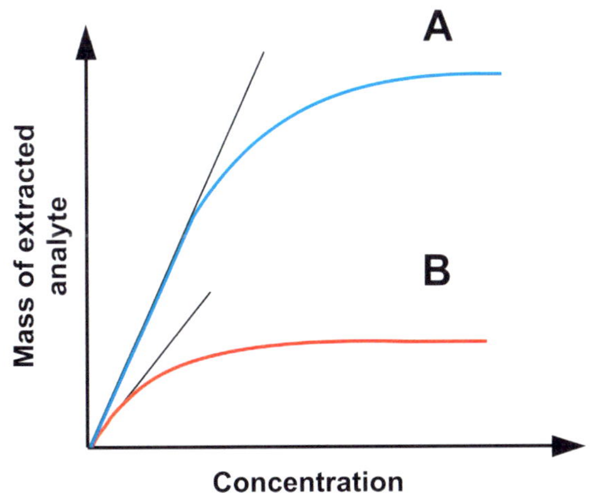

Fig. (9). Influence of the matrix composition on both extraction efficiency and linear dynamic range when using adsorption-based extraction in absence (**A**) and presence (**B**) of interfering compounds.

In practice, it is not possible to avoid the previously described effects because the conducted analyses always involve multicomponent systems. From the practical point of view, the most difficult situation definitely occurs when the analyte determinations are made in samples characterized by similar polarity, *e.g.* isolation of polar analytes from aqueous samples. In such case, due to huge excess of water molecules, even a slight affinity of water for the extractant (the latter should have polar active sites) may significantly diminish the surface available to a given analyte. In this context, we should also bear in mind the kinetics of extraction process. The temporal profile may be different than that presented in Fig. (**3**). Due to differing diffusion rates of analytes, the elimination process takes place, *i.e.* the already retained analytes are substituted by slower but better interacting molecules of other sample components, which results in a decreased number of the former on the fiber. The described effects can be avoided, to a certain degree, by using very short extraction times that fall in the kinetically controlled range due to the analyte transport towards the surface of extractant. In most cases, this solution does not secure enough retained analytes to conduct their detection therefore the fiber exposure time has to be extended. Under such conditions, the quantitative measurements should be performed based on the calibration curves that had been obtained under conditions most closely imitating the real conditions with regard to the composition of sample matrix. It also should be noted that a change in sorption properties applies to extractants based on the partition mechanism. However, such situation occurs very rarely, when the composition of extractant is changed by the absorption process at the level of at least few percent.

5.1.8. Optimization of the Sample Preparation Step in SPME

The fundamentals of thermodynamics and kinetics described earlier for the analyte extraction process have obvious implications with regard to analytical procedures. At the same time, it has to be stressed that the optimal choice of extraction conditions encompasses numerous factors, which are presented schematically in Fig. (**10**). From among the listed factors, some should be singled out as playing a key role, *e.g.* the choice of extractant; sample composition and its temperature; and the extraction mode. Therefore, more attention has been paid to the aforementioned factors in the next part of this chapter.

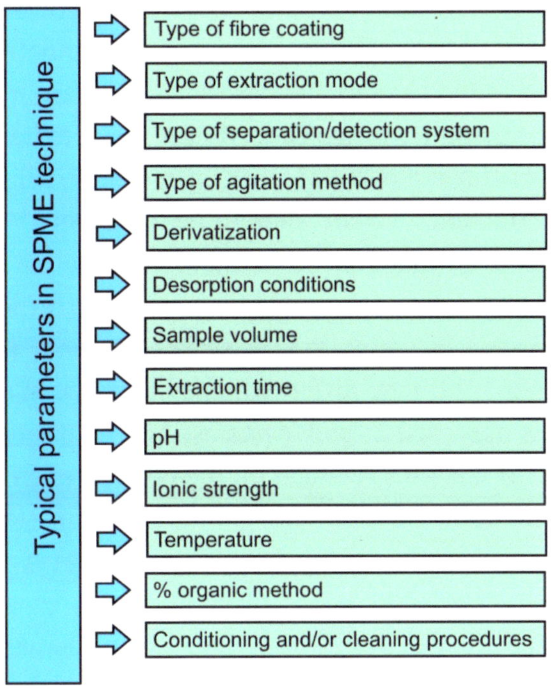

Fig. (10). Experimental parameters to be considered in SPME technique.

5.1.9. Sample Temperature

Temperature at which the fiber is exposed to the sample influences both thermodynamics and kinetics of analyte isolation. The thermodynamic effect is, in general, defined under the assumption that the variability of partition coefficient, as the equilibrium constant, can be expressed by van't Hoff isobar equation:

$$\frac{\partial \ln K_{e\ s}}{\partial T} = \frac{\Delta H_e}{\mathrm{RT}^2} \quad \textbf{(28)}$$

where ΔH_e is the enthalpy of analyte sorption and R is universal gas constant

The integral of the above equation, under the assumption that ΔH_e is constant within a given temperature range, indicates the linear relationship between the partition coefficient and reciprocal of temperature, with the slope equal to $\Delta H_e/R$:

$$\ln K_{e\ s} = \frac{\Delta H_e}{R}\frac{1}{T} + \mathrm{const} \quad \textbf{(29)}$$

Because the extraction of analytes is an exothermic process ($\Delta H_e > 0$) an increase in temperature must lead to a decrease in the partition coefficient. When the fiber is directly exposed to the sample, we observe a decrease in extraction efficiency with increasing temperature. In the case of headspace extraction, the influence of temperature on the values of partition coefficients in the sample-headspace system and headspace-extractant system should be considered in order to perform quantitative analysis. In general, the enthalpy of analyte transfer into gaseous phase is approximated by the enthalpies of vaporization corrected for the interaction with matrix in which the phase is dissolved. As a consequence, the values of slope in Eq. (29) may differ, while an increase in analyte affinity for the headspace phase is observed in both systems. The latter phenomenon results in a decreased amount of analytes retained on the coating of SPME fiber. Obviously, the above considerations concern only the equilibrium state. When the analyte isolation is performed in the kinetically controlled area (see areas A and B in Fig. **3**), the temperature increase can result in an increased amount of retained analytes for a given fiber exposure time due to the increased value of diffusion coefficients. This, in accordance with Eq. (19), increases the rate of mass transport in the system [42]. Due to the aforementioned reasons, heating of the sample is usually applied in the case of solid samples only to elevate the rate of analyte release into the headspace [43].

5.1.10. Fiber Exposure Time

The basic parameter targeted during the optimization of extraction time is the amount of analytes retained on the fiber that would be sufficient for qualitative or quantitative determination with the use of selected detector type. The second significant parameter is the calibration method of analytical procedure. Both these parameters depend on each other to a certain degree. When the calibration can be conducted based on the known values of partition coefficients then choosing equilibrium conditions is fully justified. On the other hand, when the time necessary for reaching the equilibrium is very long and significantly limits the throughput of analytical procedure, it is worthwhile to consider the possibility of conducting the kinetically controlled extraction. This idea is applicable only if the amount of retained analytes – compared to that obtained under equilibrium conditions - is adequate for performing determinations. In conclusion, it has to be noted that the optimal fiber exposure time depends on the majority of factors listed in Fig. (**10**). Bearing that in mind, one should pay more attention to the possible optimization of this specific parameter by application of more advanced techniques for experiment planning than those based on a sorption profile, the latter being most frequently found in the published literature.

5.1.11. Sample Agitation

The sample mixing is used to increase the rate of mass transport in the system, which results in a decrease in time necessary to achieve the equilibrium state as well as in an increase in extraction efficiency per unit time. From the practical point of view, it is necessary to secure homogenous mixing of the sample in order to achieve repeatability of the results in a diffusion-controlled process. A number of agitation techniques is available in SPME, *e.g.* mixing with a magnetic stirrer, sonication, needle vibration, orbital shaking, vortex stirring or the fiber exposed to the sample flow. From among the aforementioned techniques, the first two are used most frequently. In the case of mixing method employing a magnetic stirrer, its simplicity and the cost of necessary equipment are the main advantages. Nevertheless, we should discuss a couple of other factors. Firstly, in order to achieve high repeatability of measurements, the stirrer's dimensions should remain constant (besides the obvious parameter such as revolutions per unit time), while the fiber should be exposed in the same position relative to the axis of the vial so the constant linear flow rate of the mixed sample is maintained. A frequently discussed aspect of this mixing mode is the necessity of using a stirring element which can take part in the analyte extraction by acting as a competitive extractant or may be the source of sample contamination. Some authors also point out that the thermal properties of a stirring plate may result in the temperature change of the sample. The use of sonication as a mixing technique usually allows for shortening the extraction time compared to the extraction with the use of magnetic stirrer. This results from the fact that ultrasound interacts with both the sample phase and the extractant. As a consequence, sonication increases the speed of analyte molecules in all phases of the system, while the magnetic stirring device interacts, only, diminishing the static diffusion layer around the fiber. The temperature control is the basic problem of sonication method. A comparison between sonication and the use of magnetic stirrer shows that the former has a negative impact on the sample temperature because the released energy increases it. This, in turn, leads to lower extraction efficiency and decreased repeatability of analyte isolation, as discussed earlier. Mixing can be achieved by applying a vortex technique. The vortexing limits the possibility of sample contamination. The company Varian (Zug, Switzerland) introduced a commercially available SPME device with a vibrating needle which can be used for sampling and the automated injection of extracted analytes directly into a gas chromatographer [44]. The effect of different mixing techniques on the sorption profile of analytes is presented in Fig. (**11**).

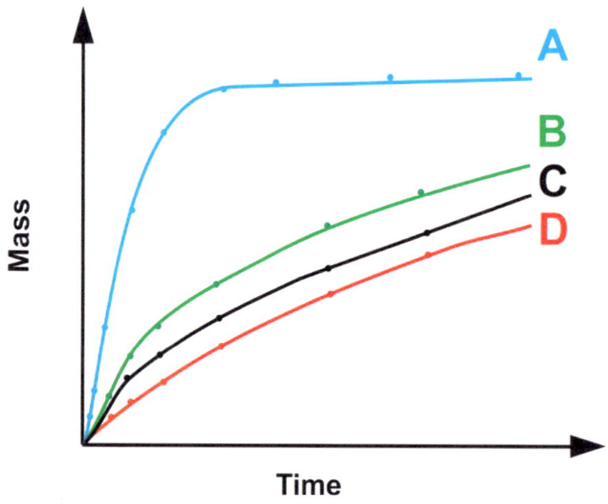

Fig. (11). Impact of different mixing techniques on the sorption of analytes as a function of time: (A) 2500 rpm magnetic stirring, (B) 100 W sonication, (C) repeated fiber insertion, and (D) no stirring.

5.1.12. The Change in pH and Matrix Effect

The influence of sample pH on the amount of retained analytes is only significant in the case of compounds with acid-base properties [45, 46] *e.g.*, amines and phenols. The mechanism of pH impact on the extraction process is related to the fact that only charge-free analyte molecules undergo partition. Therefore the proper choice of pH increases the concentration of neutral form of analytes and thus leads to better extraction efficiency. In Fig. (**12**) the relative amounts of dissociated and non-dissociated forms of propranolol are shown as an example. At pH 9.53, which is equal to the negative value of decadic logarithm ($\log_{10}$) of the dissociation constant pK_a, the neutral and ionized forms of a given compound are present at equal molar amounts. The amount of neutral form of the analyte increases with increasing pH, which enables its sorption onto the coating material. This obviously results in the increased extraction efficiency. In practice, the assumption is made that the sample pH should differ from the value of pK_a by two units (lower for acids and higher for bases). Moreover, the sample acidity is a significant parameter because of the resistance of coating; some sorption materials show poor resistance to strongly basic or strongly acidic environments [3].

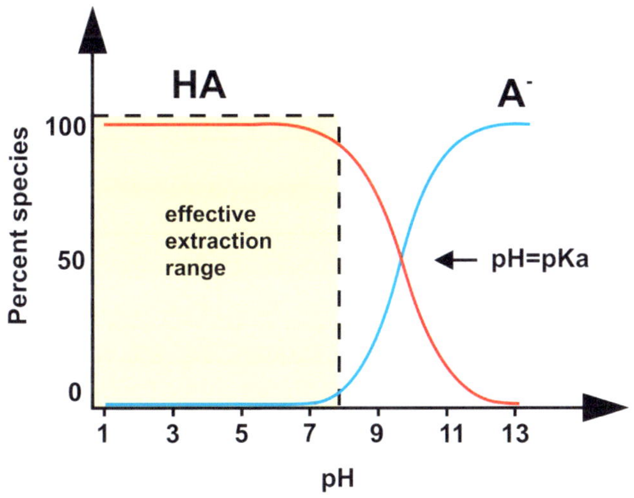

Fig. (12). Relative amounts of dissociated and non-dissociated forms of propranolol as a function of sample pH.

5.1.13. Salting-out Effect

Another parameter influencing the efficiency of extraction is the addition of salting-out agents. The addition of very small amounts of salting-out agent to liquid sample results in the increased extraction efficiency due to the higher ionic strength of the obtained solution. Organic analytes, particularly those with high polarity, become less soluble in water which results in the larger amount of analyte molecules adsorbed onto the fiber coating. An increase in extraction efficiency by salting out of sample is dependent, to a large extent, on the analyte concentration and salt concentration in the sample. Sodium chloride [47] and sodium sulfate are the most frequently used salt-out agents. However, the addition of salt-out agents highly increases the risk of sample contamination with the interfering substances, lowers selectivity and makes it necessary to thoroughly rinse the fiber as the latter becomes more prone to physical damage.

5.1.14. Derivatization

The low efficiency of SPME technique is particularly visible when sampling analytes present at trace levels as well as during the extraction of analytes from samples with complex matrices. For some compounds, the conversion of analytes into their derivatives characterized by different chemical structure and properties, followed by the extraction process, is a good solution for efficiency improvement. The chemical conversion of an analyte can be performed at different stages of extraction, *i.e.* prior to sample collection (conversion of analytes in the matrix); during the absorption of analyte onto the SPME fiber; or during the desorption of analytes in the chromatographic column [48, 49]. The derivatization of analytes also extends the range of various analyte groups which can be extracted by means of SPME fiber. The conversion of analytes into their derivatives is most frequently used in the case of polar and thermally unstable compounds; the procedure improves those compound's properties that are important with regard to chromatographic separation. Most frequently employed ways of derivatization are esterification of strongly acidic compounds such as acetic acids [50, 51], the conversion of amines and amphetamines into their less polar and more volatile derivatives [52], the conversion of aldehydes and ketones into more stable oximes [53], and phenols into acetates [54] as well as the conversion of metalorganic compounds into their more volatile derivatives [55]. Derivatization also allows the selective isolation of analytes and enables the conversion of analytes into their derivatives which are easier to extract, are more volatile or have better thermal stability.

5.1.15. Coating Material of SPME Fiber

A search for novel types of coating is the most intensely researched aspect of efficiency improvement in relation to analyte extraction from a given medium by means of SPME technique [26]. In order to achieve better efficiency of extractions performed with the use of SPME device, highly specific sorbents are applied which are characterized by high affinity for the analytes being determined. This translates into high values of partition coefficients, which allow for achieving high efficiency of the extraction procedure despite the small amount of sorption material used in SPME. The main reason for conducting the aforementioned research is the limited choice of commercially available coatings, particularly when specific applications are concerned [19]. Moreover, the commercial SPME fibers display a number of limitations, *i.e.* they have low thermal and mechanical resistance, low selectivity, and they do not fulfill user's expectations in relation to high extraction efficiency of polar analytes from samples characterized by polar matrix components. The commercial coatings are made from a number of extractants, *i.e.* polydimethylsiloxane (PDMS), divinylbenzene (DVB), polyacryl (PA), Carboxen (CAR) and polyethylene glycol (PEG, Carbowax, CW). Different combinations of fiber thickness and coating material are available [56]. Mixed coatings, consisting of couple types of extractant such as PDMS/DVB, PDMS/CAR, CW/DVB, demonstrate better efficiency during the extraction of polar organic analytes from samples with polar matrix components, which is due to combining polar and non-polar coating materials [57, 58]. A simple but in most cases sufficient rule for choosing the coating material is based on the principle that polar sorbents display high affinity for polar analytes, while non-polar sorbents for non-polar analytes. According to this rule, polar coatings, *e.g.* PA are used to collect samples of polar analytes, *i.e.* phenols [59] and some pesticides [60]. More polar coatings (PDMS/DVB, PDMS/CAR, CW/DVB) are applied in order to extract highly polar compounds such as, alcohols, amines and ethers [61, 62]. By analogy, hydrophobic PDMS is mainly used to extract non-polar compounds such as, BTEX (benzene, toluene, ethylbenzene and xylenes), PAH (polyaromatic hydrocarbons), VOC (volatile organic compounds) [63].

Due to the limited number of commercially available fibers and their drawbacks, it has become apparent that novel sorption materials have to be synthesized to fulfill the needs of SPME technique. Despite the fact that single-handed preparation of fibers is rather complicated and difficult to reproduce, a large number of publications dedicated to novel methods of fiber preparation, the conditions of their synthesis and possible uses of specific sorption materials has been noted in recent years. The materials most frequently used in the production of fiber coating in SPME technique are collated in Fig. (**13**).

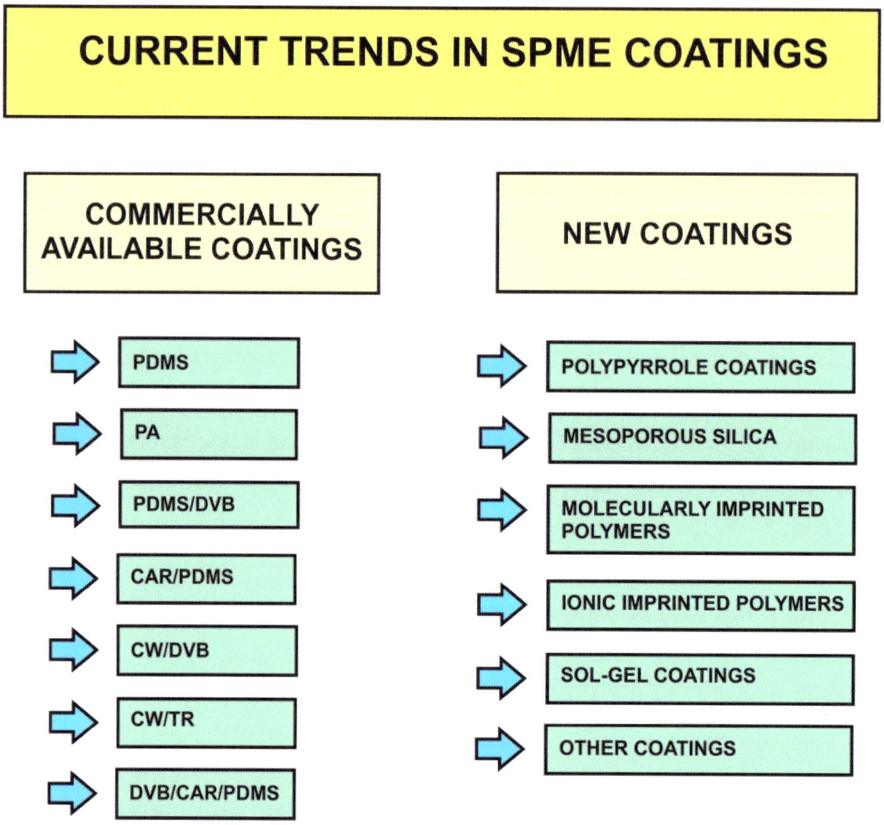

Fig. (13). Current trends in the preparation of SPME coating materials.

New technologies enable the synthesis of coatings for SPME fibers from relatively inexpensive polymeric materials. The obtained layers of extractant are stable in the entire pH range and at high temperatures, display high mechanical resistance, and can be re-used. Molecularly imprinted polymers (MIP) [64] and the materials produced using sol-gel technique [65] are sorption materials that are most frequently described in the published literature in the context of their application in SPME technique.

Coatings made from conductive polymers have been used to overcome the obstacles associated with the isolation of polar analytes from samples characterized by polar matrix composition. In this case, polypyrrole (Ppy) and its derivatives are most often applied polymers [66]. The polymers belonging to this group are relatively easy to obtain from chemical or electrochemical synthesis from monomers that are commercially available [67]. The fiber coating made from polypyrroles is characterized by high adhesion to the fiber core, the latter usually being a metal rod made from platinum, gold or stainless steel. Thanks to the application of metal cores, the mechanical resistance of such fibers is very high [68].

MIP are used as fiber coating in SPME devices because of their specific properties resulting from the synthesis type used during MIP production. In general, MIP are obtained by polymerization of monomers in the solution that contains template molecules. After polymerization, the template molecules are removed from the polymer structure with the use of appropriate solvents. As a result, cavities formed in the polymer structure have identical shape, dimensions and molecular recognition sites that are specific for a given analyte. MIPs are characterized by highly specific recognition of compounds with the structure identical or analogous to their template molecule [69]. These materials are stable in the wide pH range and have high thermal resistance [70]. MIPs are used, *inter alia*, to conduct highly efficient extractions of pesticides [71], phenols, heavy metals, antibiotics [72] and many other compounds characterized by complex matrix such as, blood [73], urine [74] as well as in food analysis [75].

Some obstacles related to low mechanical, thermal and chemical resistance have been overcome due to chemical binding of coating to the glass material from which the SPME fiber is made, as in the sol-gel technique [76]. The coating process is based on the hydrolysis of organic precursor (metal alkyl oxide; mainly tetraalkoxysilane) followed by its alcoholic or aqueous condensation. Next, the obtained product is subjected to hydrolysis and condensation. Due to the formation of bonds among the silica chains, the chains aggregate which results in an increase in solution viscosity. As a consequence, the original physical structure of sol turns into a three-dimensional network, *i.e.* rigid gel. By choosing appropriate reagents, a polymeric stationary phase with the desired surface properties and structure can be produced [77, 78]. Such coatings have chemical resistance and mechanical durability which allow for extending their application range, *e.g.* thermal desorption of high boiling point compounds. Moreover, the porosity of coating increases the specific surface area, which allows the use of polymeric thin films characterized by high extraction efficiency. Nevertheless, the sol-gel technique also has certain disadvantages and limitations. The complexity of synthesis, which requires the optimization of many parameters and, consequently, leads to difficulties with obtaining reproducible results is the main problem. The sol-gel technique enables the introduction of various compounds in the polymeric gel and therefore the creation of coatings with specific properties [79]. In order to satisfy the needs of SPME technique, the coatings containing, *inter alia*, polyvinyl alcohol (PVA) [80, 81] phenyltrimethoxysilane [82], hydroxyfullerenes [83], and n-octyltriethoxysilane (C8-TEOS). By applying the sol-gel technique, coatings containing calix [4]arenes in their structure were also produced. These materials are used to extract phenols [84, 85], aromatic amines [86, 87], BTEX as well as pesticides [88]. Other frequently used extractants are coatings containing crown ethers; they are applied to sample phenols [89, 90], aromatic amines [91], BTEX and chlorobenzenes [92] and pesticides [93].

Selected applications of SPME in environmental, food and clinical analysis are provided on Tables **1-3**, and the strengths, weaknesses, opportunities and threats of SPME are provided in Fig. (**14**).

STRENGTHS	**WEAKNESSES**
- Solvent-less technique. - High enrichment factors can be achieved. - Commercially available equipment. - Compatible with several analytical techniques. - Inexpensive, simple and easy to operate.	- Non-exhaustive technique. - Lack of robustness. - Low reproducibility. - Reduced fiber lifetime.
OPPORTUNITIES	**THREATS**
- Valid for *in situ* and *in vivo* sampling. - Applicable with scarce samples. - Novel materials can expand the aplicability of the technique.	- Lower selectivity than membrane- and three phase-based microextraction modes. - Reduced extraction kinetics compared with DLLME.

Fig. (14). Strengths, Weaknesses, Opportunities and Threats of SPME.

Table 1. Applications of SPME in environmental analysis.

Target compounds	SPME mode	SPME fiber	Analytical technique	LOD (µg/L)	Precision (RSD, %)	Samples analyzed	Ref.
organotin, organolead and organomercury compounds	HS-SPME	PDMS (100 µm)	GC-MS	0.3-1.4	1.2-3.2	water samples	[4]
alkanes and toluene	SPME (TWA passive sampling)	CAR/PDMS (75 µm)	GC-FID	--	6-15	indoor and outdoor air	[40]
phenol derivatives	HS-SPME	CAR/PDMS (75 µm) PDMS (100 µm)	GC-MS	0.001-0.012 0.001-0.020	0.7-11.6 0.3-16.2	sewage water	[54]
PAHs	HS-SPME	dodecylsulfate-doped polypyrrole (16 µm)	GC-FID and GC-MS	0.05-0.16	2.6-7.6	river and Caspian sea water	[67]
organochlorine pesticides	HS-SPME	PDMS (100 µm)	GC-ECD	0.06-0.65 ng/g	3.2-26.3	soil	[94]
aldehydes and ketones	HS-SPME DI-SPME	PDMS (100 µm)	GC-ECD	0.006-0.5 0.008-0.2	5.4-14.8 5.3-13.1	rain and drinking water	[47]
chlorophenols	DI-SPME	sol-gel-coated calix[4]arene	GC-FID	0.005-0.276	0.7-6.8	river water and soil	[84]
chlorophenols	DI-SPME	PA (85 µm)	GC-MS	0.0005-0.03	5-9	landfill leaches and soil	[95]
herbicides	DI-SPME	CW/DVB (65 µm)	GC-MS	0.01-0.06	5-31	soil and surface water	[96]
Explosives	DI-SPME	CW/DVB (65 µm)	GC-ITMS	0.005-0.325	6-20	seawater	[97]

Table 2. Applications of SPME in food analysis.

Target compounds	SPME mode	SPME fiber	Analytical technique	LOD (µg/L)	Precision (RSD, %)	Samples analyzed	Ref.
free fatty acids	HS-SPME	PA (85 µm)	GC-MS	1-5 µg/g	2.5-7.9	ewe cheese	[98]
2,4,6-trichloroanisole	HS-SPME	PDMS (100 µm)	GC-MS	0.002	5.1-13.1	wine	[99]

(Table 2) contd.....

Target compounds	SPME mode	SPME fiber	Analytical technique	LOD (μg/L)	Precision (RSD, %)	Samples analyzed	Ref.
volatile sulfides and disulfides	HS-SPME	PDMS (100 μm) and PA (85 μm)	GC-FPD	0.05-3.00	0.5-7	wine	[100]
volatile and semivolatile sulfur compounds	HS-SPME	CAR/PDMS (75 μm)	GC-FPD	0.001-0.060	< 18.3	beer	[101]
organophosphorus pesticides	HS-SPME	PDMS (100 μm)	GC-FTD and GC-MS	0.025-0.050	5.2-17.1	strawberry and cherry juice	[102]
organophosphorus pesticides	DI-SPME	sol-gel derived bisbenzo crown ether/hydroxyl-terminated silicone oil	GC-FPD	0.003-1 ng/g	2.0-15.0	honey, juice, oranche pakchoi	[93]
acidic pesticides	DI-SPME	CW/template resins (50 μm)	CE-MS	0.007-1.6	3-16	fruit	[103]
triazines	DI-SPME	atrazine-MIP	GC-FID and GC-MS	20-90	6.5-11.6	water, rice and onion	[104]
carbamate and phenylurea pesticides	DI-SPME	CW/template resins (50 μm)	LC-MS and LC-MS-MS	0.005-0.01 μg/g	1-17	fruit juice	[105]

Table 3. Applications of SPME in clinical analysis.

Target compounds	SPME mode	SPME fiber	Analytical technique	LOD (μg/L)	Precision (RSD, %)	Samples analyzed	Ref.
fatty acid ethyl esters	HS-SPME	PDMS/DVB (65 μm)	GC-MS	0.01-0.04 μg/g	3.5-16	hair	[106]
amphetamines	HS-SPME	PDMS (100 μm)	GC-FID	1.4-1.5	5.2-8.7	urine	[107]
amphetamine-like drugs	HS-SPME	PDMS (100 μm)	GC-MS	0.43-3.14 μg/g	1.5-3.6	hair and saliva	[108]
alkanes and aromatic hydrocarbons	HS-SPME	PDMS (100 μm)	GC-FID	0.012-1.26	3.7-9.8	human breath	[109]
volatile organic compounds	HS-SPME	CAR/PDMS (75 μm)	GC-TOF-MS	0.31-0.75	3.4-9.5	human breath	[110]
γ-hydroxybutyric acid	HS-SPME	CW/template resins (50 μm)	GC-MS	50-100	2.9-3.9	plasma and urine	[111]
benzodiazepines	DI-SDME	RP-amide C16-coated silica particle	LC-MS-MS	0.8-3.0	9-17	whole blood	[6]

(Table 3) contd.....

Target compounds	SPME mode	SPME fiber	Analytical technique	LOD (μg/L)	Precision (RSD, %)	Samples analyzed	Ref.
carbamazepine, propranolol, pseudoephedrine, ranitidine and diazepam	DI-SPME	C18 and RP-amide C16 (45 μm)	LC-MS-MS	<1-10	3.2-29	urine and plasma	[112]
methadone and EDDP	DI-SPME	PDMS (100 μm)	GC-MS	0.15-2.48 μg/g	7.3-13.3	hair	[113]
antidepressants	DI-SPME	polypyrrole (20 μm)	HPLC-UV	5-8	10-14	plasma	[114]

5.1.16. Other SPME Devices

Another important area of research related to SPME technique is the search for novel equipment designs, with the following main aims: achieving better efficiency of extraction, shortening the fiber exposure time, and investigating the possible use of new stationary phases. In the next sections, the most commonly applied alternative equipment designs and technical solutions in SPME have been described. Internally cooled SPME (IC-SPME).

In order to achieve the highest extraction efficiency, one aims at obtaining the equilibrium state between the sorbent and the sample matrix in the fastest possible way. However the time necessary to reach the equilibrium state may vary in dependence on the coating thickness, sorbent type, analyte concentration and its volatility, and the extraction conditions. Although raising the sample temperature has a positive effect on the kinetics of analyte isolation, it also results in lowering the partition coefficient and, consequently, in the lower retention of analytes on the fiber coating. In order to achieve both aims, *i.e.* shorter extraction time and enhanced extraction efficiency, the sample has to be heated and the SPME fiber needs to be concomitantly cooled. In this case, the SPME fiber is a capillary, not a piece of glass tubing, the latter being a typical design. The lowering of fiber temperature is obtained by the introduction of the proper cooling agent inside the capillary. Since now, liquid CO_2 [115] and alcohol cooled in a cryostat [116] have been used as cooling agents. The first internally-cooled fibers were applied to perform the efficient extractions of VOC and BTEX from the air [117]. In subsequent publications, the use of IC-SPME device for isolating VOCs from tropical fruits [118], chloroanisoles in cork stoppers from wine bottles [119], and for sampling aerosols in the atmospheric air [120] was described.

5.1.17. In-tube SPME

The low mechanical resistance of SPME fiber and the difficulties associated with coupling SPME to HPLC became the starting point for elaborating an in-tube SPME technique. The in-tube SPME design was introduced into preparative analytics in 1997 [5] and since then has been widely used for collecting a wide spectrum of analytes from environmental, biological and food samples as well as for sampling polar and thermally unstable analytes [121]. The extractant phase in the in-tube SPME device forms the internal layer of the needle or open capillary. The analytes are retained in the extractant during a number of cycles consisting of drawing in and pushing out the sample from the syringe. Next, the analytes are directly desorbed in the HPLC injector with a mobile phase, and then directly injected into the HPLC column. The desorption of analytes is also conducted in an off-line mode by washing the in-tube SPME device with a small amount of solvent. Besides the standard materials used as coatings in capillary columns, the internal coating of the needle in the in-tube SPME device can also be made from MIPs [122], conductive polymers [123, 124] and immunosorbents [125]. Moreover, monolithic capillary columns modified with octadecyl groups C_{18} and made from the material of improved porosity have been used in the in-tube SPME technique, which resulted in a very small drop in pressure for the high flow values [126]. Various polymeric materials obtained using sol-gel technique are also employed in the in-tube SPME devices [78]. Thanks to the application of sol-gel technique, both inorganic and hybrid (organic-inorganic) polymers are obtained. The in-tube SPME technique can also be coupled to gas chromatography [127, 128]. Exemplary applications of different sorption materials in the in-tube SPME technique are presented in Table **4**.

5.1.18. Membrane SPME (M-SPME)

As described earlier, the SPME of polar analytes from samples which contain polar components still requires improvement with regard to many factors. The difficulties connected to sampling of polar analytes concern, inter alia, low affinity of coatings for polar analytes. However an increase in sorbent polarity also increases its affinity for the sample matrix, which can be particularly disadvantageous when adsorption is the underlying mechanism of analyte isolation. The partition-based mechanism of analyte isolation implies that liquid or pseudo-liquid extractants have to be used. For the adsorption-based extraction, the loss of extractant can easily occur due to its dissolution in the sample matrix. One of the possible solutions to this problem may be the use of a physical barrier between the polar extractant and the matrix, namely, a semipermeable membrane. In such a membrane system, polar analytes migrate from the sample matrix through the hydrophobic, stable and thermally stable membrane to be retained in an appropriate extractant. In the M-SPME technique, the SPME fiber is coated with two layers of sorbent, i.e. the internal layer of highly polar PEG which is the proper extractant, and the external layer of hydrophobic and non-polar PDMS that forms a specific barrier between the extractant and the sample matrix. A schematic representation of the SPME fiber with a two-layer coating is shown in Fig. (**15**).

Table 4. Selected applications of different stationary phases in the in-tube SPME technique.

Type of stationary phase	Type of analyte	Type of matrix	Type of determination technique	Ref.
molecularly imprinted polymers (MIPs)	propranolol	biological fluids samples	HPLC	[5]
conducting polymers (PPy)	b-blockers	urine and serum samples	LC-ESI-MS	[121]
	aromatic compounds	aqueous samples	HPLC	[129]
	organoarsenic compounds	aqueous samples	LC-ESI-MS	[122]
monolithic capillary columns	PAH, BTEX, pesticides	aqueous samples	HPLC	[130]
	PAH	aqueous samples	HPLC	[131]
	non-steroidal drugs, phenols, endocrine disrupting chemicals	aqueous samples	HPLC	[132]
	ketamine	urine samples	HPLC	[126]

(Table 4) contd.....

Type of stationary phase	Type of analyte	Type of matrix	Type of determination technique	Ref.
sol-gel sorbents	PAH, aldehydes, ketones, phenols, amines	aqueous samples	GC	[133]
	PAH, aldehydes, ketones	aqueous samples	CME-GC	[76]
	non-steroidal drugs, ketoprofen, fenbufen, ibuprofen	urine samples	HPLC	[134]

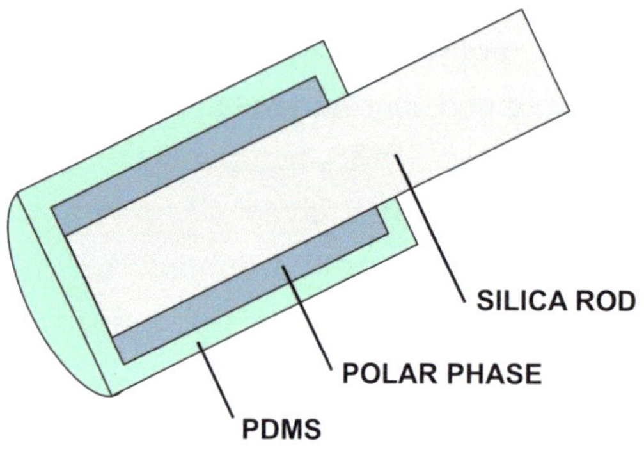

Fig. (15). Schematic representation of M-SPME.

The obtained fiber with a two-layer coating is characterized by good mechanical properties and high thermal stability, which allows for conducting thermal desorption of analytes directly in the injector of gas chromatographer. Until now, the M-SPME technique was used to isolate polar organic compounds from the aqueous samples and a headspace [135, 136].

SUMMARY

The inception of SPME in the early 90's represents a major contribution to the search for a simple and rapid technique that enabled the integration of sampling, extraction, preconcentration and sample introduction into a single step. SPME provides several benefits when compared with conventional sample preparation approaches. Specifically, its solvent-less nature involves an outstanding advantage in terms of solvent consumption, waste generation and occupational exposure. Besides, SPME allows for the convenient automation of the sample preparation step. Furthermore, the small size of the SPME device makes it being an excellent option in combination with portable analytical instrumentation for on-site analysis. Certain drawbacks have, however, being identified for SPME, including its limited sorption capacity and precision, as well as the fragility, reduced lifetime and limited choice of SPME coating fibers. In this sense, remarkable improvements of the SPME technology have been steadily made in the last years, thus yielding highly versatile and reliable sample preparation technique. Further developments are, nevertheless, expected in the coming years.

SPME is nowadays considered a well-established sample preparation technique that allows the extraction and enrichment of relevant compounds from a wide variety of samples. The applicability of this miniaturized technique has been abundantly demonstrated in several research areas, including applications in environmental, food, pharmaceutical and medical fields. The constant evolution of the technique makes us believe that SPME and related approaches will continue solving current and future challenges.

CONFLICT OF INTEREST

The authors confirm that they have no conflict of interest to declare for this publication.

ACKNOWLEDGEMENTS

F. Pena-Pereira thanks Xunta de Galicia for financial support as a post-doctoral researcher of the I2C program.

REFERENCES

[1] Arthur, C.L.; Pawliszyn, J. Solid phase microextraction with thermal desorption using fused silica optical fibers. *Anal. Chem.,* **1990**, *62*, 2145-2148. [http://dx.doi.org/10.1021/ac00218a019]

[2] Bojko, B.; Cudjoe, E.; Gómez-Ríos, G.A.; Gorynski, K.; Jiang, R.; Reyes-Garcés, N.; Risticevic, S.; Silva, É.A.; Togunde, O.; Vuckovic, D.; Pawliszyn, J. SPMEquo vadis? *Anal. Chim. Acta,* **2012**, *750*, 132-151. [http://dx.doi.org/10.1016/j.aca.2012.06.052] [PMID: 23062435]

[3] Arthur, C.L.; Killam, L.M.; Buchholz, K.D.; Pawliszyn, J.; Berg, J.R. Automation and optimization of solid-phase microextraction. *Anal. Chem.,* **1992**, *64*, 1960-1966. [http://dx.doi.org/10.1021/ac00041a034]

[4] Parkinson, D.R.; Bruheim, I.; Christ, I.; Pawliszyn, J. Full automation of derivatizationsolid-phase

microextraction-gas chromatography-mass spectrometry with a dual-arm system for the determination of organometallic compounds in aqueous samples. *J. Chromatogr. A,* **2004**, *1025*(1), 77-84. [http://dx.doi.org/10.1016/j.chroma.2003.10.061] [PMID: 14753674]

[5] Eisert, R.; Pawliszyn, J. Automated in-tube solid-phase microextraction coupled to high-performance liquid chromatography. *Anal. Chem.,* **1997**, *2700*, 3140-3147. [http://dx.doi.org/10.1021/ac970319a]

[6] Vuckovic, D.; Cudjoe, E.; Hein, D.; Pawliszyn, J. Automation of solid-phase microextraction in high-throughput format and applications to drug analysis. *Anal. Chem.,* **2008**, *80*(18), 6870-6880. [http://dx.doi.org/10.1021/ac800936r] [PMID: 18712934]

[7] Pawliszyn, J.; Vuckovic, D.; Mirnaghi, F.; Risticevic, S. *Handbook of Solid Phase Microextraction*; Pawliszyn, J., Ed.; Chemical Industry Press: Beijing, **2009**, pp. 135-165.

[8] O'Reilly, J.; Wang, Q.; Setkowa, L.; Hutchinson, J.P.; Chen, Y.; Lord, H.L.; Linton, C.M.; Pawliszyn, J. Automation of solid-phase microextraction. *J. Sep. Sci.,* **2005**, *28*, 2010-2022. [http://dx.doi.org/10.1002/jssc.200500244]

[9] Vuckovic, D. High-throughput solid-phase microextraction in multi-well-plate format. *TrAC. Trends Analyt. Chem.,* **2013**, *45*, 136-153. [http://dx.doi.org/10.1016/j.trac.2013.01.004]

[10] Anastas, P.T.; Warner, J.C. *Green chemistry: Theory and Practice*; Oxford Univerity Press: New York, **1998**.

[11] Tobiszewski, M.; Mechlińska, A.; Namieśnik, J. Green analytical chemistry theory and practice. *Chem. Soc. Rev.,* **2010**, *39*(8), 2869-2878. [http://dx.doi.org/10.1039/b926439f] [PMID: 20502819]

[12] Keith, L.H.; Gron, L.U.; Young, J.L. Green analytical methodologies. *Chem. Rev.,* **2007**, *107*(6), 2695-2708. [http://dx.doi.org/10.1021/cr068359e] [PMID: 17521200]

[13] Anastas, P.T. Green Chemistry and the role of analytical methodology development. *Crit. Rev. Anal. Chem.,* **1999**, *29*, 167-175. [http://dx.doi.org/10.1080/10408349891199356]

[14] Handley, J.; Harris, C.M. Great ideas of a decade. *Anal. Chem.,* **2001**, *73*(23), 660A-666A. [http://dx.doi.org/10.1021/ac012642q] [PMID: 11774924]

[15] Pawliszyn, J. *Solid Phase Microextraction. Theory and Practice*; Wiley: New York, **1997**.

[16] Pawliszyn, J. *Applications of Solid Phase Microextraction*; Pawliszyn, J., Ed.; Royal Society of Chemistry: Cambridge, **1999**.

[17] Pawliszyn, J. *Handbook of Solid Phase Microextraction*; Pawliszyn, J., Ed.; Chemical Industry Press: Beijing, **2009**.

[18] Pen, A. Trends in solid-phase microextraction for determining organic pollutants in environmental samples. *TrAC. Trends Analyt. Chem.,* **1999**, *18*, 557-568. [http://dx.doi.org/10.1016/S0165-9936(99)00145-4]

[19] Dietz, C.; Sanz, J.; Cámara, C. Recent developments in solid-phase microextraction coatings and

related techniques. *J. Chromatogr. A,* **2006**, *1103*(2), 183-192.
[http://dx.doi.org/10.1016/j.chroma.2005.11.041] [PMID: 16337213]

[20] Risticevic, S.; Niri, V.H.; Vuckovic, D.; Pawliszyn, J. Recent developments in solid-phase microextraction. *Anal. Bioanal. Chem.,* **2009**, *393*(3), 781-795.
[http://dx.doi.org/10.1007/s00216-008-2375-3] [PMID: 18836706]

[21] Ouyang, G.; Pawliszyn, J. Recent developments in SPME for on-site analysis and monitoring. *TrAC. Trends Analyt. Chem.,* **2006**, *25*, 692-703.
[http://dx.doi.org/10.1016/j.trac.2006.05.005]

[22] Vuckovic, D.; Risticevic, S.; Pawliszyn, J. *In vivo* solid-phase microextraction in metabolomics: opportunities for the direct investigation of biological systems. *Angew. Chem. Int. Ed. Engl.,* **2011**, *50*(25), 5618-5628.
[http://dx.doi.org/10.1002/anie.201006896] [PMID: 21608090]

[23] Ouyang, G.; Vuckovic, D.; Pawliszyn, J. Nondestructive sampling of living systems using *in vivo* solid-phase microextraction. *Chem. Rev.,* **2011**, *111*(4), 2784-2814.
[http://dx.doi.org/10.1021/cr100203t] [PMID: 21271696]

[24] Pawliszyn, J. Integration and miniaturization facilitate on-site and in-vivo analysis. *TrAC. Trends Analyt. Chem.,* **2011**, *30*, 1363-1364.
[http://dx.doi.org/10.1016/j.trac.2011.08.002]

[25] Souza Silva, E.A.; Risticevic, S.; Pawliszyn, J. Recent trends in SPME concerning sorbent materials, configurations and *in vivo* applications. *TrAC. Trends Analyt. Chem.,* **2013**, *43*, 24-36.
[http://dx.doi.org/10.1016/j.trac.2012.10.006]

[26] Spietelun, A.; Pilarczyk, M.; Kloskowski, A.; Namieśnik, J. Current trends in solid-phase microextraction (SPME) fibre coatings. *Chem. Soc. Rev.,* **2010**, *39*(11), 4524-4537.
[http://dx.doi.org/10.1039/c003335a] [PMID: 20882243]

[27] Namieśnik, J.; Zygmunt, B.; Jastrzebska, A. Application of solid-phase microextraction for determination of organic vapours in gaseous matrices. *J. Chromatogr. A,* **2000**, *885*(1-2), 405-418.
[http://dx.doi.org/10.1016/S0021-9673(99)01157-7] [PMID: 10941687]

[28] Spietelun, A.; Kloskowski, A.; Chrzanowski, W.; Namieśnik, J. Understanding solid-phase microextraction: key factors influencing the extraction process and trends in improving the technique. *Chem. Rev.,* **2013**, *113*(3), 1667-1685.
[http://dx.doi.org/10.1021/cr300148j] [PMID: 23273266]

[29] Spietelun, A.; Marcinkowski, Ł.; de la Guardia, M.; Namieśnik, J. Recent developments and future trends in solid phase microextraction techniques towards green analytical chemistry. *J. Chromatogr. A,* **2013**, *1321*, 1-13.
[http://dx.doi.org/10.1016/j.chroma.2013.10.030] [PMID: 24238710]

[30] Poling, B.E.; Prausnitz, J.M.; O'Connell, J.P. *Properties of Gases and Liquids,* 5th ed; McGraw-Hill Education: New York, **2000**.

[31] Atkins, P.W. *Physical Chemistry,* 4th ed; Freeman: New York, **1978**.

[32] Berthod, A.; Carda-Broch, S. Determination of liquid-liquid partition coefficients by separation methods. *J. Chromatogr. A,* **2004**, *1037*(1-2), 3-14.

[http://dx.doi.org/10.1016/j.chroma.2004.01.001] [PMID: 15214657]

[33] Pecsok, R.L. *Principles and Practice of Gas Chromatography*; Wiley: New York, **1959**.

[34] Kloskowski, A.; Chrzanowski, W.; Pilarczyk, M.; Namieśnik, J. Partition coefficients of selected environmentally important volatile organic compounds determined by gas-liquid chromatography with polydimethylsiloxane stationary phase. *J. Chem. Thermodyn.,* **2005**, *37*, 21-29. [http://dx.doi.org/10.1016/j.jct.2004.07.020]

[35] Liu, Z.; Wene, M.J. Measurement of gas-liquid partition coefficient and headspace concentration profiles of perfume materials by solid-phase microextraction and capillary gas chromatography-mass spectrometry. *J. Chromatogr. Sci.,* **2000**, *38*(9), 377-382. [http://dx.doi.org/10.1093/chromsci/38.9.377] [PMID: 11011720]

[36] Louch, D.; Motlagh, S.; Pawliszyn, J. Dynamics of organic compound extraction from water using liquid-coated fused silica fibers. *Anal. Chem.,* **1992**, *64*, 1187-1199. [http://dx.doi.org/10.1021/ac00034a020]

[37] Chen, Y.; Pawliszyn, J. Kinetics and the on-site application of standards in a solid-phase microextraction fiber. *Anal. Chem.,* **2004**, *76*(19), 5807-5815. [http://dx.doi.org/10.1021/ac0495081] [PMID: 15456301]

[38] Pawliszyn, J. Theory of solid-phase microextraction. *J. Chromatogr. Sci.,* **2000**, *38*(7), 270-278. [http://dx.doi.org/10.1093/chromsci/38.7.270] [PMID: 10901412]

[39] Martos, P.A.; Pawliszyn, J. Time-weighted average sampling with solid-phase microextraction device: implications for enhanced personal exposure monitoring to airborne pollutants. *Anal. Chem.,* **1999**, *71*(8), 1513-1520. [http://dx.doi.org/10.1021/ac981028k] [PMID: 10221072]

[40] Chen, Y.; Pawliszyn, J. Time-weighted average passive sampling with a solid-phase microextraction device. *Anal. Chem.,* **2003**, *75*(9), 2004-2010. [http://dx.doi.org/10.1021/ac026315+] [PMID: 12720333]

[41] Górecki, T.; Yu, X.; Pawliszyn, J. Theory of analyte extraction by selected porous polymer SPME fibres. *Analyst (Lond.),* **1999**, *124*, 643-649. [http://dx.doi.org/10.1039/a808487d]

[42] Nerín, C.; Salafranca, J.; Aznar, M.; Batlle, R. Critical review on recent developments in solventless techniques for extraction of analytes. *Anal. Bioanal. Chem.,* **2009**, *393*(3), 809-833. [http://dx.doi.org/10.1007/s00216-008-2437-6] [PMID: 18931996]

[43] Prosen, H.; Zupančič-Kralj, L. Solid-phase microextraction. *Trends Analyt. Chem.,* **1999**, *18*, 272-282. [http://dx.doi.org/10.1016/S0165-9936(98)00109-5]

[44] Geppert, H. Solid-phase microextraction with rotation of the microfiber. *Anal. Chem.,* **1998**, *70*, 3981-3982. [http://dx.doi.org/10.1021/ac970419g]

[45] van Doorn, H.; Grabanski, C.B.; Miller, D.J.; Hawthorne, S.B. Solid-phase microextraction with pH adjustment for the determination of aromatic acids and bases in water. *J. Chromatogr. A,* **1998**, *829*(1-2), 223-233. [http://dx.doi.org/10.1016/S0021-9673(98)00760-2] [PMID: 9923082]

[46] Pan, L.; Adams, M.; Pawliszyn, J. Determination of fatty acids using solid phase microextraction. *Anal. Chem.,* **1995**, *67*, 4396-4406.
[http://dx.doi.org/10.1021/ac00119a031]

[47] Bao, M.L.; Pantani, F.; Griffini, O.; Burrini, D.; Santianni, D.; Barbieri, K. Determination of carbonyl compounds in water by derivatization-solid-phase microextraction and gas chromatographic analysis. *J. Chromatogr. A,* **1998**, *809*(1-2), 75-87.
[http://dx.doi.org/10.1016/S0021-9673(98)00188-5] [PMID: 9677712]

[48] Alzaga, R.; Bayona, J.M. Determination of perfluorocarboxylic acids in aqueous matrices by ion-pair solid-phase microextraction-in-port derivatization-gas chromatography-negative ion chemical ionization mass spectrometry. *J. Chromatogr. A,* **2004**, *1042*(1-2), 155-162.
[http://dx.doi.org/10.1016/j.chroma.2004.05.015] [PMID: 15296400]

[49] Alzaga, R.; Peña, A.; Ortiz, L.; Bayona, J.M. Determination of linear alkylbenzensulfonates in aqueous matrices by ion-pair solid-phase microextraction-in-port derivatization-gas chromatography-mass spectrometry. *J. Chromatogr. A,* **2003**, *999*(1-2), 51-60.
[http://dx.doi.org/10.1016/S0021-9673(03)00493-X] [PMID: 12885051]

[50] Wittman, G.Y.; Van Langenhove, H.; Dewulf, J. Determination of acetic acid in aqueous samples, by water-phase derivatisation, solid-phase microextraction and gas chromatography. *J. Chromatogr. A,* **2000**, *874*(2), 225-234.
[http://dx.doi.org/10.1016/S0021-9673(00)00114-X] [PMID: 10817361]

[51] Sarrión, M.N.; Santos, F.J.; Galceran, M.T. Solid-phase microextraction coupled with gas chromatography-ion trap mass spectrometry for the analysis of haloacetic acids in water. *J. Chromatogr. A,* **1999**, *859*(2), 159-171.
[http://dx.doi.org/10.1016/S0021-9673(99)00873-0] [PMID: 10574209]

[52] Zimmermann, T.; Ensinger, W.J.; Schmidt, T.C. *In situ* derivatization/solid-phase microextraction: determination of polar aromatic amines. *Anal. Chem.,* **2004**, *76*(4), 1028-1038.
[http://dx.doi.org/10.1021/ac035098p] [PMID: 14961735]

[53] Cancho, B.; Ventura, F.; Galceran, M.T. Determination of aldehydes in drinking water using pentafluorobenzylhydroxylamine derivatization and solid-phase microextraction. *J. Chromatogr. A,* **2002**, *943*(1), 1-13.
[http://dx.doi.org/10.1016/S0021-9673(01)01437-6] [PMID: 11820270]

[54] Llompart, M.; Lourido, M.; Landín, P.; García-Jares, C.; Cela, R. Optimization of a derivatization-solid-phase microextraction method for the analysis of thirty phenolic pollutants in water samples. *J. Chromatogr. A,* **2002**, *963*(1-2), 137-148.
[http://dx.doi.org/10.1016/S0021-9673(02)00646-5] [PMID: 12187964]

[55] Davis, W.C.; Vander, P.S.; Schantz, M.M.; Long, S.E.; Day, R.D.; Christopher, S.J. An accurate and sensitive method for the determination of methylmercury in biological specimens using GC-ICP-MS with solid phase microextraction. *J. Anal. At. Spectrom.,* **2004**, *19*, 1546-1551.
[http://dx.doi.org/10.1039/b412668h]

[56] Fontanals, N.; Marcé, R.M.; Borrull, F. New materials in sorptive extraction techniques for polar compounds. *J. Chromatogr. A,* **2007**, *1152*(1-2), 14-31.
[http://dx.doi.org/10.1016/j.chroma.2006.11.077] [PMID: 17187808]

[57] Mani, V. *Applications Of Solid-Phase Microextraction*; Pawliszyn, J., Ed.; Royal Society of Chemistry: Cambridge, **1999**, pp. 57-72.

[58] Barro, R.; Ares, S.; Garcia-Jares, C.; Llompart, M.; Cela, R. Development of a sensitive methodology for the analysis of chlorobenzenes in air by combination of solid-phase extraction and headspace solid-phase microextraction. *J. Chromatogr. A,* **2004**, *1045*(1-2), 189-196. [http://dx.doi.org/10.1016/j.chroma.2004.06.033] [PMID: 15378894]

[59] Buchholz, K.D.; Pawliszyn, J. Determination of phenols by solid-phase microextraction and gas chromatographic analysis. *Environ. Sci. Technol.,* **1993**, *27*, 2844-2848. [http://dx.doi.org/10.1021/es00049a026]

[60] Boyd-Boland, A.; Pawliszyn, J. Solid-phase microextraction of nitrogen-containing herbicides. *J. Chromatogr. A,* **1995**, *704*, 163-172. [http://dx.doi.org/10.1016/0021-9673(95)00151-C]

[61] Miller, M.E.; Stuart, J.D. Comparison of gas-sampled and SPME-sampled static headspace for the determination of volatile flavor components. *Anal. Chem.,* **1999**, *71*(1), 23-27. [http://dx.doi.org/10.1021/ac980576v] [PMID: 21662922]

[62] Ábalos, M.; Bayona, J.M.; Ventura, F. Development of a solid-phase microextraction GC-NPD procedure for the determination of free volatile amines in wastewater and sewage-polluted waters. *Anal. Chem.,* **1999**, *71*(16), 3531-3537. [http://dx.doi.org/10.1021/ac990197h] [PMID: 10464481]

[63] Valor, I.; Cortada, C.; Moltó, J.C. Direct solid phase microextraction for the determination of BTEX in water and wastewater. *J. High Resolut. Chromatogr.,* **1996**, *19*, 472-474. [http://dx.doi.org/10.1002/jhrc.1240190813]

[64] Martín-Esteban, A. Molecularly imprinted polymers: new molecular recognition materials for selective solid-phase extraction of organic compounds. *Fresenius J. Anal. Chem.,* **2001**, *370*(7), 795-802. [http://dx.doi.org/10.1007/s002160100854] [PMID: 11569855]

[65] Kumar, A.; Gaurav, K.; Malik, A.K.; Tewary, D.K.; Singh, B. A review on development of solid phase microextraction fibers by sol-gel methods and their applications. *Anal. Chim. Acta,* **2008**, *610*(1), 1-14. [http://dx.doi.org/10.1016/j.aca.2008.01.028] [PMID: 18267134]

[66] Wu, J.; Pawliszyn, J. Preparation and applications of polypyrrole films in solid-phase microextraction. *J. Chromatogr. A,* **2001**, *909*(1), 37-52. [http://dx.doi.org/10.1016/S0021-9673(00)01025-6] [PMID: 11218140]

[67] Mohammadi, A.; Yamini, Y.; Alizadeh, N. Dodecylsulfate-doped polypyrrole film prepared by electrochemical fiber coating technique for headspace solid-phase microextraction of polycyclic aromatic hydrocarbons. *J. Chromatogr. A,* **2005**, *1063*(1-2), 1-8. [http://dx.doi.org/10.1016/j.chroma.2004.11.087] [PMID: 15700451]

[68] Alizadeh, N.; Zarabadipour, H.; Mohammadi, A. Headspace solid-phase microextraction using an electrochemically deposited dodecylsulfate-doped polypyrrole film to determine of phenolic compounds in water. *Anal. Chim. Acta,* **2007**, *605*(2), 159-165. [http://dx.doi.org/10.1016/j.aca.2007.10.039] [PMID: 18036379]

[69] Pichon, V. Selective sample treatment using molecularly imprinted polymers. *J. Chromatogr. A,* **2007**, *1152*(1-2), 41-53.
[http://dx.doi.org/10.1016/j.chroma.2007.02.109] [PMID: 17412351]

[70] Kloskowski, A.; Pilarczyk, M.; Przyjazny, A.; Namieśnik, J. Progress in development of molecularly imprinted polymers as sorbents for sample preparation. *Crit. Rev. Anal. Chem.,* **2009**, *39*, 43-58.
[http://dx.doi.org/10.1080/10408340802570223]

[71] Siemann, M.; Andersson, L.I.; Mosbach, K. Selective recognition of the herbicide atrazine by noncovalent molecularly imprinted polymers. *J. Agric. Food Chem.,* **1996**, *44*, 141-145.
[http://dx.doi.org/10.1021/jf950233n]

[72] Urraca, J.L.; Moreno-Bondi, M.C.; Hall, A.J.; Sellergren, B. Direct extraction of penicillin G and derivatives from aqueous samples using a stoichiometrically imprinted polymer. *Anal. Chem.,* **2007**, *79*(2), 695-701.
[http://dx.doi.org/10.1021/ac061622r] [PMID: 17222039]

[73] Koster, E.H.; Crescenzi, C.; den Hoedt, W.; Ensing, K.; de Jong, G.J. Fibers coated with molecularly imprinted polymers for solid-phase microextraction. *Anal. Chem.,* **2001**, *73*(13), 3140-3145.
[http://dx.doi.org/10.1021/ac001331x] [PMID: 11467565]

[74] Yan, H.; Row, K.H.; Yang, G. Water-compatible molecularly imprinted polymers for selective extraction of ciprofloxacin from human urine. *Talanta,* **2008**, *75*(1), 227-232.
[PMID: 18371872]

[75] Shi, X.; Wu, A.; Zheng, S.; Li, R.; Zhang, D. Molecularly imprinted polymer microspheres for solid-phase extraction of chloramphenicol residues in foods. *J. Chromatogr. B Analyt. Technol. Biomed. Life Sci.,* **2007**, *850*(1-2), 24-30.
[http://dx.doi.org/10.1016/j.jchromb.2006.10.057] [PMID: 17126085]

[76] Kim, T-Y.; Alhooshani, K.; Kabir, A.; Fries, D.P.; Malik, A. High pH-resistant, surface-bonded sol-gel titania hybrid organic-inorganic coating for effective on-line hyphenation of capillary microextraction (in-tube solid-phase microextraction) with high-performance liquid chromatography. *J. Chromatogr. A,* **2004**, *1047*(2), 165-174.
[http://dx.doi.org/10.1016/S0021-9673(04)01027-1] [PMID: 15460245]

[77] Popp, P.; Paschke, A. Solid phase microextraction of volatile organic compounds using carboxen-polydimethylsiloxane fibers. *Chromatographia,* **1997**, *46*, 419-424.
[http://dx.doi.org/10.1007/BF02490880]

[78] Pérez, S.; Ferrer, I.; Hennion, M-C.; Barceló, D. Isolation of priority polycyclic aromatic hydrocarbons from natural sediments and sludge reference materials by an anti-fluorene immunosorbent followed by liquid chromatography and diode array detection. *Anal. Chem.,* **1998**, *70*(23), 4996-5001.
[http://dx.doi.org/10.1021/ac980533e] [PMID: 9852782]

[79] Basheer, C.; Jegadesan, S.; Valiyaveettil, S.; Lee, H.K. Sol-gel-coated oligomers as novel stationary phases for solid-phase microextraction. *J. Chromatogr. A,* **2005**, *1087*(1-2), 252-258.
[http://dx.doi.org/10.1016/j.chroma.2005.02.024] [PMID: 16130721]

[80] Lopes, A.L.; Augusto, F. Preparation and characterization of polydimethylsiloxane/poly(vinylalcohol) coated solid phase microextraction fibers using sol-gel technology. *J. Chromatogr. A,* **2004**, *1056*(1-

2), 13-19.
[http://dx.doi.org/10.1016/j.chroma.2004.05.047] [PMID: 15595527]

[81] Zuin, V.G.; Lopes, A.L.; Yariwake, J.H.; Augusto, F. Application of a novel sol-gel polydimethylsiloxane-poly(vinyl alcohol) solid-phase microextraction fiber for gas chromatographic determination of pesticide residues in herbal infusions. *J. Chromatogr. A,* **2004**, *1056*(1-2), 21-26.
[http://dx.doi.org/10.1016/S0021-9673(04)01278-6] [PMID: 15595528]

[82] Azenha, M.; Malheiro, C.; Silva, A.F. Ultrathin phenyl-functionalized solid phase microextraction fiber coating developed by sol-gel deposition. *J. Chromatogr. A,* **2005**, *1069*(2), 163-172.
[http://dx.doi.org/10.1016/j.chroma.2005.02.013] [PMID: 15830942]

[83] Yu, J.; Dong, L.; Wu, C.; Wu, L.; Xing, J. Hydroxyfullerene as a novel coating for solid-phase microextraction fiber with sol-gel technology. *J. Chromatogr. A,* **2002**, *978*(1-2), 37-48.
[http://dx.doi.org/10.1016/S0021-9673(02)01347-X] [PMID: 12458943]

[84] Li, X.; Zeng, Z.; Zhou, J. High thermal-stable sol-gel-coated calix[4]arene fiber for solid-phase microextraction of chlorophenols. *Anal. Chim. Acta,* **2004**, *509*, 27-37.
[http://dx.doi.org/10.1016/j.aca.2003.12.008]

[85] Zhou, F.; Li, X.; Zeng, Z. Determination of phenolic compounds in wastewater samples using a novel fiber by solid-phase microextraction coupled to gas chromatography. *Anal. Chim. Acta,* **2005**, *538*, 63-70.
[http://dx.doi.org/10.1016/j.aca.2005.02.009]

[86] Li, X-j.; Xiao, R-m.; Li, H-b.; Zeng, Z-r. Sol-gel-coated calix[4]arene fiber for solid-phase microextraction. *Wuhan Univ. J. Natur. Sci.,* **2003**, *8*, 1153-1155.
[http://dx.doi.org/10.1007/BF02903690]

[87] Li, X.; Zeng, Z.; Gao, S.; Li, H. Preparation and characteristics of sol-gel-coated calix[4]arene fiber for solid-phase microextraction. *J. Chromatogr. A,* **2004**, *1023*(1), 15-25.
[http://dx.doi.org/10.1016/j.chroma.2003.09.042] [PMID: 14760845]

[88] Dong, C.; Zeng, Z.; Li, X. Determination of organochlorine pesticides and their metabolites in radish after headspace solid-phase microextraction using calix[4]arene fiber. *Talanta,* **2005**, *66*(3), 721-727.
[http://dx.doi.org/10.1016/j.talanta.2004.12.020] [PMID: 18970044]

[89] Zeng, Z.; Qiu, W.; Huang, Z. Solid-phase microextraction using fused-silica fibers coated with sol-ge--derived hydroxy-crown ether. *Anal. Chem.,* **2001**, *73*(11), 2429-2436.
[http://dx.doi.org/10.1021/ac0012750] [PMID: 11403282]

[90] Wang, D.; Xing, J.; Peng, J.; Wu, C. Novel benzo-15-crown-5 sol-gel coating for solid-phase microextraction. *J. Chromatogr. A,* **2003**, *1005*(1-2), 1-12.
[http://dx.doi.org/10.1016/S0021-9673(03)00884-7] [PMID: 12924778]

[91] Zeng, Z.; Qiu, W.; Yang, M.; Wei, X.; Huang, Z.; Li, F. Solid-phase microextraction of monocyclic aromatic amines using novel fibers coated with crown ether. *J. Chromatogr. A,* **2001**, *934*(1-2), 51-57.
[http://dx.doi.org/10.1016/S0021-9673(01)01303-6] [PMID: 11762763]

[92] Yun, L. High extraction efficiency solid-phase microextraction fibers coated with open crown ether stationary phase using sol-gel technique. *Anal. Chim. Acta,* **2003**, *486*, 63-72.
[http://dx.doi.org/10.1016/S0003-2670(03)00491-4]

[93] Yu, J.; Wu, C.; Xing, J. Development of new solid-phase microextraction fibers by sol-gel technology for the determination of organophosphorus pesticide multiresidues in food. *J. Chromatogr. A,* **2004**, *1036*(2), 101-111.
[http://dx.doi.org/10.1016/j.chroma.2004.02.081] [PMID: 15146910]

[94] Doong, R-A.; Liao, P.L. Determination of organochlorine pesticides and their metabolites in soil samples using headspace solid-phase microextraction. *J. Chromatogr. A,* **2001**, *918*(1), 177-188.
[http://dx.doi.org/10.1016/S0021-9673(01)00740-3] [PMID: 11403446]

[95] Lee, M-R.; Yeh, Y-C.; Hsiang, W-S.; Hwang, B-H. Solid-phase microextraction and gas chromatography–mass spectrometry for determining chlorophenols from landfill leaches and soil. *J. Chromatogr. A,* **1998**, *806*, 317-324.
[http://dx.doi.org/10.1016/S0021-9673(98)00075-2]

[96] Hernandez, F.; Beltran, J.; Lopez, F.J.; Gaspar, J.V. Use of solid-phase microextraction for the quantitative determination of herbicides in soil and water samples. *Anal. Chem.,* **2000**, *72*(10), 2313-2322.
[http://dx.doi.org/10.1021/ac991115s] [PMID: 10845380]

[97] Pinho, O.; Ferreira, I.M.; Ferreira, M.A. Solid-phase microextraction in combination with GC/MS for quantification of the major volatile free fatty acids in ewe cheese. *Anal. Chem.,* **2002**, *74*(20), 5199-5204.
[http://dx.doi.org/10.1021/ac020296m] [PMID: 12403571]

[99] Evans, T.J.; Butzke, C.E.; Ebeler, S.E. Analysis of 2,4,6-trichloroanisole in wines using solid-phase microextraction coupled to gas chromatography-mass spectrometry. *J. Chromatogr. A,* **1997**, *786*(2), 293-298.
[http://dx.doi.org/10.1016/S0021-9673(97)00581-5] [PMID: 9408990]

[100] Mestres, M.; Busto, O.; Guasch, J. Headspace solid-phase microextraction analysis of volatile sulphides and disulphides in wine aroma. *J. Chromatogr. A,* **1998**, *808*(1-2), 211-218.
[http://dx.doi.org/10.1016/S0021-9673(98)00100-9] [PMID: 9652119]

[101] Hill, P.G.; Smith, R.M. Determination of sulphur compounds in beer using headspace solid-phase microextraction and gas chromatographic analysis with pulsed flame photometric detection. *J. Chromatogr. A,* **2000**, *872*(1-2), 203-213.
[http://dx.doi.org/10.1016/S0021-9673(99)01307-2] [PMID: 10749498]

[102] Lambropoulou, D.A.; Albanis, T.A. Headspace solid phase microextraction applied to the analysis of organophosphorus insecticides in strawberry and cherry juices. *J. Agric. Food Chem.,* **2002**, *50*(12), 3359-3365.
[http://dx.doi.org/10.1021/jf0112771] [PMID: 12033796]

[103] Rodríguez, R.; Mañes, J.; Picó, Y. Off-line solid-phase microextraction and capillary electrophoresis mass spectrometry to determine acidic pesticides in fruits. *Anal. Chem.,* **2003**, *75*(3), 452-459.
[http://dx.doi.org/10.1021/ac025884e] [PMID: 12585470]

[104] Djozan, D.; Ebrahimi, B. Preparation of new solid phase micro extraction fiber on the basis of atrazine-molecular imprinted polymer: application for GC and GC/MS screening of triazine herbicides in water, rice and onion. *Anal. Chim. Acta,* **2008**, *616*(2), 152-159.
[http://dx.doi.org/10.1016/j.aca.2008.04.037] [PMID: 18482598]

[105] Sagratini, G.; Mañes, J.; Giardiná, D.; Damiani, P.; Picó, Y. Analysis of carbamate and phenylurea pesticide residues in fruit juices by solid-phase microextraction and liquid chromatography-mass spectrometry. *J. Chromatogr. A,* **2007**, *1147*(2), 135-143. [http://dx.doi.org/10.1016/j.chroma.2007.02.066] [PMID: 17362971]

[106] Pragst, F.; Auwaerter, V.; Sporkert, F.; Spiegel, K. Analysis of fatty acid ethyl esters in hair as possible markers of chronically elevated alcohol consumption by headspace solid-phase microextraction (HS-SPME) and gas chromatography-mass spectrometry (GC-MS). *Forensic Sci. Int.,* **2001**, *121*(1-2), 76-88. [http://dx.doi.org/10.1016/S0379-0738(01)00456-X] [PMID: 11516891]

[107] Lord, H.L.; Pawliszyn, J. Method optimization for the analysis of amphetamines in urine by solid-phase microextraction. *Anal. Chem.,* **1997**, *69*(19), 3899-3906. [http://dx.doi.org/10.1021/ac970375b] [PMID: 9322428]

[108] Gentili, S.; Torresi, A.; Marsili, R.; Chiarotti, M.; Macchia, T. Simultaneous detection of amphetamine-like drugs with headspace solid-phase microextraction and gas chromatography-mass spectrometry. *J. Chromatogr. B Analyt. Technol. Biomed. Life Sci.,* **2002**, *780*(1), 183-192. [http://dx.doi.org/10.1016/S1570-0232(02)00522-6] [PMID: 12383494]

[109] Yu, H.; Xu, L.; Wang, P. Solid phase microextraction for analysis of alkanes and aromatic hydrocarbons in human breath. *J. Chromatogr. B Analyt. Technol. Biomed. Life Sci.,* **2005**, *826*(1-2), 69-74. [http://dx.doi.org/10.1016/j.jchromb.2005.08.013] [PMID: 16154390]

[110] Rudnicka, J.; Kowalkowski, T.; Ligor, T.; Buszewski, B. Determination of volatile organic compounds as biomarkers of lung cancer by SPME-GC-TOF/MS and chemometrics. *J. Chromatogr. B Analyt. Technol. Biomed. Life Sci.,* **2011**, *879*(30), 3360-3366. [http://dx.doi.org/10.1016/j.jchromb.2011.09.001] [PMID: 21982505]

[111] Frison, G.; Tedeschi, L.; Maietti, S.; Ferrara, S.D. Determination of γ-hydroxybutyric acid (GHB) in plasma and urine by headspace solid-phase microextraction and gas chromatography/positive ion chemical ionization mass spectrometry. *Rapid Commun. Mass Spectrom.,* **2000**, *14*(24), 2401-2407. [http://dx.doi.org/10.1002/1097-0231(20001230)14:24<2401::AID-RCM179>3.0.CO;2-I] [PMID: 11114057]

[112] Vuckovic, D.; Shirey, R.; Chen, Y.; Sidisky, L.; Aurand, C.; Stenerson, K.; Pawliszyn, J. *In vitro* evaluation of new biocompatible coatings for solid-phase microextraction: implications for drug analysis and *in vivo* sampling applications. *Anal. Chim. Acta,* **2009**, *638*(2), 175-185. [http://dx.doi.org/10.1016/j.aca.2009.02.049] [PMID: 19327457]

[113] Lucas, A.C.; Bermejo, A.M.; Tabernero, M.J.; Fernández, P.; Strano-Rossi, S. Use of solid-phase microextraction (SPME) for the determination of methadone and EDDP in human hair by GC-MS. *Forensic Sci. Int.,* **2000**, *107*(1-3), 225-232. [http://dx.doi.org/10.1016/S0379-0738(99)00165-6] [PMID: 10689574]

[114] Chaves, A.R.; Chiericato Júnior, G.; Queiroz, M.E. Solid-phase microextraction using poly(pyrrole) film and liquid chromatography with UV detection for analysis of antidepressants in plasma samples. *J. Chromatogr. B Analyt. Technol. Biomed. Life Sci.,* **2009**, *877*(7), 587-593. [http://dx.doi.org/10.1016/j.jchromb.2008.12.070] [PMID: 19185550]

[115] Zhang, Z.; Pawliszyn, J. Quantitative extraction using an internally cooled solid phase microextraction device. *Anal. Chem.,* **1995**, *67*, 34-43.
[http://dx.doi.org/10.1021/ac00097a007]

[116] Chia, K-J.; Lee, T-Y.; Huang, S-D. Simple device for the solid-phase microextraction screening of polychlorodibenzo-*p*-dioxins and polychlorodibenzofurans in heavily contaminated soil samples. *Anal. Chim. Acta,* **2004**, *527*, 157-162.
[http://dx.doi.org/10.1016/j.aca.2004.09.049]

[117] Kaur, V.; Malik, A.K.; Verma, N. Applications of solid phase microextraction for the determination of metallic and organometallic species. *J. Sep. Sci.,* **2006**, *29*(3), 333-345.
[http://dx.doi.org/10.1002/jssc.200500319] [PMID: 16544873]

[118] Carasek, E.; Pawliszyn, J. Screening of tropical fruit volatile compounds using solid-phase microextraction (SPME) fibers and internally cooled SPME fiber. *J. Agric. Food Chem.,* **2006**, *54*(23), 8688-8696.
[http://dx.doi.org/10.1021/jf0613942] [PMID: 17090108]

[119] Carasek, E.; Cudjoe, E.; Pawliszyn, J. Fast and sensitive method to determine chloroanisoles in cork using an internally cooled solid-phase microextraction fiber. *J. Chromatogr. A,* **2007**, *1138*(1-2), 10-17.
[http://dx.doi.org/10.1016/j.chroma.2006.10.092] [PMID: 17134712]

[120] Koziel, J.A.; Haddadi, S.H.; Koch, W.; Pawliszyn, J. Sampling and analysis of nanoparticles with cold fibre SPME device. *J. Sep. Sci.,* **2009**, *32*(11), 1975-1980.
[http://dx.doi.org/10.1002/jssc.200900023] [PMID: 19425018]

[121] Kataoka, H. Automated sample preparation using in-tube solid-phase microextraction and its application a review. *Anal. Bioanal. Chem.,* **2002**, *373*(1-2), 31-45.
[http://dx.doi.org/10.1007/s00216-002-1269-z] [PMID: 12012170]

[122] Mullett, W.M.; Martin, P.; Pawliszyn, J. In-tube moleculary imprinted polymer solid-phase microextraction for the selective determination of propranolol. *Anal. Chem.,* **2001**, *73*(11), 2383-2389.
[http://dx.doi.org/10.1021/ac0100502] [PMID: 11403276]

[123] Wu, J.C.; Lord, H.L.; Pawliszyn, J.; Kataoka, H. Polypyrrole-coated capillary in-tube solid phase microextraction coupled with liquid chromatography-electrospray ionization mass spectrometry for the determination of beta-blockers in urine and serum samples. *J. Microcol.,* **2000**, *12*, 255-266.
[http://dx.doi.org/10.1002/(SICI)1520-667X(2000)12:4<255::AID-MCS10>3.0.CO;2-J]

[124] Wu, J.; Pawliszyn, J. Polypyrrole-coated capillary coupled to HPLC for in-tube solid-phase microextraction and analysis of aromatic compounds in aqueous samples. *Anal. Chem.,* **2001**, *73*(1), 55-63.
[http://dx.doi.org/10.1021/ac000885x] [PMID: 11195512]

[125] Queiroz, M.E.; Lancas, F.M. Analysis of drugs in biological samples: Automated "in tube" solid-phase microextraction and high performance liquid chromatography. *Quim. Nova,* **2005**, *28*, 880-886.
[http://dx.doi.org/10.1590/S0100-40422005000500027]

[126] Shintani, Y.; Zhou, X.; Furuno, M.; Minakuchi, H.; Nakanishi, K. Monolithic silica column for in-tube solid-phase microextraction coupled to high-performance liquid chromatography. *J. Chromatogr. A,*

2003, *985*(1-2), 351-357.
[http://dx.doi.org/10.1016/S0021-9673(02)01447-4] [PMID: 12580503]

[127] Carasek, E.; Cudjoe, E.; Pawliszyn, J. Fast and sensitive method to determine chloroanisoles in cork using an internally cooled solid-phase microextraction fiber. *J. Chromatogr. A,* **2007**, *1138*(1-2), 10-17.
[http://dx.doi.org/10.1016/j.chroma.2006.10.092] [PMID: 17134712]

[128] Picó, Y.; Fernández, M.; Ruiz, M.J.; Font, G. Current trends in solid-phase-based extraction techniques for the determination of pesticides in food and environment. *J. Biochem. Biophys. Methods,* **2007**, *70*(2), 117-131.
[http://dx.doi.org/10.1016/j.jbbm.2006.10.010] [PMID: 17175029]

[129] Basheer, C.; Lee, H.K. Hollow fiber membrane-protected solid-phase microextraction of triazine herbicides in bovine milk and sewage sludge samples. *J. Chromatogr. A,* **2004**, *1047*(2), 189-194.
[http://dx.doi.org/10.1016/S0021-9673(04)01140-9] [PMID: 15460248]

[130] Kataoka, H.; Narimatsu, S.; Lord, H.L.; Pawliszyn, J. Automated in-tube solid-phase microextraction coupled with liquid chromatography/electrospray ionization mass spectrometry for the determination of beta-blockers and metabolites in urine and serum samples. *Anal. Chem.,* **1999**, *71*(19), 4237-4244.
[http://dx.doi.org/10.1021/ac990356x] [PMID: 10517146]

[131] Kataoka, H.; Lord, H.L.; Yamamoto, S.; Narimatsu, S.; Pawliszyn, J. Development of automated in-tube SPME/LC/MS method for drug analysis. *J. Microcol.,* **2000**, *12*, 493-500.
[http://dx.doi.org/10.1002/1520-667X(2000)12:9<493::AID-MCS3>3.0.CO;2-C]

[132] Tan, B.C.; Marriott, P.J.; Lee, H.K.; Morrison, P.D. In-tube solid phase micro-extraction-gas chromatography of volatile compounds in aqueous solution. *Analyst (Lond.),* **1999**, *124*, 651-655.
[http://dx.doi.org/10.1039/a902567g]

[133] Fan, Y.; Feng, Y-Q.; Da, S-L.; Wang, Z.H. In-tube solid phase microextraction using a β-cyclodextrin coated capillary coupled to high performance liquid chromatography for determination of non-steroidal anti-inflammatory drugs in urine samples. *Talanta,* **2005**, *65*(1), 111-117.
[PMID: 18969772]

[134] Zheng, M.M.; Lin, B.; Feng, Y.Q. Hybrid organic-inorganic octyl monolithic column for in-tube solid-phase microextraction coupled to capillary high-performance liquid chromatography. *J. Chromatogr. A,* **2007**, *1164*(1-2), 48-55.
[http://dx.doi.org/10.1016/j.chroma.2007.07.020] [PMID: 17675044]

[135] Kloskowski, A.; Pilarczyk, M.; Namieśnik, J. Membrane solid-phase microextractiona new concept of sorbent preparation. *Anal. Chem.,* **2009**, *81*(17), 7363-7367.
[http://dx.doi.org/10.1021/ac9007978] [PMID: 19653632]

[136] Spietelun, A.; Marcinkowski, Ł.; Kloskowski, A.; Namieśnik, J. Determination of volatile organic compounds in water samples using membrane-solid phase microextraction (M-SPME) (headspace version). *Analyst (Lond.),* **2013**, *138*(17), 5099-5106.
[http://dx.doi.org/10.1039/c3an36851c] [PMID: 23807267]

CHAPTER 6

Stir Bar Sorptive Extraction and Related Techniques

José Manuel Florêncio Nogueira*

University of Lisbon, Faculty of Sciences, Chemistry and Biochemistry Department and Centre of Chemistry and Biochemistry, Campo Grande, Ed. C8, 1749-016 Lisbon, Portugal

Abstract: From the sorption-based methods available nowadays, stir bar sorptive extraction (SBSE) became a well-established analytical technique for sample preparation, in which hundreds of applications in almost all scientific areas have already been proposed in the literature. This remarkable analytical tool shows great capacity for static microextraction and outstanding performance to operate at the ultra-trace level, in particular for the analysis of complex systems. Furthermore, is very effective, present easy manipulation in comparison to other alternative techniques and great reproducibility for the analysis of priority and emerging organic compounds. Recently, related static microextraction techniques were introduced, with particular emphasis to bar adsorptive microextraction (BAμE) that operates under the floating sampling technology, in which has demonstrated high analytical capacity and remarkable performance. This novel concept has also proved great effectiveness for ultra-trace analysis of organic compounds with polar characteristics, in particular from complex systems. The present contribution describes the fundamental principles, the experimental methodology, the main applications, as well as, the analytical potential of these novel microextraction techniques.

Keywords: Activated carbons, BAμE, Enrichment methods, Floating sampling, GC-MS, HS, LC-MS, LD, Microextraction techniques, Nanostructured materials, PDMS, Polymeric coatings, Real matrices, Sample preparation, SBSE, Sorbent phases, Sorption-based methods, Static microextraction TD, Trace analysis.

* **Corresponding author José Manuel Florêncio Nogueira:** University of Lisbon, Faculty of Sciences, Chemistry and Biochemistry Department and Centre of Chemistry and Biochemistry, Campo Grande, Ed. C8, 1749-016 Lisbon, Portugal; Tel: +351-217500899; Fax: +351-217500088; E-mail: nogueira@fc.ul.pt

M. Valcárcel, S. Cárdenas, R. Lucena (Eds.)

6.1. INTRODUCTION

Due to the very high complexity presented by many matrices, the modern sample enrichment techniques run towards the great simplification, miniaturization, easy manipulation of the analytical devices, strong reduction or absence of organic toxic solvents, as well as, low sample volume requirements in agreement with the green analytical chemistry principles [1]. For trace analysis in particular, the sorption-based methods have demonstrated to be the best choice enabling the direct microextraction, mainly of volatile and semi-volatile compounds, from almost all types of aqueous matrices. Moreover, these solvent-free approaches have demonstrating high-throughput to enhance selectivity and sensitivity prior to the application of chromatographic or hyphenated techniques. Although, some of the currently well-established methodologies use active or dynamic sampling mode, such as solid phase extraction (SPE), the passive or static sampling mode has gained more acceptances in almost all the scientific areas, due to the cost-effectiveness, much easier manipulation with emphasis to solid-phase microextraction (SPME) [2] and stir bar sorptive extraction (SBSE) [3]. These enrichment techniques combine the extraction and concentration of the analytes simultaneously, having the possibility to use immersion or headspace (HS) sampling modes in only a single step. On the other hand, they reduce the manipulation, the overall time required for sample preparation, and are indicated to be combined with the great sensitivity of the modern analytical instrumentation. SBSE in particular, was introduced as a novel sample preparation method based on the same principles of SPME, which has been successfully applied to monitor traces of priority organic compounds in many complex matrices, due to the very high sensitivity exhibited. A good indicator of the greatest advantages over other sorbent-based techniques is the number of publications related with SBSE that strongly increased in the last years, reaching up to several hundreds of scientific articles. Furthermore, many review articles had already been reported covering a wide range of features concerning SBSE principles, method development and applications [4 - 9]. Nevertheless, once commercial stirbars still had presented some analytical limitations, in particular for the enrichment of the polar organic compounds, several analytical solutions and related techniques have been proposed. For this reason, bar adsorptive microextraction (BAμE) was introduced as an alternative and complementary static microextraction technique for ultra-trace analysis [9 - 11].

6.2. STIR BAR SORPTIVE EXTRACTION (SBSE)

6.2.1. Fundamentals

The SBSE technique belongs to the silicone-type sorbents since uses just the polydimethylsiloxane (PDMS) polymeric phase and is commercially known with the trademark Twister®. Fig. (**1.a**) depicts the SBSE analytical device, which is constituted by magnetic stir bars incorporated into a glass jacket (10 to 20 mm in length) typically coated with 24-126 μL (0.3 to 1.0 in thickness) of PDMS, a nonpolar polymeric phase characterized to promote hydrophobic interactions with target molecules, where the retention mechanism occurs mainly through Van-der-Waals forces but also hydrogen bonds can be formed with oxygen atoms of the sorbent network, depending on the molecular structure of the analytes. Furthermore, this polymer has interesting diffusion and thermo-stable properties allowing to operate in a wide range of temperatures and, because of that, this solventless approach was firstly proposed to be associated with thermal desorption (TD) following gas chromatography (GC) analysis. The substantial volume of PDMS involved, in relation to the SPME fibbers (up to 0.5 μL for100 μm film thickness), promotes a lower phase ratio between the sorbent phase and the sample media, which provides an increasing capacity, and quantitative recoveries can be reached by SBSE, especially for nonpolar solutes. This approach enables to increase the sensitivity by a factor within 50 and 250, in comparison to SPME, decreasing the detection limits to the ultra-trace level (parts-per-trillion) many times. In the SBSE theory [3, 9], the efficiency of the analyte partitioning between the PDMS phase of the stir bar and the water sample (W), presents a similar behaviour as the distribution described by the octanol-water partition coefficients ($K_{PDMS/W} \approx K_{O/W}$) during the static equilibrium. In general, nonpolar analytes are characterized to have values of log $K_{O/W}$ higher or equal to 3, whereas for the polar are lower. Thus, the $K_{O/W}$ and the phase ratio ($\beta = V_W/V_{PDMS}$, in which V_W is the water sample volume and V_{PDMS} is the PDMS volume), are important parameters to predict the theoretical recovery by using the following expression:

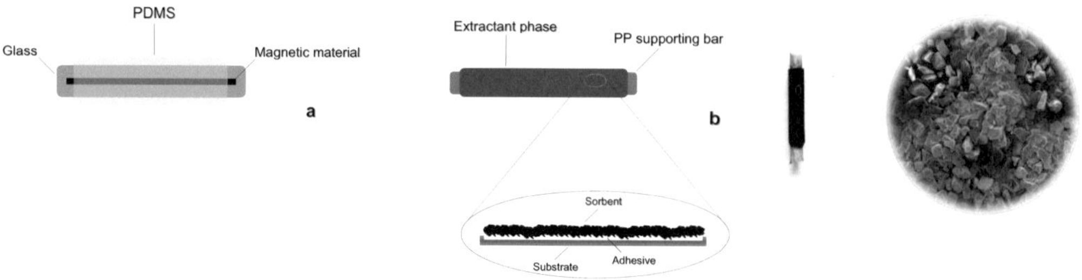

Fig. (1). Schematic representations, image and SEM micrograph of the SBSE **(a)** and BAμE **(b)** devices, respectively [8, 9].

$$\text{Recovery}(\%) = \frac{K_{O/W}/\beta}{1 + K_{O/W}/\beta} \times 100\%$$

From this expression, we can see that the higher hydrophobicity has the target compounds or the PDMS volume (lower β), the higher extraction yields are expected by SBSE. This expression shows us the great advantage to predict the theoretical recovery, just know β and $K_{O/W}$, in which the latter can be obtained from theoretical tables in the literature or calculated through available software. For instance, if a 50 mL sample having a solute (*e.g.* log $K_{O/W} = 4$) is analysed by SBSE using a stir bar with 50 μL of PDMS, *i.e.* $\beta = 1{,}000$, an average theoretical recovery of 90.9% will be expected. We can also understand that for SBSE processes that present $K_{O/W}/\beta = 1$, the theoretical recovery will be 50.0%, as well as for log $K_{O/W}$ values higher than 5, the extraction will be quantitative or complete. Once the $K_{O/W}$ value is constant for each solute, β could be optimized for each particular application, as well as other experimental parameters, which can influence the kinetics or thermodynamics microextraction process. Fig. (**2**) depicts the plots profile of the theoretical recoveries against log $K_{O/W}$, where it can be observed that the higher the hydrophobic degree of the target analytes the greater extraction yields will be predicted through SBSE(PDMS). It can be also observed the influence of β in the theoretical recovery by SBSE, in which for lower values (low sample volume) the plots moves towards the left side and higher extraction capacity are expected.

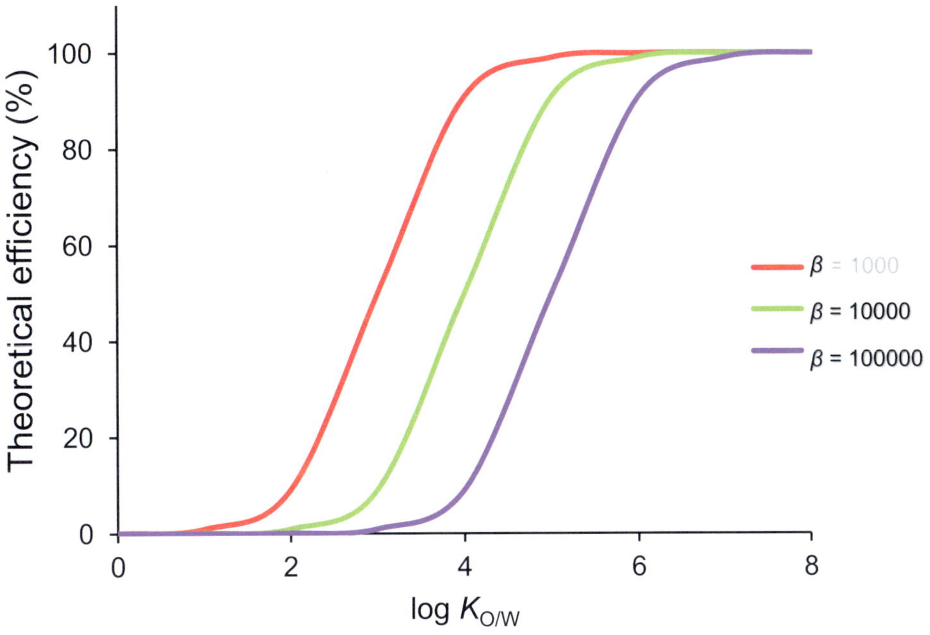

Fig. (2). Influence of the phase ratio (*β*) on the theoretical efficiency of SBSE(PDMS) against log $K_{O/W}$.

6.2.2. Method Development

The SBSE operating principle consists in two major steps, namely, the extraction or enrichment of the analytes from the sample bulk to the PDMS polymer and then, the back-extraction or desorption of the analytes from the sorbent phase to the chromatographic system. In any SBSE analytical scheme, the method development should start by optimizing the instrumental system, *i.e.* the detection and separation conditions, followed by the back-extraction step and then, the extraction most important experimental parameters. The SBSE method development should always be designed through systematic assays using univariate or multivariate optimization strategies in order to achieve the best experimental parameters that affect those analytical systems making possible the evaluation of main interactions among the different variables [9, 13]. Subsequently, takes place the validation procedure and the application to real matrices in order to evaluate the performance and efficiency of the analytical methodology developed.

6.2.2.1. Extraction Step

During the extraction process, the stirbars can be put into contact with the solutes using either immersion or HS sampling modes, as exemplified in Fig. (**3.a**), the latter one indicated to analyse volatile compounds. In the HS-SBSE mode, the sampling is performed by suspending the stir bar at the top of the flask, where the polymer is in static contact with the vapour phase from liquid or solid matrices, avoiding therefore possible contamination with non-volatile interferences. The extraction step is usually carried out under steady state conditions in order to obtain maximum efficiency, where the control of several experimental parameters is always compulsory. Very often, systematic assays must be performed to optimise the most important variables that influence the extraction process, namely equilibrium time and agitation speed but also, pH adjustment, polarity and ionic strength of the matrix [9 - 14]. Other parameters such as sample and PDMS volumes, dilution factor, *etc.* are also very important, in particular to meet enough sensitivity during ultra-trace analysis. Some of these variables affect the extraction kinetics (*e.g.* equilibrium time and stirring speed), as well as, the interaction among the analytes and the PDMS phase (*e.g.* sample pH, polarity and ionic strength); matrix pH influences the analytes extraction that have dissociation characteristics, where non-dissociation conditions is always a must; the addition of an electrolyte (*e.g.* sodium chloride) favours the "salting-out" or "oil-effect" phenomena and therefore, is very effective for the enrichment the more hydrophilic (log $K_{O/W}$< 3) compounds, in particular when HS sampling mode is adopted; finally, the addition of a modifier (*e.g.* methanol) minimizes the "wall-effect" phenomena that can occur for the more hydrophobic or nonpolar solutes (log $K_{O/W}$≥ 3). Additionally, the PDMS volume also influences the recovery yields, in particular the more polar targets, whereas temperature can have a great influence on the vapour pressure of the compounds involved especially during the HS mode. However, in order to minimize the analytical time, the sensitivity and precision can be sacrificed by working under non-equilibrium conditions also with enough performance. It must be emphasized that the development of any SBSE method should start by performing assays in ultra-pure water spiked with the desired model or target compounds, to assess the analytical behaviour.

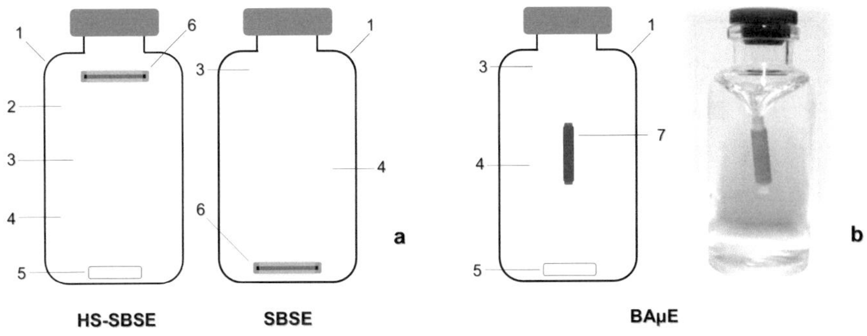

Fig. (3). Schematic representations and image exemplifying the microextraction modes through HS and immersion by SBSE (a) and floatation by BAμE (b), respectively [8, 9]. **Legend**: 1 - Sampling flask; 2 - HS; 3 - Vortex; 4 - Sample; 5 - Teflon magnetic stir bar; 6 - SBSE device; 7 - BAμE device.

6.2.2.2. Back-extraction Step

After the extraction step, the stirbars are removed, rinsed with distilled water to clean-up from any potential interference (*e.g.* salts, sugars, proteins or other undesirable sample constituents), dipped on a clean paper tissue to remove water, and submitted to the back-extraction process, through TD or liquid desorption (LD) modes, as depicted in Fig. (**4**). Although TD is the most straightforward back-extraction mode, it is limited to thermally stable volatile to semi-volatile solutes, whereas LD is of wide application, but usually more indicated for semi-volatile to non-volatile and thermo-labile compounds. The TD approach needs a glass tube and a dedicated unit for the bake-out operation (up to 350 °C), being only possible for GC analysis, as is illustrated in Fig. (**4.a**)

During TD several instrumental parameters should be taken into consideration that influences the back-extraction operation, in particular, the desorption temperature, the purge flow, the GC injector temperature, in which a programmed temperature vaporisation (PTV) inlet is compulsive. The PTV cryo focusing the analytes under low temperatures, using liquid nitrogen, following vaporisation through fast heating towards the chromatographic column, to avoiding band broadening. As alternative, the solvent type (*e.g.* methanol, acetonitrile, mixtures, *etc.*), soak time and number of desorption steps are important variables during LD, in which the complete immersion of the stirbar on the solvent inside a vial or glass insert, under mechanical or sonication treatment, is a must to improve efficiency, as depicted in Fig. (**4.b**) [12]. In the back-extraction point of view, the TD is an *on-line* approach since makes possible the direct and complete sample introduction into the GC system, promoting a very high sensitivity and the possibility of automation by using a dedicated auto-sampler for routine analysis. On the other hand, the LD is an *off-line* approach, offering the opportunity of reanalysis and besides very cost-effective, is able to operate in combination with GC, high performance liquid chromatography (HPLC) [15, 16] or even capillary electrophoresis (CE) [17, 18] systems.

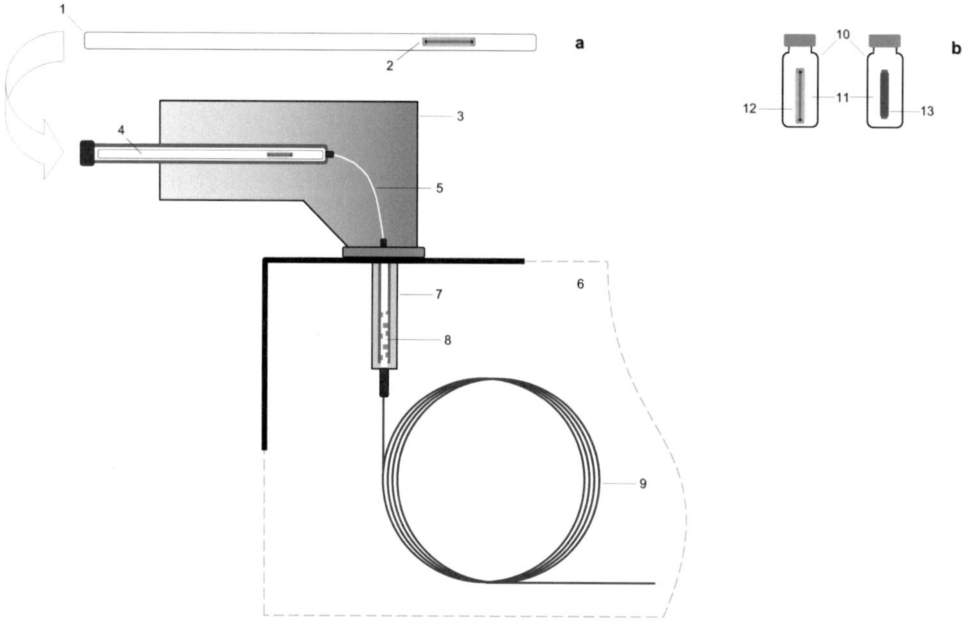

Fig. (4). Schematic representation of the TD (a) and LD (b) modes used for back-extraction by SBSE or BAμE techniques [8, 9]. **Legend**: 1 - Glass tube for TD; 2 - SBSE device; 3 - TD unit; 4 - TD tube with the SBSE device; 5 - Hot transfer capillary; 6 - Gas chromatograph; 7 - PTV injector; 8 - *Liner* with glass wool; 9 - Capillary column; 10 - *Vial*; 11 - LD solvent; 12 - SBSE device; 13 - BAμE device.

6.2.3. Validation Stage

After the analytical optimisation stage, takes place the validation procedure, including the determination of parameters such as analytical limits, linear dynamic ranges, accuracy and precision of the developed methodology. For ultra-trace analysis in particular, the validation parameters should meet reproducibility in compliance with the requirements of Directive 98/83/EC [19]. After validation, the application to real samples should be evaluate in order to demonstrate all analytical advantages, as alternative and reliable technique, in comparison to other conventional methodologies, especially if complex matrices are involved. The efficiency of SBSE can be strongly affected by the complexity of the matrices involved since potential interferences can influence the recovery yields. Therefore, the application of any optimized and validated SBSE methodology to a particular real sample should be performed in order to verify the analytical behaviour, as well as, the occurrence of possible interferences that can cause matrix effects. The easiest way to compensate for matrix effects is based on the standard addition methodology (SAM). One of the most interesting features of SBSE techniques is that each stir bar can be used hundreds of times without showing any physical degradation of the PDMS coating. The only practical condition to be implemented is that before being reused, the stir bars should be cleaned-up with suitable solvents (*e.g.* acetonitrile) or through TD treatment (*e.g.* 320ºC). Another advantage comparatively to other enrichment approaches (*e.g.* dynamic techniques) is that SBSE technique allows the possibility to operate overnight, if much longer equilibrium times are demanded to reach the steady state conditions, without any special requirements.

6.2.4. Examples of Application

In general, the great number of applications has been focusing on the replacement of other conventional techniques since SBSE presents an easier practical manipulation, is a solvent less approach, allows great selectivity and sensitivity in particular as a multi residue methodology for complex matrices. According to literature, most of the SBSE applications have been performed using both HS and immersion sampling modes involving trace level analysis of emerging priority targets, either with natural or anthropogenic sources, in areas with recognized impact in the society such as environment, food and flavour, biomedical, forensic, pharmaceutical, petrochemical, *etc*. Some good examples of application are the determination of volatile and semi-volatile organic compounds in air [20], persistent organic pollutants (*e.g.* pesticides) in water matrices [12 - 18, 21 - 24], contaminants and aroma compounds in beverages [25 - 27], fruits and vegetables

[28, 29], as well as, metabolites (*e.g.* sexual hormones), drugs, pharmaceuticals and personal care products (PPCPs) in biological fluids such as saliva, urine samples, *etc* [16, 30 - 32]. In short, SBSE(PDMS) is a remarkable sorption-based technique indicated to monitor trace or ultra-trace levels of nonpolar to medium polar solutes having volatile to semi-volatile characteristics, mainly from aqueous media. Fig. (**5.a**) exemplifies the application of SBSE/TD-GC-MS to a sample of Madeira wine, where it is possible to observe the complex aromatic profile constituted by several classes of organic compounds at the trace level; in Fig. (**5.b**) can be observed the presence of triclosan, a disinfectant commonly used in toothpastes, in saliva by SBSE-LD/HPLC-DAD.

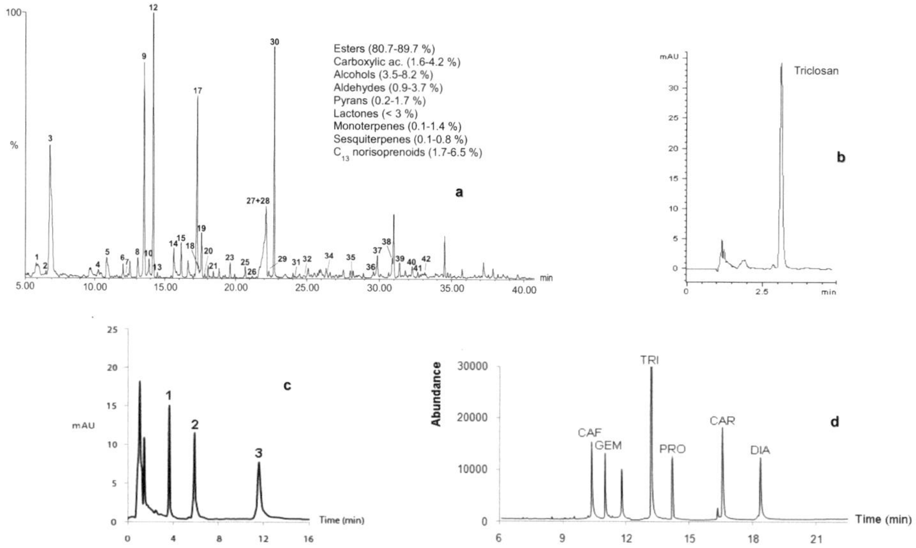

Fig. (5). Chromatograms exemplifying applications in real matrices.

a) Aromatic profile from a Madeira wine sample (Verdelho-99) by SBSE/TD-GC-MS [25].

b) Determination of triclosan from a saliva sample by SBSE-LD/HPLC-DAD [32].

c) Triazinic herbicide analysis from a fortified superficial water by BAμE(AC)-LD/HPLC-DAD [34].

1 - Simazine; 2 - Atrazine; 3 - Terbuthylazine.

d) Monitoring several PPCPs from an wastewater sample by BAμE(PS-DVB)-LD/LVI-GC-MS [37].

CAF - Caffeine; GEM - Gemfibrosil; TRI - Triclosan; PRO - Propranolol; CAR - Carbamazepine; DIA - Diazepam.

6.2.5. Main Limitations

Although the recent analytical developments by SBSE(PDMS) have proved interesting features as a novel sorption-based technique, in many cases, it presents great limitations in the microextraction of a particular type of solutes involved. Therefore, if we focus our attention just on analytes with polar characteristics, *i.e.* log $K_{O/W} < 3$, the SBSE(PDMS) technique has demonstrated the lack of efficiency due to the weak hydrophobic interactions involved. For this reason, novel analytical strategies or concepts have been proposed and improved to overcome this drawback including multi-mode assays, derivatisation procedures, new polymeric phases or alternative and related sorption-based approaches such as the adsorptive microextraction techniques. For more detailed explanations concern the most important practical issues readers are advised to consult dedicated bibliography [8 - 10, 14, 33].

6.3. BAR ADSORPTIVE MICROEXTRACTION (BAME)

6.3.1. Principles and Advantages

Solid materials are known to easily adsorb polar molecules once presenting large specific areas and nanotexture porous structures with suitable active sites, where the "adsorption-desorption" mechanisms take place [9 - 11]. Like activated carbons (ACs), most of these nanoparticles are characterized to have microporous (pore width < 2 nm) nature, associated to mesoporous (2 nm < pore width < 50 nm) structures, with volumes estimated up to approximately 60% of the total. Under powder form, the average micropore widths, most of the times around 1.50 nm, allows the target molecules to access the porous structure and therefore, be retained, as they not suffer steric restrictions. Moreover, the mesopore size distribution used to be very broad, revealing that AC materials present pore with openings in all the mesopores range, which favours the diffusivity of the target molecules to the inner porosity. The large specific areas exhibited (up to 1,000 m^2 g^{-1}) by these materials present a remarkable adsorptive capacity (≈ 100-500 µg mg^{-1}), just depending many times on pH_{pzc} (pH at the point of zero charge) and texture. As a matter of fact, pH_{pzc} has a great influence in almost all types of interaction mechanisms involved, where the acidic or basic characteristics of the surface area of those materials conditioning the microextraction phenomena. The acidic/basic nature of the AC surface depends mainly on the content of heteroatoms as, for instance, oxygen, hydrogen and nitrogen, corresponding usually to a small percentage of the total AC composition. These heteroatoms form surface species that strongly influence the material adsorptive properties in liquid phase, which is often ideal to retain organic compounds with much higher polar characteristics. Furthermore, because ACs are amorphous solids, composed essentially by carbon and presenting high adsorption capacities, they can be prepared from carbon-rich raw materials. Depending on the preparation methodology used and on the precursor, it is possible to tailor the pore size distribution and surface chemistry to meet the requirements for a given application. For trace analysis in particular, the ACs are ideal sorbents for microextraction purposes, since the great enrichment capacity is definitely below the isothermal plateau (saturation of the sorbent) and therefore, the Langmuir and Freundlich theoretical considerations are not applicable [9 - 11]. Meanwhile, other powdered phases such as polymeric-based materials present also strong sorbent properties and can be potential alternatives whenever ACs do not present enough selectivity or capacity to retain particular target solutes. Even so, powdered sorbents are constituted by strongly divided materials composed through thousands of microscopic particles (particle size < 30 μm), which make them difficult to be manipulated from the practical point of view. Nevertheless, if those powered materials could be fixed or supported to suitable substrates through "sticking-based technologies", the new devices can represent a powerful microextraction alternative to overcome the limitations of the SBSE(PDMS) technique to retain the more polar solutes. In short, the greatest advantage of this analytical concept is that these microextraction devices or micro-collectors can be home-made prepared using the most convenient sorbent to each particular type of application at low cost without wasting too much time.

6.3.2. Sorbent Coatings and Configurations

As stated before, the microextraction devices can be prepared in-house by using sorbent coating materials fixed to convenient substrates through sticking-based technologies. Therefore, for the preparation of the micro-collectors, we should start to choose the most selective sorbent phase to retain the target analytes of interest. Several materials are commercially available, including ACs, alumina, silica, as well as, polystyrene-divinylbenzene (PS-DVB), modified pyrrolidone, alumina, silica-based polymers and mixtures among many others, most of them commonly used in SPE, which are known to have strong sorptive properties. For instance, some sorbent materials can also be easily prepared and characterized in-house, like those already proposed from cork waste that showed remarkable sorptive properties when compared with others commercially available [34, 35]. Depending on the type of phases involved, the AC-based materials present mainly electrostatic and/or dispersive mechanisms, whereas the polymeric-based materials are almost of reversed phase type, by retaining the analytes according to π-π, dipole-dipole, hydrogen bonding and ionic interactions. In general, particle size, specific surface, microporosity and pH are important properties known to have a great influence on the interaction mechanisms involved between the target molecules and the sorbents network. Moreover, some sorbents present much higher efficiency once offering multiple mechanisms of retention when compared with other materials, which most of the times are recommended, improving the overall selectivity and capacity. Two geometrical configurations were already designed using substrates having bar and sphere shape. According to the configuration of the devices, the BAμE technique presents a bar-shaped geometry, where the sorbent phases are fixed with convenient adhesive films on PP cylindrical substrates (adhesive supporting) [11]. Fig. (**1.b**) depicts schematic representations and images of BAμE, where a scanning electron microscopy (SEM) micrograph of the device coated with AC powder, shows that the grains are fixed to the substrate in a very homogeneous coating. Additionally, BAμE devices are very easy to prepare independently of the sorbent coating involved and the great advantage of these novel analytical tools over other sorptive extraction approaches is the ability to choose the most convenient and specific phase for each particular target or class of compounds. For BAμE, the amount of sorbent phase involved (up to 5 mg) is usually limited to the available surface of the adhesive supporting film, according to the expected content level of the target compounds. Before being applied to particular samples, the microextraction devices must be evaluated in terms of stability and robustness mainly concerning the sorbents fixation. Therefore, appropriate physico-chemical tests should be performed, including assays to assess their resistance to organic solvents, pH, and temperature, as well as, mechanical and regeneration treatments. As a matter of fact, the temperature tests performed on the BAμE devices proved that the sorbent phases stayed fixed and stable until 40 °C, whereas above this value started to disaggregate from the supporting films. Re-utilization is also possible, since the microextraction devices can be re-used several times depending on the type of sample matrix complexity, solutes and experimental conditions, as well as, the sorbent phases involved. In general, it was clearly observed that if the sorbent phases are based on ACs, the microextraction devices can be used more times because this type of material present a much stronger fixation stability and mechanical robustness, when compared with other solid or polymeric powdered coatings. From the experimental point of view, AC and PS-DVB phases have showed the best efficiency performance through BAμE, giving a very good indication as potential sorbent phases to retain trace levels of polar compounds from aqueous media. Apart of the much better stability fixation demonstrated through BAμE, ACs also present interesting features such as convenient textures (*e.g.* mesopores) conjugated with suitable strong adsorptive properties, which are critical for the retention process. On the other hand, PS-DVB sorbents can present multiple functionality since they can join reverse phase with ionic exchange properties. Once these sorbents have presented very interesting interactions features, the adsorptive microextraction techniques also demonstrated very good performance even for compounds with nonpolar characteristics. Nevertheless, it must be emphasized that the enrichment performance of the sorbent phases for a specific solute depends on a case by case assessment, *i.e.* the material should be careful selected for each particular type of application.

6.3.3. Floating Sampling Technology

After preparing and testing the stability and robustness of the microextraction devices, we can start to focus our attention on the BAμE analytical performance. For operating purposes, the BAμE devices coated with convenient sorbent materials are introduced into the sampling flasks (*e.g.* 20 mL) to operate under the floating technology mode [9 - 11]. This new microextraction concept is based on the use of an analytical device, light in weight, simultaneously with a conventional Teflon magnetic stirring bar at the bottom of the sampling flask. When the sample matrix is rapidly spinning around, due to centripetal force promoted by the magnetic bar, the analytical device is left under free-floating just below the vortex centre promoted the steady state rotational motion of the liquid matrix with very high stability and reproducibility. During a static process, the analytes migrating by diffusion from the sample bulk towards a selective sorbent phase where the microextraction takes place. This novel approach has the great advantage of avoiding the direct contact of the micro-collector with the walls or the bottom of the sampling flask, thus minimizing the mechanical disaggregation of the sorbents, increasing the lifetime. Fig. (**3.b**) depicts a schematic representation and an image exemplifying the floating sampling process mode for BAμE devices during the static operation. Although presenting a bit different enrichment philosophy, when compared to SBSE technique, the optimization of the experimental conditions for BAμE is quite similar and compulsory for each type of application. To maximize the efficiency of the BAμE technique, beyond the sorbent materials type and amount involved, systematic assays on parameters such as equilibrium time, agitation speed and matrix characteristics, *i.e.* pH, polarity and ionic strength, should also be performed using both univariate and multivariate optimization strategies. Most often, longer hours of equilibrium and lower agitation speed are needed, in order to obtain steady state conditions comparable to SBSE. Although the ionic strength and organic modifiers have showed not to be very important variables, once this analytical technology is indicated to polar molecules, the pH is often critical either for the target compounds or sorbents surface involved, especially if acidic or basic characteristics are present. Once most of the medium polar to polar solutes are semi-volatile to non-volatile compounds, in which some of them present thermolabile properties, LD followed by HPLC or hyphenated to mass spectrometry (LC-MS) systems are definitely the subsequent instrumental association of choice for identification and quantification purposes. Meanwhile, for back-extraction conditions, *i.e.* the solutes desorption from the sorbent coatings, the use of an appropriate solvent under optimised experimental conditions needs also to be assessed. Like SBSE-LD (Fig. **4.b**), the optimisation of the back-extraction parameters such as desorption solvent type and time, as well as, the number of desorption steps is very important, usually processed under sonification treatment, in which the pH often plays a very important role. For the BAμE operating process, the strength of the interaction in between the sorbent phase and the target molecules also determines the desorption process required, in order to obtain the best analytical performance. In general, the overall optimization is always a must to establish the most convenient validation conditions for any particular application, in which at the ultra-trace level the reproducibility should also meet the requirements of Directive 98/83/EC. In short, BAμE that uses the floating sampling approach is very easy to work-up and allow the possibility to choose the best sorbent phase for each particular type of application.

6.3.4. Potential Applications

After establishing the best experimental conditions and validation procedures, the BAμE devices are ready for application to monitor polar solutes in real samples. The data obtained, using the SAM to avoid matrix effects, have shown remarkable performance for the determination of several classes of priority compounds in many types of matrices, where SBSE technique have demonstrated lack of effectiveness. BAμE have been applied to monitor emerging polar compounds in important areas such as environment, food, pharmaceutical, forensic, biomedical, *etc.* Under optimized experimental conditions, trace levels (parts-per-trillion range) of priority compounds such as pesticides, herbicides (triazinic type), fungicides, insecticide repellents, parabens, irgarol, benzotriazoles, disinfection by-products (short chain carbonyl type), PPCPs, pharmaceuticals, sexual hormones, anabolic steroids, antibiotics, ibuprofen, clofibric acid, acetaminophen, caffeine, drugs of abuse (morphine and codeine), benzophenones, phenols, phenolic acids, *etc.*, have been successfully determined in real matrices including drinking water, surface water, wastewater, estuarine water, biological fluids (*e.g.* urine), food and beverages (*e.g.* wine), commercial products and many other complex samples [9, 10, 34-45]. Fig. (**5.c**) exemplifies the application of BAμE(AC)-LD/HPLC-DAD to a superficial water sample fortified with triazinic herbicides demonstrating the great performance achieved at the trace level; (Fig. **5d**.) exemplifies the monitoring of several PPPCs in a wastewater sample by BAμE(PS-DVB)-LD/LVI-GC-MS, where it is notice great selectivity and sensitivity at the ultra-trace level. From the data obtained, BAμE have demonstrated remarkable selectivity, high sensitivity and outstanding robustness for trace level analysis. In almost all cases, the recovery yields of polar targets used to be above 75% showing the great effectiveness of the BAμE methodology comparatively to SBSE. Furthermore, the floating sampling technology has showed great effectiveness for trace level analysis of polar organic compounds and in many cases, it can be a remarkable alternative over other approaches.The potential applications presented herein clearly demonstrate that BAμE technique containing appropriate sorbent phases and operating under optimized experimental conditions, are undoubtedly indicated to monitor trace levels of polar organic compounds in many type of aqueous media and must be understood as a novel advantageous sorption-based tools for analytical purposes.

CONCLUSION

Among the sorption-based microextraction methods, SBSE(PDMS) is nowadays, one of the most well-established static sampling tools, showing reliability in terms of the enrichment capacity, outstanding performance, as well as, great sensitivity and selectivity for trace analysis of complex matrices. Besides all the advantages discussed, as a sample preparation method to be associated with chromatographic and hyphenated systems, this remarkable analytical approach cannot be so effective for some complex matrices, in particular to monitor the large group of polar organic compounds. Although other analytical strategies can be implemented, this limitation can be much better overcome through the recent introduced BAμE that use the properties of the nano structurated materials and operates under the novel floating sampling technology mode, showing great performance and reproducibility. Furthermore, BAμE is cost-effective and easy to work-up, demonstrating to be a very promising analytical tool in the near future to overcome many other dedicated methodologies and analytical challenges from emerging areas in the life sciences (*e.g.* metabolomics).

CONFLICT OF INTEREST

The authors confirm that they have no conflict of interest to declare for this publication.

ACKNOWLEDGEMENTS

The author thanks all research collaborators of the Chromatography & Capillary

Electrophoresis Lab. from the Separation Science & Technology group of the Centre of Chemistry and Biochemistry of the Faculty of Sciences of the University of Lisbon (Portugal) and Fundação para a Ciência e a Tecnologia for funding (Portugal; Pest-OE/QUI/UI0612/2014).

DISCLOSURE

Part of this chapter has been previously published in Nogueira, J.M "Novel sorption-based methodologies for static microextraction analysis: A review on SBSE and related techniques", Analytica Chimica Acta, 757, 1-10 (2012).

REFERENCES

[1] Raynie, D.E. Modern extraction techniques. *Anal. Chem.,* **2010**, *82*(12), 4911-4916. [http://dx.doi.org/10.1021/ac101223c] [PMID: 20481433]

[2] Arthur, C.L.; Pawlizyn, J. Solid phase microextraction with thermal desorption using fused silica optical fibers. *Anal. Chem.,* **1990**, *62*, 2145. [http://dx.doi.org/10.1021/ac00218a019]

[3] Baltussen, E.; Sandra, P.; David, F.; Cramers, C.A. Stir bar sorptive extraction (SBSE), a novel extraction technique for aqueous samples: Theory and principles. *J. Microcol.,* **1999**, *11*, 737. [http://dx.doi.org/10.1002/(SICI)1520-667X(1999)11:10<737::AID-MCS7>3.0.CO;2-4]

[4] Baltussen, E.; Cramers, C.A.; Sandra, P.J. Sorptive sample preparation a review. *Anal. Bioanal. Chem.,* **2002**, *373*(1-2), 3-22. [http://dx.doi.org/10.1007/s00216-002-1266-2] [PMID: 12012168]

[5] David, F.; Sandra, P. Stir bar sorptive extraction for trace analysis. *J. Chromatogr. A,* **2007**, *1152*(1-2), 54-69. [http://dx.doi.org/10.1016/j.chroma.2007.01.032] [PMID: 17239895]

[6] Sánchez-Rojas, F. F.; Bosch-Ojeda, C.; Cano-Pavón, J.M. A review of Stir Bar sorptive extraction. *Chromatographia,* **2008**, *69*, S79. [http://dx.doi.org/10.1365/s10337-008-0687-2]

[7] Lanças, F.M.; Queiroz, M.E.; Grossi, P.; Olivares, I.R. Recent developments and applications of stir bar sorptive extraction. *J. Sep. Sci.,* **2009**, *32*(5-6), 813-824. [http://dx.doi.org/10.1002/jssc.200800669] [PMID: 19278005]

[8] Nogueira, J.M. Extração sortiva em barra de agitação (SBSE): uma metodologia inovadora para microextração estática. *Sci. Chromatogr.,* **2012**, *4*, 1-11. [http://dx.doi.org/10.4322/sc.2012.019]

[9] Nogueira, J.M. Novel sorption-based methodologies for static microextraction analysis: A review on SBSE and related techniques. *Anal. Chim. Acta,* **2012**, *757*, 1-10. [http://dx.doi.org/10.1016/j.aca.2012.10.033] [PMID: 23206390]

[10] Nogueira, J.M. Microextração adsortiva em barra (BAμE): Um conceito analítico inovador para

microextração estática. *Sci. Chromatogr.,* **2013**, *5*, 275-283.
[http://dx.doi.org/10.4322/sc.2014.010]

[11] Neng, N.R.; Silva, A.R.; Nogueira, J.M. Adsorptive micro-extraction techniques novel analytical tools for trace levels of polar solutes in aqueous media. *J. Chromatogr. A,* **2010**, *1217*(47), 7303-7310.
[http://dx.doi.org/10.1016/j.chroma.2010.09.048] [PMID: 20950811]

[12] Serôdio, P.; Nogueira, J.M. Multi-residue screening of endocrine disrupters chemicals in water samples by stir bar sorptive extraction-liquid desorption-capillary gas chromatography-mass spectrometry detection. *Anal. Chim. Acta,* **2004**, *517*, 21.
[http://dx.doi.org/10.1016/j.aca.2004.04.045]

[13] Serôdio, P.; Cabral, M.S.; Nogueira, J.M. Use of experimental design in the optimization of stir bar sorptive extraction for the determination of polybrominated diphenyl ethers in environmental matrices. *J. Chromatogr. A,* **2007**, *1141*(2), 259-270.
[http://dx.doi.org/10.1016/j.chroma.2006.12.011] [PMID: 17196215]

[14] Prieto, A.; Basauri, O.; Rodil, R.; Usobiaga, A.; Fernández, L.A.; Etxebarria, N.; Zuloaga, O. Stir-bar sorptive extraction: A view on method optimisation, novel applications, limitations and potential solutions. *J. Chromatogr. A,* **2010**, *1217*(16), 2642-2666.
[http://dx.doi.org/10.1016/j.chroma.2009.12.051] [PMID: 20083248]

[15] Popp, P.; Bauer, C.; Wennrich, L. Application of stir bar sorptive extraction in combination with column liquid chromatography for the determination of polycyclic aromatic hydrocarbons in water samples. *Anal. Chim. Acta,* **2001**, *436*, 1.
[http://dx.doi.org/10.1016/S0003-2670(01)00895-9]

[16] Almeida, C.; Nogueira, J.M. Determination of steroid sex hormones in water and urine matrices by stir bar sorptive extraction and liquid chromatography with diode array detection. *J. Pharm. Biomed. Anal.,* **2006**, *41*(4), 1303-1311.
[http://dx.doi.org/10.1016/j.jpba.2006.02.037] [PMID: 16581218]

[17] Juan-García, A.; Picó, Y.; Font, G. Capillary electrophoresis for analyzing pesticides in fruits and vegetables using solid-phase extraction and stir-bar sorptive extraction. *J. Chromatogr. A,* **2005**, *1073*(1-2), 229-236.
[http://dx.doi.org/10.1016/j.chroma.2004.09.028] [PMID: 15909524]

[18] do Rosário, P.M.; Nogueira, J.M. Combining stir bar sorptive extraction and MEKC for the determination of polynuclear aromatic hydrocarbons in environmental and biological matrices. *Electrophoresis,* **2006**, *27*(23), 4694-4702.
[http://dx.doi.org/10.1002/elps.200600210] [PMID: 17136717]

[19] European Commission. Off J 608. *EurCommun,* **1998**, *L330*, 32.

[20] Woolfenden, E. Sorbent-based sampling methods for volatile and semi-volatile organic compounds in air Part 1: Sorbent-based air monitoring options. *J. Chromatogr. A,* **2010**, *1217*(16), 2674-2684.
[http://dx.doi.org/10.1016/j.chroma.2009.12.042] [PMID: 20106481]

[21] Serôdio, P.; Nogueira, J.M. Development of a stir-bar-sorptive extraction-liquid desorption-large-volume injection capillary gas chromatographic-mass spectrometric method for pyrethroid pesticides in water samples. *Anal. Bioanal. Chem.,* **2005**, *382*(4), 1141-1151.
[http://dx.doi.org/10.1007/s00216-005-3210-8] [PMID: 15937663]

[22] Peñalver, A.; García, V.; Pocurull, E.; Borrull, F.; Marce, R.M. Stir bar sorptive extraction and large volume injection gas chromatography to determine a group of endocrine disrupters in water samples. *J. Chromatogr. A,* **2003**, *1007*(1-2), 1-9. [http://dx.doi.org/http://dx.doi. org/10.1016/S0021-9673(03)00933-6] [PMID: 12924545]

[23] Lacorte, S.; Quintana, J.; Tauler, R.; Ventura, F.; Tovar-Sánchez, A.; Duarte, C.M. Ultra-trace determination of persistent organic pollutants in arctic ice using stir bar sorptive extraction and gas chromatography coupled to mass spectrometry. *J. Chromatogr. A,* **2009**, *1216*(49), 8581-8589. [http://dx.doi.org/10.1016/j.chroma.2009.10.029] [PMID: 19879587]

[24] Neng, N.R.; Santalla, R.P.; Nogueira, J.M. Determination of tributyltin in environmental water matrices using stir bar sorptive extraction with in-situ derivatisation and large volume injection-gas chromatography-mass spectrometry. *Talanta,* **2014**, *126*, 8-11. [http://dx.doi.org/10.1016/j.talanta.2014.03.021] [PMID: 24881528]

[25] Alves, R.F.; Nascimento, A.M.; Nogueira, J.M. Characterization of the aroma profile of Madeira wine by sorptive extraction techniques. *Anal. Chim. Acta,* **2005**, *546*(1), 11-21. [http://dx.doi.org/10.1016/j.aca.2005.05.012]

[26] Weldegergis, B.T.; Tredoux, A.G.; Crouch, A.M. Application of a headspace sorptive extraction method for the analysis of volatile components in South African wines. *J. Agric. Food Chem.,* **2007**, *55*(21), 8696-8702. [http://dx.doi.org/10.1021/jf071554p] [PMID: 17927151]

[27] Barletta, J.Y.; Gomes, P.C.; Dos Santos-Neto, A.J.; Lanças, F.M. Development of a new stir bar sorptive extraction coating and its application for the determination of six pesticides in sugarcane juice. *J. Sep. Sci.,* **2011**, *34*(11), 1317-1325. [http://dx.doi.org/10.1002/jssc.201100096] [PMID: 21548086]

[28] Pereira, M.B.; Serôdio, P.; Castro, M.J.; Nogueira, J.M. Determination of organochlorine pesticides in vegetable matrices by stir bar sorptive extraction with liquid desorption and large volume injection-gas chromatography-mass spectrometry towards compliance with European Union directives. *J. Chromatogr. A,* **2010**, *1217*(1), 119-126. [http://dx.doi.org/10.1016/j.chroma.2009.10.076] [PMID: 19931087]

[29] Madrera, R.R.; Valles, B.S. Determination of volatile compounds in apple pomace by stir bar sorptive extraction and gas chromatography-mass spectrometry (SBSE-GC-MS). *J. Food Sci.,* **2011**, *76*(9), C1326-C1334. [http://dx.doi.org/10.1111/j.1750-3841.2011.02406.x] [PMID: 22416695]

[30] Soini, H.A.; Bruce, K.E.; Wiesler, D.; David, F.; Sandra, P.; Novotny, M.V. Stir bar sorptive extraction: a new quantitative and comprehensive sampling technique for determination of chemical signal profiles from biological media. *J. Chem. Ecol.,* **2005**, *31*(2), 377-392. [http://dx.doi.org/10.1007/s10886-005-1347-8] [PMID: 15856790]

[31] Chaves, A.R.; Queiroz, M.E. Extração sortiva em barra de agitação para análise de fármacos em fluidos biológicos. *Quim. Nova,* **2008**, *31*, 1814. [http://dx.doi.org/10.1590/S0100-40422008000700037]

[32] Silva, A.R.; Nogueira, J.M. New approach on trace analysis of triclosan in personal care products, biological and environmental matrices. *Talanta,* **2008**, *74*(5), 1498-1504.

[http://dx.doi.org/10.1016/j.talanta.2007.09.040] [PMID: 18371809]

[33] Neng, N.R.; Pinto, M.L.; Pires, J.; Marcos, P.M.; Nogueira, J.M. Development, optimisation and application of polyurethane foams as new polymeric phases for stir bar sorptive extraction. *J. Chromatogr. A,* **2007**, *1171*(1-2), 8-14. [http://dx.doi.org/10.1016/j.chroma.2007.09.033] [PMID: 17927992]

[34] Neng, N.R.; Mestre, A.S.; Carvalho, A.P.; Nogueira, J.M. Powdered activated carbons as effective phases for bar adsorptive micro-extraction (BAμE) to monitor levels of triazinic herbicides in environmental water matrices. *Talanta,* **2011**, *83*(5), 1643-1649. [http://dx.doi.org/10.1016/j.talanta.2010.11.049] [PMID: 21238763]

[35] Neng, N.R.; Mestre, A.S.; Carvalho, A.P.; Nogueira, J.M. Cork-based activated carbons as supported adsorbent materials for trace level analysis of ibuprofen and clofibric acid in environmental and biological matrices. *J. Chromatogr. A,* **2011**, *1218*(37), 6263-6270. [http://dx.doi.org/10.1016/j.chroma.2011.07.025] [PMID: 21820664]

[36] Neng, N.R.; Nogueira, J.M. Determination of short-chain carbonyl compounds in drinking water matrices by bar adsorptive micro-extraction (BAμE) with *in situ* derivatization. *Anal. Bioanal. Chem.,* **2010**, *398*(7-8), 3155-3163. [http://dx.doi.org/10.1007/s00216-010-4256-9] [PMID: 20936263]

[37] Neng, N.R.; Nogueira, J.M. Development of a bar adsorptive micro-extraction-large-volume injection-gas chromatography-mass spectrometric method for pharmaceuticals and personal care products in environmental water matrices. *Anal. Bioanal. Chem.,* **2012**, *402*(3), 1355-1364. [http://dx.doi.org/10.1007/s00216-011-5515-0] [PMID: 22076312]

[38] Almeida, C.; Nogueira, J.M. Comparison of the selectivity of different sorbent phases for bar adsorptive microextractionapplication to trace level analysis of fungicides in real matrices. *J. Chromatogr. A,* **2012**, *1265*, 7-16. [http://dx.doi.org/10.1016/j.chroma.2012.09.047] [PMID: 23084822]

[39] Gonçalves, A.F.; Neng, N.R.; Mestre, A.S.; Carvalho, A.P.; Nogueira, J.M. Development of a powdered activated carbon in bar adsorptive micro-extraction for the analysis of morphine and codeine in human urine. *J. Chromatogr. Sci.,* **2012**, *50*(7), 574-581. [http://dx.doi.org/10.1093/chromsci/bms051] [PMID: 22562817]

[40] Almeida, C.; Stępkowska, A.; Alegre, A.; Nogueira, J.M. Determination of trace levels of benzophenone-type ultra-violet filters in real matrices by bar adsorptive micro-extraction using selective sorbent phases. *J. Chromatogr. A,* **2013**, *1311*, 1-10. [http://dx.doi.org/10.1016/j.chroma.2013.08.007] [PMID: 24011723]

[41] Almeida, C.; Strzelczyk, R.; Nogueira, J.M. Improvements on bar adsorptive microextraction (BAμE) technique application for the determination of insecticide repellents in environmental water matrices. *Talanta,* **2014**, *120*, 126-134. [http://dx.doi.org/10.1016/j.talanta.2013.11.031] [PMID: 24468351]

[42] Almeida, C.; Nogueira, J.M. Determination of trace levels of parabens in real matrices by bar adsorptive microextraction using selective sorbent phases. *J. Chromatogr. A,* **2014**, *1348*, 17-26. [http://dx.doi.org/10.1016/j.chroma.2014.04.057] [PMID: 24835595]

[43] Neng, N.R.; Nogueira, J.M. Determination of phenol compounds in surface water matrices by bar

[43] Neng, N.R.; Nogueira, J.M. Determination of phenol compounds in surface water matrices by bar adsorptive microextraction-high performance liquid chromatography-diode array detection. *Molecules,* **2014**, *19*(7), 9369-9379.
[http://dx.doi.org/10.3390/molecules19079369] [PMID: 24995922]

CHAPTER 7

Microextraction by Packed Sorbent (MEPS): Theory, Developments and Applications

Mohammad Mahdi Moein, Abbi Abdel-Rehim and **Mohamed Abdel-Rehim***

Department of Analytical Chemistry, Stockholm University, SE10691 Stockholm, Sweden

Abstract: Sample preparation is a critical issue in bioanalysis since the matrix is often complex. A good sample preparation technique will remove the possible interferences from the sample matrix pre-concentrate the analyte and to be reproducible independent of the sample matrix. Recent developments of sample preparation techniques are directed toward automation, the smaller sample volumes and online coupling of sample preparation units and detection systems. Microextraction in packed syringe (MEPS) is a new type of solid-phase extraction (SPE) technique that is miniaturized and can be fully automated. In MEPS the solid bed is integrated in the injection syringe. The MEPS syringe can be used online for both extraction and injection of samples. The present chapter gives an overview of MEPS technique, including the MEPS description, formats, sorbents, experimental and protocols and factors that affect the MEPS performance. We also summarize MEPS recent applications in bioanalysis, environmental and food analysis.

Keywords: Applications, Experimental protocols, Microextraction by packed sorbent (MEPS), Sorbents.

7.1. INTRODUCTION

Owing to the need of daily rotation of results from huge numbers of biological samples in big pharmaceutical companies, it is essential to apply sample preparation methods with high throughput properties. Sample preparation as the first step in the analysis procedure has an important role in many different fields

* **Corresponding author Mohamed Abdel-Rehim:** Department of Analytical Chemistry, Stockholm University, SE10691 Stockholm, Sweden; Tel: +46 (0) 707108122; Fax: +46 709569894; E-mail: mohamed.abdel-rehim @anchem.su.se

M. Valcárcel, S. Cárdenas, R. Lucena (Eds.)

such as bioanalysis and environmental analysis. Additionally sample preparation methods utilizing small sample volumes (10-50 μL) should be useful for clinical and preclinical studies. This can give new possibilities for handling of samples from children. As well, it allows serial collection of more than one time point from the same animal, such as mice or rat. Such sample preparation methods will save money and time in drug discovery and development. To obtain this purpose the integration of the sample preparation with the analysis apparatus and miniaturization were significantly required.

Microextraction by packed sorbent (MEPS) is a new technique for miniaturized solid phase extraction (SPE) that can be connected online to GC or LC without any modifications and can handle small sample volumes. MEPS technique is easy to use, fully automated, of low cost and rapid. In comparison with previously used methods, MEPS significantly decreases the sample handling time, decreases the sample volumes and the solvent required.

This chapter provides an overview on the emerging technique microextraction by packed sorbent (MEPS). The recent developments in MEPS device and finding ways for automated application are presented. MEPS provides differentiation from the conventional SPE and in some applications it could replace the need of the analytical column in LC/MS/MS (*i.e.*, directly used online with the MS detector). In addition, factors affecting the MEPS performance and extraction procedures will be discussed.

7.2. MEPS DESCRIPTION AND THEORY

MEPS is a self-acting kind of solid phase extraction (SPE) technique. MEPS was emerged in 2004 by Dr. Abdel-Rehim [1]. The MEPS technique was developed to meet the request of automated and high-throughput sample preparation methods. The MEPS technique is fully automated, reduces the handle time and handles small sample volumes, it is simple and inexpensive. The fast process and the ease of the sample preparation process are advantages of MEPS. The main attributes of MEPS is the decrease of the sample and solvent used (10-50 μL) and the elution solution can be injected directly into GC or LC instruments.

In MEPS 1–4 mg of sorbent is packed in the barrel of a gas-tight syringe

(100–250 μL), or in a special container positioned between the needle and the barrel as a cartridge (Fig. **1**).

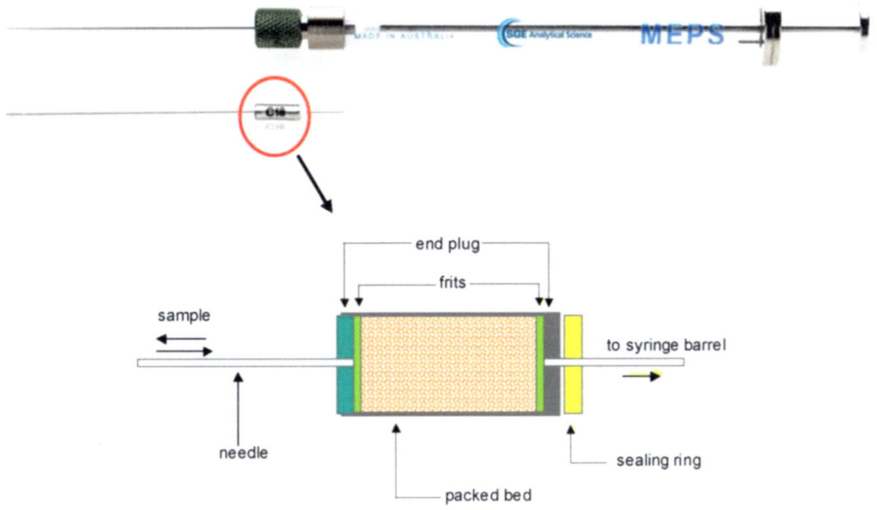

Fig. (1). MEPS syringe [2].

In MEPS the sample solution is withdrawn through the syringe by an autosampler and when the sample solution has passed through the solid phase, the analytes are adsorbed to the solid sorbent. Then the solid phase is washed once or more with water to remove the proteins and other interfering materials. Finally, for desorption of analyte organic solvent, such as methanol or the LC mobile phase (20–50 μL) can used and directly through the analytical instrument's injector.

The MEPS processing, extraction and injection steps are performed online using the same syringe. The main concept of MEPS and SPE techniques is approximately similar and, both techniques build on the same principles and therefore the transferring of a method from traditional SPE to MEPS would be quite straight forward. Typical MEPS is designed in the syringe format and a gas-tight glass syringe is used (the syringe volume is 100, 250 or 500 μL). The solid sorbent may be embedded in a small container situated between the barrel and the needle, as mentioned above. Miniaturization is critical point in MEPS technique, thereby only approximately low volume of solvents is required for elution of the analytes from the adsorbent, a quantity that is suitable for direct introduction into the analysis instrument.

The MEPS solid sorbent can be packed or coated to create selective and efficient sampling conditions. To obtain clean sample and good recovery choosing appropriate sorbent is serious issue. Moreover, creating stable and effective extraction situation considering the analyte properties (non-polar, polar or ionized) is an important parameter. In MEPS, large variety of sorbents material such silica based (C2, C8, C18), strong cation exchanger (SCX) using sulfonic acid bonded silica, restricted access material (RAM), hilic, carbon, polystyrene–divinylbenzene copolymer (PS-DVB), molecular imprinted polymers (MIPs) and organic monolithic sorbent can be used. Typically, silica particles have been irregular, 50 μm, 60 Å, and ENV+ polymer particles have also been irregular, 90 μm, 800 Å. In addition, we have shown that a highly viscous polymer, a silicone-silarylene copolymer, could be used as sorbent for MEPS [3]. In addition most important thing to get high recovery is the choice of sorbent. C2–C18 phases are suitable for lipophilic analytes (non-polar) and polymeric phases such as polystyrene–divinylbenzene or mixed mod phases (anion–cation exchange mode) are suitable for polar analytes such as acidic and basic compounds.

MEPS present much more optimization opportunities that we could anticipate from a so simple design and in fact there are many factors affecting MEPS and has four main stages which are highlighted in below:

1. Conditioning-Regeneration; the type and the volume of the solvents should be optimized to avoid carryover of the previous extractions and extend the sorbent utilization. Usually two volumes of the total syringe capacity are recommended. Commonly, organic solvent such as methanol or acetonitrile should be used as first solution and water as the second solution.
2. Sample loading; one or more loading cycles can be performed according to the concentration of target analytes. Sample can be used directly (water saples) or diluted if they are too viscous (plasma, blood). Several cycles of loading/ dispensing can be done to enrich the analytes on MEPS sorbent.
3. Sample washing; water or 0.1% formic acid in water are the often used solutions. Washing should be tentatively optimized to remove as much as possible of the interferences and keep the target analytes. Therefore low

organic solvent amount such as methanol (5-10%) in water with suitable pH should be investigated.

4. Sample elution; performed with an appropriate solvent to release the retained analytes from the sorbent. Solvents such as methanol or acetonitrile are often used. This is also a critical step to recover the target analytes. It is important to select a solvent compatible with the LC mobile phase to avoid supplementary procedures. In addition elution solution pH has an important role to achieve good elution capacity. The schematic processes of MEPS extraction and some protocols for blood and plasma samples with different sorbents are shown in Fig. (**2**).

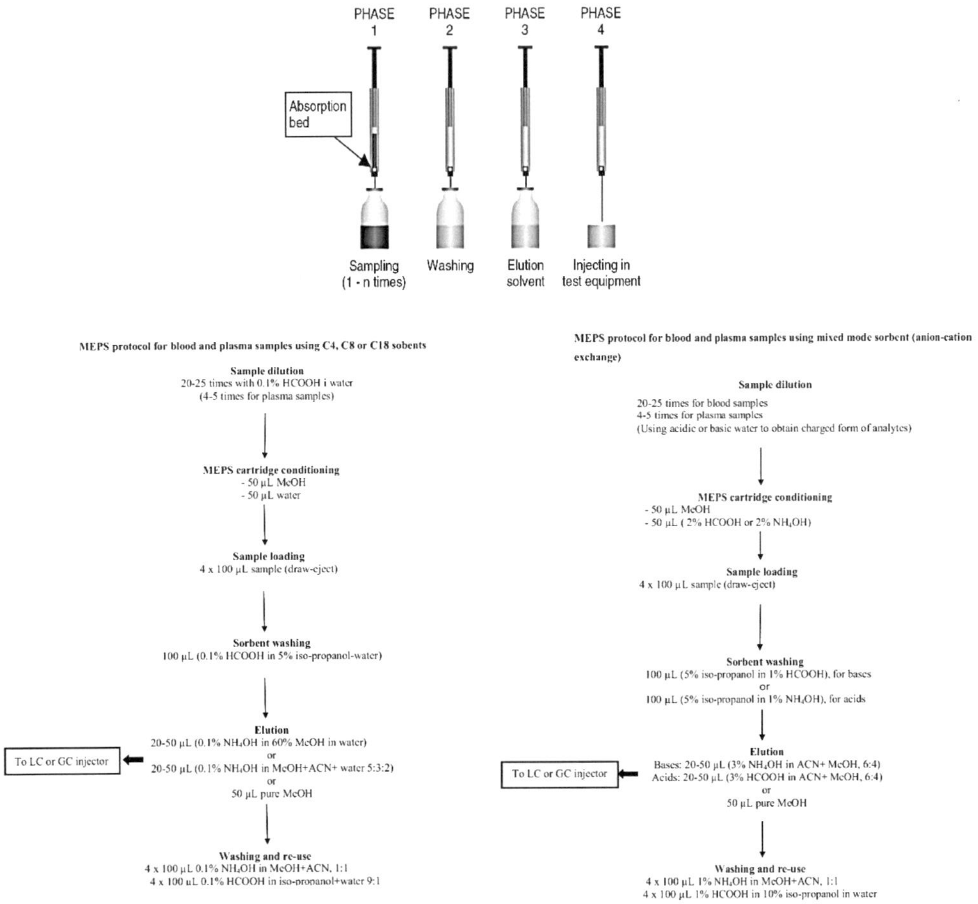

Fig. (2). MEPS main steps and extraction protocols [4, 5].

In continue part the new automated types of MEPS instruments will be considered and discussed.

7.3. MEPS DEVELOPMENT FOR ON LINE EXTRACTION

The importance of MEPS approach is the fully integration of the sample preparation with the analytical system. Thereby, automation of the entire analysis system becomes straightforward. MEPS can be connected with GC, LC, or MS without any modification of the instrument.

There are three types of MEPS ; manually, semi-automated or fully automated as are shown in Fig. (**3**). The sample is drawn once or more through the sorbent manually or by an autosampler (draw–eject in same vial or draw from the vial and eject into waste). When the sample has passed through the solid phase, the analytes have been adsorbed to the solid phase. The solid phase is then washed once by 50-100 μL water to remove the interferences. The analytes are then eluted with a solvent (20–50 μL) directly into the instrument's injector.

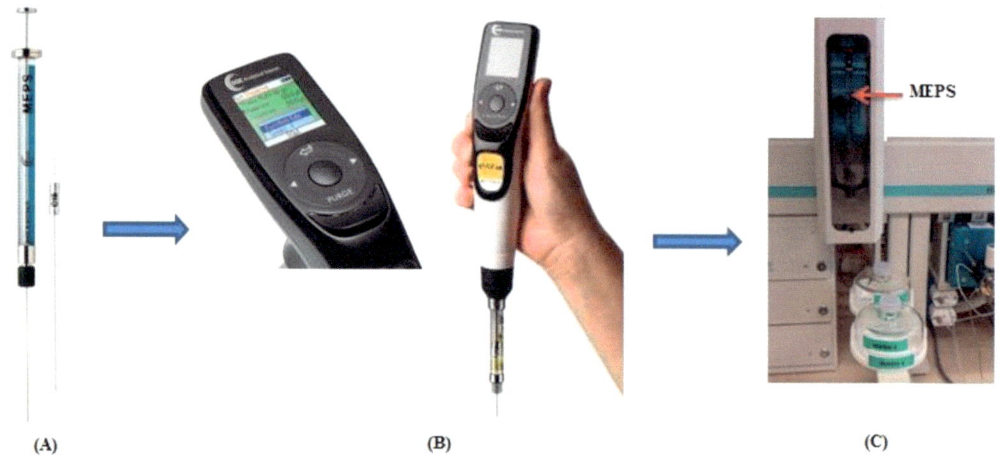

Fig. (3). MEPS Formats: (**A**) Manually; (**B**) Semi-automated (MEPS-eVol) and (**C**) Fully-automated (on-line) [Source: www.sge.com].

7.4. MEPS HIGHLIGHTS

MEPS is simple sample preparation technique, presenting several advantages when compared with other sample preparation techniques. MEPS not only retained the major successful features from SPE, namely its broad applications and simplicity, as improved the analytical performance that can be obtained with several analytes and matrices [2]. This was achieved through the miniaturization of all procedures, particularly the solvent requirements and the sorbent amount and presentation, which is limited to a few milligrams (1-4 mg) and packed in a small cylinder through which the sample is drawn several times. MEPS is therefore much more environment-friendly then SPE as well as other extraction techniques as LLE. Additionally, although sample requirements for a proper extraction are highly variable, with MEPS, these volumes are strongly reduced [2], which can be extremely important in precious samples, as the biological ones whose collection is invasive and painful and produce a limited volume of sample. Unlike other techniques, MEPS sorbents can be reused many times, up to one hundred or more, depending on the complexity of the matrix being processed and thus the cost by analysis is much lower that we could obtain using SPE, for instance. MEPS utilization is very easy and straightforward, making the technique also very economic and user-friendly. Finally, MEPS is also faster than other sample extraction techniques as each step is performed faster, most times using an automatic autosampler, and target analytes are eluted in a suitable volume without the need for pre-concentration steps that are usually associated to analyte losses.

7.5. MEPS APPLICATIONS IN BIOANALYSIS

The performance of MEPS was illustrated by online LC–MS and GC–MS assays of drugs and metabolites in blood, plasma and urine samples. A huge number of bioanalysis research works showed the useful of MEPS in bioanalysis [6-11]. The online connection of MEPS with LC/MS/MS reduces sample handling time and also can improve the limit of detection, for detection of methadone from human urine samples [12]. Several research papers reported the use of MEPS for extraction of a wide range of analytes from biological samples, using different matrices. Hence, several drugs such as local anaesthetics and their metabolites [1, 3, 4, 13, 32], the anti-cancer drugs roscovitine, olomoucine, busulphan, cyclophosphamide, and AZD3409 were extracted from plasma, urine and blood utilizing MEPS online with LC/MS [14 - 17]. The beta-blockers acebutolol and metoprolol in urine and plasma samples [18], the neurotransmitters dopamine and serotonine in human urine samples [19], cocaine and its metabolites in urine samples [20] the sensory neuron-specific receptors agonist peptide BAM8-22 in plasma [21], have been extracted utilizing MEPS technique. In addition online MEPS-GC/MS was used for the determination of amphetamine in human hair and monoterpene metabolites in human urine [22, 23]. Recently, MEPS-LC/MS/MS has been used for the determination of lidocaine in human saliva samples [24]. Furthermore, MEPS was used for the determination of remifentanil in human plasma using small sample volume ($\leq$ 20 μL) utilizing LC/MS/MS [29]. Said *et al.* reported on-line MEPS as sample preparation method with LC/MS/MS for the quantification of immunosuppressive drugs: cyclosporine, everolimus, sirolimus and tacrolimus in human whole blood and plasma samples [30].

7.6. MEPS APPLICATIONS IN ENVIRONMENTAL ANALYSIS

MEPS have been applied in several research works for environmental analysis such as; extraction of polycyclic aromatic hydrocarbons [25, 26, 31] and chlorobenzenes [27] from water samples utilizing GC/MS. Also, MIPs in MEPS were used for the extraction of fluoro-quinolone related compounds from water samples and detection by LC/MS/MS [28]. In addition, the determination of parabens, triclosan and its derivative in wastewater was carried out using MEPS online with GC/MS [37]. The determination of local anaesthetics in plasma samples using a cation-exchange sorbent in MEPS and LC-MS/MS was reported [38].

7.7. MEPS APPLICATIONS IN FOOD ANALYSIS

MEPS was used for the analysis of wine samples and for the determination of various compounds such as haloanisol, phenolic compounds, flavonols, resveratrol *etc*. The determination of haloanisol in wine was published by Pinto *et al.* [33]. The determination of melatonin and other indolic and phenolic antioxidants in grape- related foodstuffs and beverages was reported utilizing MEPS and HPLC, the method was validated and applied into the real samples [34]. The determination of vitamin C in beverages utilizing MEPS and LC-UV was reported [35]. Due to the wide use of pesticides in the farming the suitable analytical method for the multi-residual analysis of pesticides is very desirable. In published work the MEPS-GC-MS was used for the analysis of 22 pesticides in honey samples [36]. The determination of hydroxybenzoic and hydroxycinnamic acids in wine was reported by Goncalves *et al.* using MEPS and UHPLC-PDA [39].

CONCLUSION

MEPS is a miniaturized sort of SPE that is attractive in the field of bioanalysis and other fields. Compared to SPE, MEPS uses 10-100 times less sample volumes and reagents. The small volumes also significantly reduce extraction time from 10-20 mins down to 1-2 minutes. The MEPS format allows routine SPE of samples as small as 10 μL and up to 250 μL in a single cycle. Very small elution volumes and the integration of the sorbent into the sampling syringe and needle allow MEPS to be used for direct injection into LC and GC systems. The format is also suitable for complete automation with online sample handling robots and autosamplers. MEPS has been commercialised by SGE Analytical Science and is available in a wide range of silica and polymeric chemistries. An extensive range of applications for drugs, metabolites, alkaloids, pollutants and other substances in the most diverse matrices were developed by different research groups and the number is continuously growing.

CONFLICT OF INTEREST

The authors confirm that they have no conflict of interest to declare for this publication.

ACKNOWLEDGEMENTS

Declared none.

REFERENCES

[1] Abdel-Rehim, M. New trend in sample preparation: on-line microextraction in packed syringe for liquid and gas chromatography applications. I. Determination of local anaesthetics in human plasma samples using gas chromatography-mass spectrometry. *J. Chromatogr. B Analyt. Technol. Biomed. Life Sci.,* **2004**, *801*(2), 317-321. [http://dx.doi.org/10.1016/j.jchromb.2003.11.042] [PMID: 14751801]

[2] Abdel-Rehim, M. Recent advances in microextraction by packed sorbent for bioanalysis. *J. Chromatogr. A,* **2010**, *1217*(16), 2569-2580. [http://dx.doi.org/10.1016/j.chroma.2009.09.053] [PMID: 19811788]

[3] Abdel-Rehim, M.; Dahlgren, M.; Blomberg, L.; Claude, S.; Tabacchi, R. Microextraction in packed syringe (MEPS) utilizing methylcyanopropyl–Silarylene as coating polymer for extraction of drugs in biological samples. *J. Liq. Chrom. Rel. Tech.,* **2006**, *29*(17), 2537-2544. [http://dx.doi.org/10.1080/10826070600915023]

[4] Abdel-Rehim, M.; Altun, Z.; Blomberg, L. Microextraction in packed syringe (MEPS) for liquid and gas chromatographic applications. Part II Determination of ropivacaine and its metabolites in human plasma samples using MEPS with liquid chromatography/tandem mass spectrometry. *J. Mass Spectrom.,* **2004**, *39*(12), 1488-1493. [http://dx.doi.org/10.1002/jms.731] [PMID: 15578759]

[5] Abdel-Rehim, M. Microextraction by packed sorbent (MEPS): a tutorial. *Anal. Chim. Acta,* **2011**, *701*(2), 119-128. [http://dx.doi.org/10.1016/j.aca.2011.05.037] [PMID: 21801877]

[6] Alves, G.; Rodrigues, M.; Fortuna, A.; Falcão, A.; Queiroz, J. A critical review of microextraction by packed sorbent as a sample preparation approach in drug bioanalysis. *Bioanalysis,* **2013**, *5*(11), 1409-1442. [http://dx.doi.org/10.4155/bio.13.92] [PMID: 23742310]

[7] Pereira, J.; Gonçalves, J.; Alves, V.; Câmara, J.S. Microextraction using packed sorbent as an effective and high-throughput sample extraction technique: Recent applications and future trends. *Sample Preparation,* **2013**, 38-53.

[8] Vita, M.; Abdel-Rehim, M.; Nilsson, C.; Hassan, Z.; Skansen, P.; Wan, H.; Meurling, L.; Hassan, M. Stability, pKa and plasma protein binding of roscovitine. *J. Chromatogr. B Analyt. Technol. Biomed. Life Sci.,* **2005**, *821*(1), 75-80. [http://dx.doi.org/10.1016/j.jchromb.2005.04.014] [PMID: 15894521]

[12] El-Beqqali, A.; Abdel-Rehim, M. Quantitative analysis of methadone in human urine samples by microextraction in packed syringe-gas chromatography-mass spectrometry (MEPS-GC-MS). *J. Sep. Sci.,* **2007**, *30*(15), 2501-2505. [http://dx.doi.org/10.1002/jssc.200700067] [PMID: 17763516]

[13] Abdel-Rehim, M.; Askemark, Y.; Norsten-Höög, C.; Pettersson, K.J.; Halldin, M. Quantification of 4-OH-2,6-Xylidine and its conjugates in human urine samples utilising microextraction in packed syringe on-line with liquid chromatography and electrospray tandem mass spectrometry (MEPS-LC/MS/MS). *J. Liq. Chrom. Rel. Tech.,* **2006**, *29*(16), 2413-2424. [http://dx.doi.org/10.1080/10826070600864882]

[14] Abdel-Rehim, M.; Skansen, P.; Vita, M.; Hassan, Z.; Blomberg, L.; Hassan, M. Microextraction in packed syringe/liquid chromatography/electrospray tandem mass spectrometry for quantification of olomoucine in human plasma samples. *Anal. Chim. Acta,* **2005**, *539*(1-2), 35-39. [http://dx.doi.org/10.1016/j.aca.2005.02.061]

[15] Abdel-Rehim, M.; Hassan, Z.; Skansem, P.; Hassan, M. Simultaneous determination of busulphan in plasma samples by liquid chromatography-electrospray ionization mass spectrometry utilizing microextraction in packed syringe (MEPS) as on-line sample preparation method. *J. Liq. Chrom. Rel. Tech.,* **2007**, *30*(20), 3029-3041. [http://dx.doi.org/10.1080/10826070701632337]

[16] Said, R.; Hassan, Z.; Hassan, M.; Abdel-Rehim, M. Rapid and sensitive method for determination of cyclophosphamide in patients plasma samples utilizing microextraction by packed sorbent online with liquid chromatography-tandem mass spectrometry (MEPS-LC/MS/MS). *J. Liq. Chrom. Rel. Tech.,* **2008**, *31*(5), 683-694.
[http://dx.doi.org/10.1080/10826070701853867]

[17] Vita, M.; Skansen, P.; Hassan, M.; Abdel-Rehim, M. Development and validation of a liquid chromatography and tandem mass spectrometry method for determination of roscovitine in plasma and urine samples utilizing on-line sample preparation. *J. Chromatogr. B Analyt. Technol. Biomed. Life Sci.,* **2005**, *817*(2), 303-307.
[http://dx.doi.org/10.1016/j.jchromb.2004.12.022] [PMID: 15686999]

[18] El-Beqqali, A.; Kussak, A.; Blomberg, L.; Abdel-Rehim, M. Microextraction in packed syringe/liquid chromatography/electrospray tandem mass spectrometry for quantification of acebutolol and metoprolol in human plasma and urine samples. *J. Liq. Chrom. Rel. Tech.,* **2007**, *30*(4), 575-586.
[http://dx.doi.org/10.1080/10826070601093895]

[19] El-Beqqali, A.; Kussak, A.; Abdel-Rehim, M. Determination of dopamine and serotonin in human urine samples utilizing microextraction online with liquid chromatography/electrospray tandem mass spectrometry. *J. Sep. Sci.,* **2007**, *30*(3), 421-424.
[http://dx.doi.org/10.1002/jssc.200600369] [PMID: 17396602]

[20] Jagerdeo, E.; Abdel-Rehim, M. Screening of cocaine and its metabolites in human urine samples by direct analysis in real-time source coupled to time-of-flight mass spectrometry after online preconcentration utilizing microextraction by packed sorbent. *J. Am. Soc. Mass Spectrom.,* **2009**, *20*(5), 891-899.
[http://dx.doi.org/10.1016/j.jasms.2009.01.010] [PMID: 19264505]

[21] Ashri, N.Y.; Daryanavard, M.; Abdel-Rehim, M. Microextraction by packed sorbent and liquid chromatography-tandem mass spectrometry as a tool for quantification of peptides in plasma samples: determination of sensory neuron-specific receptors agonist BAM822 and antagonist BAM228 in plasma samples. *Biomed. Chromatogr.,* **2013**, *27*(3), 396-403.
[PMID: 22987540]

[22] Matysik, S.; Matysik, F.M. Microextraction by packed sorbent coupled with gas chromatography—mass spectrometry: application to the determination of metabolites of monoterpenes in small volumes of human urine. *Mikrochim. Acta,* **2009**, *166*(1-2), 109-114.
[http://dx.doi.org/10.1007/s00604-009-0170-2]

[23] Miyaguchi, H.; Iwata, Y.T.; Kanamori, T.; Tsujikawa, K.; Kuwayama, K.; Inoue, H. Rapid identification and quantification of methamphetamine and amphetamine in hair by gas chromatography/mass spectrometry coupled with micropulverized extraction, aqueous acetylation and microextraction by packed sorbent. *J. Chromatogr. A,* **2009**, *1216*(18), 4063-4070.
[http://dx.doi.org/10.1016/j.chroma.2009.02.093] [PMID: 19286186]

[24] Abdel-Rehim, A.; Abdel-Rehim, M. Screening and determination of drugs in human saliva utilizing microextraction by packed sorbent and liquid chromatography-tandem mass spectrometry. *Biomed. Chromatogr.,* **2013**, *27*(9), 1188-1191.
[http://dx.doi.org/10.1002/bmc.2925] [PMID: 23629973]

[25] Fu, S.; Fan, J.; Hashi, Y.; Chen, Z. Determination of polycyclic aromatic hydrocarbons in water samples using online microextraction by packed sorbent coupled with gas chromatography-mass spectrometry. *Talanta,* **2012**, *94*(94), 152-157. [http://dx.doi.org/10.1016/j.talanta.2012.03.010] [PMID: 22608428]

[26] Quinto, M.; Amodio, P.; Spadaccino, G.; Centonze, D. Development of a mathematical model for online microextraction by packed sorbent under equilibrium conditions and its application for polycyclic aromatic hydrocarbon determination in water by gas chromatography-mass spectrometry. *J. Chromatogr. A,* **2012**, *1262*, 19-26. [http://dx.doi.org/10.1016/j.chroma.2012.08.098] [PMID: 22999202]

[27] Grueiro Noche, G.; Fernández Laespada, M.E.; Pérez Pavón, J.L.; Moreno Cordero, B.; Muniategui Lorenzo, S. Determination of chlorobenzenes in water samples based on fully automated microextraction by packed sorbent coupled with programmed temperature vaporization-gas chromatography-mass spectrometry. *Anal. Bioanal. Chem.,* **2013**, *405*(21), 6739-6748. [http://dx.doi.org/10.1007/s00216-013-7112-x] [PMID: 23780226]

[28] Prieto, A.; Schrader, S.; Bauer, C.; Möder, M. Synthesis of a molecularly imprinted polymer and its application for microextraction by packed sorbent for the determination of fluoroquinolone related compounds in water. *Anal. Chim. Acta,* **2011**, *685*(2), 146-152. [http://dx.doi.org/10.1016/j.aca.2010.11.038] [PMID: 21168563]

[29] Said, R.; Pohanka, A.; Andersson, M.; Beck, O.; Abdel-Rehim, M. Determination of remifentanil in human plasma by liquid chromatography-tandem mass spectrometry utilizing micro extraction in packed syringe (MEPS) as sample preparation. *J. Chromatogr. B Analyt. Technol. Biomed. Life Sci.,* **2011**, *879*(11-12), 815-818. [http://dx.doi.org/10.1016/j.jchromb.2011.02.002] [PMID: 21376673]

[30] Said, R.; Pohanka, A.; Abdel-Rehim, M.; Beck, O. Determination of four immunosuppressive drugs in whole blood using MEPS and LC-MS/MS allowing automated sample work-up and analysis. *J. Chromatogr. B Analyt. Technol. Biomed. Life Sci.,* **2012**, *897*, 42-49. [http://dx.doi.org/10.1016/j.jchromb.2012.04.006] [PMID: 22542514]

[31] El-Beqqali, A.; Kussak, A.; Abdel-Rehim, M. Fast and sensitive environmental analysis utilizing microextraction in packed syringe online with gas chromatography-mass spectrometry: determination of polycyclic aromatic hydrocarbons in water. *J. Chromatogr. A,* **2006**, *1114*(2), 234-238. [http://dx.doi.org/10.1016/j.chroma.2006.02.024] [PMID: 16515791]

[32] Said, R.; Kamel, M.; El-Beqqali, A.; Abdel-Rehim, M. Microextraction by packed sorbent for LC-MS/MS determination of drugs in whole blood samples. *Bioanalysis,* **2010**, *2*(2), 197-205. [http://dx.doi.org/10.4155/bio.09.187] [PMID: 21083303]

[33] García Pinto, C.; Pérez Antón, A.; Pérez Pavón, J.L.; Moreno Cordero, B. Coupling of microextraction by packed sorbents with gas chromatography with ionic liquid stationary phases for the determination of haloanisoles in wines. *J. Chromatogr. A,* **2012**, *1260*, 200-205. [http://dx.doi.org/10.1016/j.chroma.2012.08.078] [PMID: 22981462]

[34] Mercolini, L.; Mandrioli, R.; Raggi, M.A. Content of melatonin and other antioxidants in grape-related foodstuffs: measurement using a MEPS-HPLC-F method. *J. Pineal Res.,* **2012**, *53*(1), 21-28. [http://dx.doi.org/10.1111/j.1600-079X.2011.00967.x] [PMID: 22017461]

[35] Adam, M.; Pavlíková, P.; Cížková, A.; Bajerová, P.; Ventura, K. Microextraction by packed sorbent (MEPS) as a suitable selective method for l-ascorbic acid determination in beverages. *Food Chem.*, **2012**, *135*(3), 1613-1618.
[http://dx.doi.org/10.1016/j.foodchem.2012.06.039] [PMID: 22953901]

[36] Salami, F.H.; Queiroz, M.E. Microextraction in packed sorbent for the determination of pesticides in honey samples by gas chromatography coupled to mass spectrometry. *J. Chromatogr. Sci.*, **2013**, *51*(10), 899-904.
[http://dx.doi.org/10.1093/chromsci/bms187] [PMID: 23192738]

[37] González-Mariño, I.; Benito, Quintanaa , J.; Rodrígueza, I.; Schraderb, S.; Moeder, M. Fully automated determination of parabens, triclosan and methyl triclosan in wastewater by microextraction by packed sorbents and gaschromatography–mass spectrometry. *Anal. Chim. Acta,* **2011**, *684*(1-2), 59-66.
[http://dx.doi.org/10.1016/j.aca.2010.10.049] [PMID: 21167985]

[38] Altun, Z.; Abdel-Rehim, M.; Blomberg, L.G. New trends in sample preparation: on-line microextraction in packed syringe (MEPS) for LC and GC applications Part III: Determination and validation of local anaesthetics in human plasma samples using a cation-exchange sorbent, and MEPS-LC-MS-MS. *J. Chromatogr. B Analyt. Technol. Biomed. Life Sci.,* **2004**, *813*(1-2), 129-135.
[http://dx.doi.org/10.1016/j.jchromb.2004.09.020] [PMID: 15556525]

[39] Goncalves, J.; Silva, C.J.; Castilho, P.C.; Camara, J.S. An attractive, sensitive and high-throughput strategy based on microextraction by packed sorbent followed by UHPLC-PDA analysis for quantification of hydroxybenzoic and hydroxycinnamic acids in wines. *Microchem. J.,* **2013**, *106*, 129-138.
[http://dx.doi.org/10.1016/j.microc.2012.05.037]

CHAPTER 8

Miniaturized Dispersive Solid-phase Extraction

Guillermo Lasarte-Aragonés, Rafael Lucena, Soledad Cárdenas* and **Miguel Valcárcel**

Department of Analytical Chemistry, Institute of Fine Chemistry and Nanochemistry, Marie Curie Building (Annex), Campus de Rabanales, University of Córdoba, Córdoba, Spain

Abstract: The usefulness of dispersive extraction techniques leans on their ability to maximize the interaction between the sample and the extractant phase, thus increasing the extraction efficiency. As far as dispersive solid phase extraction is concerned, it was initially developed to increase the selectivity of the analytical process because the solid was added to retain the potential interferents from the sample matrix. In spite of its efficient sample clean-up, the sensitivity is its Achilles' heel as no preconcentration is usually achieved. Recently, the use of few milligrams of sorbent which is dispersed in a liquid sample for analytes isolation has raised a new miniaturized extraction technique, the so-called dispersive micro solid-phase extraction. This alternative is mainly focused on sensitivity enhancement. The chapter begins with a short contextualization of this extraction technique, followed by a brief description of the first approach in this context, *viz*, dispersive solid phase extraction. Next, the main contributions in the field of dispersive micro solid-phase extraction in this context will be described on the basis of the nature of the solid used. Also, the combination with dispersive liquid phase microextraction and the expected evolution of this miniaturized extraction technique are included.

Keywords: Auxiliary energies, Carbonaceous solids, Conventional sorbents, Dispersive extraction, Dispersive liquid-liquid microextraction, Dispersive solid phase microextraction, Dispersive sorbent, Enrichment factors, Green chemistry, Miniaturization, Molecularly imprinted polymers, Nanostructured sorbents, Polymeric sorbents, Silica-based sorbents, Solid phase microextraction.

* **Corresponding author Soledad Cárdenas:** Department of Analytical Chemistry, Institute of Fine Chemistry and Nanochemistry, Marie Curie Building (Annex), Campus de Rabanales, University of Córdoba, Córdoba, Spain; Tel/Fax: +34 957 218616; E-mail:qa1caarm@uco.es

M. Valcárcel, S. Cárdenas, R. Lucena (Eds.)

8.1. INTRODUCTION

The evolution of Analytical Chemistry has been lead by three trends: automation, simplification and miniaturization. While the former is clearly consolidated in any analytical laboratory, the other two are still being developed. The improvement of the sample pre-treatment operations is the target of these tendencies taking into consideration the influence they have in the basic (sensitivity, selectivity and precision) and productivity-related (cost, time, personal and environmental risks) analytical properties. Therefore, many research efforts are focused on the simplification and miniaturization of the sample pretreatment, as it is practically an unavoidable step of the analytical process.

The determination of a family of compounds in a given sample usually involves its isolation from the matrix (to increase the selectivity) and preconcentration (for sensitivity enhancement) prior to the instrumental measurement. According to the above commented tendency, the introduction of novel microextraction techniques is highly recommended in order to reduce the amount of sample and organic solvents as well as the time, cost and manipulation needed. This tendency is in good agreement with the principles of Green Chemistry often described in the literature as the three Rs: replace, reduce and recycle [1]. Solid phase microextraction and liquid phase microextraction can be considered as consolidated miniaturized sample treatment techniques, given their full acceptance in routine laboratories, particularly when automated or even fully integrated analytical instrumentation is used.

The main requirement of any microextraction technique is that the extractant media used should be highly efficient in analyte extraction taking into account the reduced amount used. In this context, dispersive-based procedures have gained relevance in the last years as they enhance the contact surface area between the sample and the extractant and thus the kinetics of the overall extraction procedure while reducing the amount/volume of the solid/liquid extractant phase needed [2].

8.2. DISPERSIVE SOLID PHASE EXTRACTION

Dispersive solid-phase extraction (DSPE) was first proposed by Anastassiades *et al.* in 2003 as a powerful tool to enhance the selectivity of the measurement. It is commonly known as QuEChERS, acronym of quick, easy, cheap, effective, rugged and safe, its main analytical features [3]. DSPE requires the presence of an organic solvent, for analyte partitioning; salts, to favour analyte transference from the sample to the organic media and a dispersive sorbent to retain the coextracts while maintaining the analytes dissolved in the organic phase. Each of the three elements plays a different role in the extraction and they must be carefully chosen to improve the efficiency of the process. They are briefly commented on. The organic solvent must be selected taking into account the analyte polarity, as it is responsible for the extraction of the target compounds. In this regards, acetonitrile is by far the most used as it permits the extraction of both polar and non-polar compounds. Its compatibility with liquid chromatography is also advantageous in this context. Magnesium sulphate and sodium chloride are used during the extraction process (i) as salting out agents and (ii) to favour phases' separation. The amounts added are determined by the analyte hydrophobicity, as a rule. The dispersive sorbent added depends on the type of interferences to be retained. Primary secondary amine, graphitized carbon black, RP-C_{18} and alumina are commonly used. In addition to the sorbent nature, the amount used is also critical, as it has to retain the interferents (selectivity improvement) while minimizing analyte adsorption (sensitivity of the measurement). The amount added is usually in the low milligram level, *viz.* 10 mg. After a centrifugation step, the organic phase can be directly used or concentrated, usually by an evaporation/redissolution step, prior to the instrumental measurement step, commonly by a chromatographic technique.

This general procedure was mainly designed to analyze horticultural samples on account of the complexity of the matrices and the low concentration of the target compounds (mainly pesticides). The presence of co-extracted compounds in the chromatogram makes difficult the unequivocal identification of the pesticides, even when detectors with high discrimination capabilities, such as mass spectrometry, are employed.

The former procedure was modified by Lehotay *et al.* in 2005 to improve the recovery of certain pesticides [4]. Recently, Anastassiades *et al.* have reported a second evolution of their initial proposal to expand the application range of the DSPE to samples with different water and acid contents [5]. DSPE has also been used to enhance the selectivity of the analysis of environmental samples (soil and compost) to determine pesticides.

The excellent performance of DSPE has also been exploited to improve sensitivity. The miniaturized approach of this dispersive extraction technique is the main focus of this chapter.

8.3. DISPERSIVE MICRO SOLID PHASE EXTRACTION

The main goal of dispersive micro solid-phase extraction focused on the isolation and preconcentration of the target analytes is to achieve the complete disaggregation of the sorbent inside the sample matrix. This fact compensates the low amount of sorbent used, usually in the low milligram level. This step is commonly assisted by an external energy (vortex, ultrasounds or microwave) or by using the appropriate solvent or chemical reagent. Depending on the nature of the solid selected, oxidation or functionalization can also be used to increase its solubility and/or interaction with the target compounds. Additionally, reversible interaction between the sorbent and the analytes is mandatory in order to elute them properly (either thermally or chemically) prior to their instrumental determination, often by a chromatographic technique.

The advances in this microextraction modality are lead by the incorporation of micro and nanomaterials as extraction phases. This fact leans on their excellent sorbent properties, which derived from their reduced size and high surface area, which maximized the interaction with the target analytes. Also, the variety of sorption mechanisms, such as van der Waals forces, electrostatic interactions, hydrogen bonds or dispersion forces, can be cited as a clear advantage. In this section, the most relevant applications of dispersive micro-solid phase extraction based on the use of micro/nanoparticles are presented, classified by the nature of the solid material (Fig. **1**). Magnetic nanoparticles, which are also used in this context, are not included as chapter 10 is fully devoted to their role in analytical microextraction techniques.

8.3.1. Polymeric Solids

Alcudia *et al.* were the first in demonstrating the usefulness of dispersive microsolid phase extraction for in sorbent surface attenuated total reflection infrared (ATR-IR) detection of sorbic and benzoic acids in fruit juices [6]. The dispersive extraction allows the selective isolation of the analytes from the sample, allowing their preconcentration prior to the IR detection. The extraction efficiency surpasses that provided by its counterparts as (i) the interaction sorbent-analyte is favored with the stirring step and (ii) the detection is directly carried out on the sorbent, so the dilution associated to the chemical elution is avoided.

In this application, the sorbent selection was critical, as it must fulfill two requirements. The first one is the favourable interaction with the analytes in order to isolate them from the sample. The second one is its compatibility with the IR detection of the analyte *i.e.* it should not absorb all the photons coming from the IR source. Two typical reversed-phase sorbents, C_{18} (silica-based) and LiChrolut-EN (polymeric) were evaluated, the latter being the most compatible with the IR detection. The sorbent amount used in the dispersive μ-SPE was also critical as the surface monitored depends on the limited penetration depth of the evanescent wave. In this case, an amount of 2 mg of LiChrolut-EN resulted in the best results. Finally, a comparison of the performance of SPE and dispersive μ-SPE demonstrated the better sensitivity of the dispersive approach by a factor of 10.1 for sorbic acid and 4.1 for benzoic acid. In 2011, Garg *et al.* applied the proposed dispersive μ-SPE-ATR-FTIR method to determine methylphosphonate in waters. In this case, the sorbent used was OASIS-HLB, a polystyrene-co-divinylbenze-e-co-polyvinylpyrrolidine-based material [7].

The homogeneous dispersion of the sorbent inside the sample matrix is usually accomplished by vortexing the mixture for a given time. Sometimes, the addition of few microliters of slurry of the sorbent in an organic solvent is selected to ensure the maximum dispersibility of the particles in the sample. Both approaches present some limitations regarding the need for an external apparatus (vortex) or the potential interference of the organic media selected for the initial dispersion of the solid, mainly increasing the solubility of the analytes in the sample, thus reducing the extraction efficiency. Our research group has proposed a green alternative for the efficient dispersion of the sorbent in the sample matrix [8]. The so-called effervescence-assisted μ-SPE requires the preparation of a tablet containing the effervescence precursors (sodium carbonate as carbon dioxide source and a phosphate salt as proton donor) as well as the sorbent, in this case, OASIS HLB. The tablet is placed inside of a glass syringe and, when the 10 mL of sample is aspirated, the effervescence process begins. The release of CO_2 yields a homogeneous dispersion of the sorbent. The composition of the tablet in both, nature and amount of precursors is critical as it determines the extraction time (*viz.* the time required for complete tablet dissolution). The novel microextraction approach was evaluated using the determination of nitroaromatic compounds in waters as model analytical problem. The effervescence assisted dispersive μ-SPE was ca. 3 times more sensitive as regards conventional SPE. Also, it was more efficient than vortex-assisted dispersion of the OASIS HLB sorbent.

8.3.2. Silica-Based Solids

The conventional, silica-based RP-C^{18}, has also been proposed as sorbent in dispersive μ-SPE [9]. In this specific application, the solid is previously dispersed in acetone to a final concentration of 10 g/L. This fact allows the appropriate spreading of the particles inside the sample, a cloud of fine particles being immediately observed. Both, the solvent and the final concentration are key factors. The solvent should fulfill the following requirements: (i) miscibility with the sample; (ii) to yield a stable suspension of the sorbent; and (iii) no interference during the extraction process. On the other hand, the concentration used must be high enough to reduce the amount of organic phase added to the sample while remaining stable and homogeneous with the time. The operational procedure is rather simple: 500 μL of the RP-C_{18} suspension in acetone were sprayed in 12.5 mL of sample, previously added to a dedicated stainless steel interface fitted with a cotton bead. The liquid phase is eliminated under vacuum, being the solid retained in the interface. Then, the interface is coupled to the injection port of a gas chromatograph and the analytes are thermally eluted by means of a focused heated air stream (300ºC, 2 min). The continuous flow of an helium stream carried the analytes to the chromatographic column and further to the mass spectrometer for unequivocal identification and quantitation. The thermal desorption of the analytes increased the method sensitivity as the dilution inherent to the chemical elution is avoided. In fact, the limits of detection (LODs) for the selected polycyclic aromatic hydrocarbons (PAHs) were in the range 0.1-0.2 μg/L with relative standard deviations lower than 8.4%.

Tsai *et al.* compared the performance of silica and polymeric-based sorbents in the dispersive μ-SPE format using tetracyclines (TCs) as target analytes [10]. The comparison includes the dispersion of the sorbents in organic and aqueous media. Up to 16 sorbents with different functional groups: primary amine, secondary amine, carbonyl or ion exchange groups were evaluated. In both cases, the extraction recoveries of TCs were higher in acetonitrile than in deionized water, with better results of silica-based sorbents in both media.

Among the sorbents studied, primary-secondary amine, NH_2, DSC-WCX (Discovery® **Weak Cation Exchange)** and SU-LC-WCX (Superclean® **Carboxylic acid Weak Cation Exchange)** provided quantitative extraction of TCs from acetonitrile extracts, being PSA selected as optimum for the dispersive microextraction configuration. The method was analytically characterized, being the limits of detection between 0.7 ng/mL and 3.5 ng/mL with recovery values in the range 96-106.6% and the precision (expressed as relative standard deviation) lower than 4.4%. In addition, no interferences from other matrix components were detected in the chromatographic analysis, which demonstrates the usefulness of the proposed dispersive μ-SPE for the given analytical problem.

8.3.3. Metallic Nanoparticles

As it has been mentioned before, the efficiency of the miniaturized extraction techniques relies on the surface area of the sorbent material available for analytes interaction. This requirement is not fulfilled by conventional sorbents but it is an outstanding property of nanostructured materials.

In the dispersive context, the first nanoparticles used were gold nanoparticles (AuNPs) although the authors named the extraction as solid phase nanoextraction (SPNE). In this case, commercial AuNPs colloids were used for the extraction of PAHs [11]. A very interesting study correlates the extraction efficiency with the AuNPs diameter. Authors observed that, although the number of molecules extracted per nanoparticle increased with the particle diameter and the total surface area of Au, the inverse tendency was observed if the extracted molecules per gram of Au and nanometer squared of surface area of Au are considered.

Therefore, it can be concluded that the smaller particles were more efficient for the extraction and the 20 nm AuNPs were selected as optimum. This fact demonstrated the relevance of the nanometric size in miniaturized extraction procedures. The analytical performance was excellent using only 500 µL of water sample and HPLC with fluorescence detection for the determination of 16 PAHs [11] or benzo[a]pyrene by laser excited time resolved Shpol'skii spectrometry [12].

8.3.3.1. Oxides and Oxides-Hybrid Nanoparticles

Titanium nanotubes (TNTs) have been used in different areas thanks to their inertness, stability, non-toxicity, low cost and corrosion resistance [13, 14]. The structure of TNTs consists of TiO_2 sheets rolled and separated by H^+ ions [15]. They are open-ended structures, which confers to the TNTs with large internal, external and interlayer interactions surfaces/spaces. They have been used as-grown bare-TNTs, functionalized for selectivity increase or combined with metallic nanoparticles [16].

In 2014, García-Valverde *et al.* synthesized TNTs which were further coated with a carbon layer following an hydrothermal process. The so-called carbon coated TNTs (c-TNTs) were evaluated using the determination of naproxen and ketoprofen in biological samples [17]. The c-TNTs were characterized by TEM (morphology) FTIR (monitoring of the superficial carbonization process) thermogravimetry and elemental analysis (carbon content) and X-ray powder diffraction (to evaluate the crystallinity of the material). In addition to their simple, low cost synthesis procedure, another favourable feature of c-TNTs is their better dispersibility in aqueous media in comparison with that of carbonaceous nanoparticles. The dispersive µ-SPE was carried out in a dedicated unit that: (i) is adapted to low sample volumes; (ii) integrates extraction and elution of analytes; and (iii) allows dispersion of the sorbent in the sample. As it can be seen in Fig. (**2**), 10 mg of the c-TNTs are placed in a pipette tip section and confined by a cotton bead. Then, a 1 mL plastic syringe is connected to the tip for sorbent conditioning, sample loading, washing and elution steps. The dispersion of the solid takes place in the syringe body. By using this unit, limited-volume biological samples such as saliva can be processed. Moreover, the nanomaterial is easily recovered after extraction. The comparison of the performance of c-TNTs and multiwalled carbon nanotubes (MWCNTs) under the optimized conditions revealed that the c-TNTs exhibited a 2-fold extraction recovery thanks to their better dispersibility in aqueous and organic media, compared to that obtained with the MWCNTs.

Khodadoust *et al.* have proposed the use of NiZnS nanoparticles loaded on activated carbon (NiZnS-AC) as sorbent for the dispersive μ-SPE of carbamates pesticides from waters [18]. In this case, the complete dispersion of the nanoparticles (15.5 mg) requires the use of a vortex-assisted platform (5.5 min) followed by ultrasonication (5.5 min). The extraction is carried out on an end part of a 3 mL glass pipette fitted with a filter to isolate the solid from the liquid phase under vacuum. Elution is carried out with acetone, a solvent changeover step to methanol being required prior to liquid chromatographic analysis. The hybrid nanoparticles were characterized in terms of surface area, pore volume and size and nanoparticle size. As far as the analytical features are concerned, the miniaturized extraction configuration allows the quantification of bendiocarb and promecarb in tap, river and mineral waters at the trace level.

8.3.4. Carbon-Based Material

The discovery of the different allotropic nanostructured forms of carbon has been a cornerstone in many scientific areas, including analytical nanoscience and nanotechnology and analytical chemistry. The outstanding properties of carbon nanomaterials have been extensively pointed out in the literature [19]. In the specific field of analytical chemistry, fullerenes, carbon nanotubes and nowadays graphene, have contributed to the development of more competitive (bio)chemical measurement processes. Additionally, the miniaturization of the whole procedures can be highlighted.

The incorporation of carbon nanostructures to miniaturized extraction devices has resulted in the development of more efficient microextraction units, thanks to the excellent sorbent properties of these nanomaterials. However, it is well-known that this property is inherent to the nanometric dimensions and, therefore, the aggregation of the nanoparticles must be avoided. This aggregation results from the insolubility of carbon nanomaterials in aqueous and organic media, being especially relevant for their reference allotropic from, carbon nanotubes. This fact limits their use in conventional extraction devices such as SPE cartridges or in miniaturized units which require the flow of sample through them as the presence of nanoparticle bundles (i) reduces the efficiency of the process; and (ii) results in system overpressure. No doubt, the use of miniaturized extraction devices, which immobilized the carbon nanoparticles on an inert or active surface, is the simplest alternative to exploit their sorbent properties in the microextraction context. Excellent reviews have been published on the topic [20, 21].

Taking into account that the bundle formation is usually exhibited by as-grown carbon nanotubes, surface modification is a core procedure to prevent carbon nanoparticles from aggregation as well as to provide additional selectivity when they are used as extracting media. In the case of carbon nanoparticles, oxidation (to generate hydroxyl, carboxyl or carbonyl groups) and more scarcely functionalization with different active groups have been proposed to facilitate the use of carbon nanoparticles in dispersive μ-SPE by increasing their solubility in aqueous media.

The first application described in the literature in this context used MWCNTs as dispersed sorbent in a conventional QuEChERS multiresidue analysis [22]. The proposed method follows the steps defined in DSPE but substituting the conventional sorbent (PSA, GBC or RP-C_{18}) by 10 mg of MWCNTs. The use of nanostructured material resulted in cleaner extracts compared to PSA even at the lower concentrations, with comparable recoveries. However, the better clean-up reduces the limits of detection up to 1 μg/g level. On the negative side, the non-reusability of the material and its relatively high cost should be noted.

Single walled carbon nanohorns (SWNHs) have been studied as sorbent in dispersive μ-SPE. They are smaller in size but tend to form stable aggregates of dahlia shape with an average diameter of 60-80 nm. These aggregates provide additional interaction sites, which increase the adsorption capacity of the SWNHs. To increase the aggregates solubility in water and polar organic solvents, they can be oxidized (o-SWNHs) by microwave irradiation [23]. For the dispersive μ-SPE, a dispersion of o-SWNHs at a concentration of 0.2 mg/mL in Milli-Q water was firstly prepared. It was sonicated for 60 min and centrifugated to remove potential non-functionalized material. The dispersion was tested using PAHs as model analytes and environmental waters as sample [24]. For extraction, 1 mL of the dispersion was added to 10 mL of water sample, being the vial stirred for 2 min at 1600 rpm, an homogeneous dispersion being immediately formed. The o-SWNHs enriched with the PAHs were retained on a 0.45 μm disposable nylon filter and the analytes were further eluted with 100 μL of hexane. The identification and quantitation of the analytes was carried out by gas chromatography-mass spectrometry. The performance of o-SWNHs was compared with that obtained with carbon nanocones and oxidized carbon nanotubes under the optimized experimental conditions. As a general conclusion, it can be said that the conical nanoparticles provided the highest recoveries thanks to their better dispersibility in water. Moreover, it was also observed the increased efficiency of the dispersive μ-SPE using carbon nanocones with increasing number of aromatic rings in the PAH molecule thanks to the larger dimensions of this nanostructure as regards SWNHs, which permits the analyte interaction in the inner part of the nanoparticle. This fact makes the elution difficult, resulting in poor reproducibility.

A green alternative to disperse carbon nanotubes in water is the previously described effervescence assisted μ-SPE. The potential of the CO_2 to avoid MWCNTs aggregation was evaluated by Lasarte-Aragonés *et al.* using triazines as target analytes [25]. Neither oxidation/functionalization nor external energy source, such as ultrasounds, microwave or vortex is needed for sorbent dispersion. The effervescent tablet is prepared by mixing the carbon nanotubes with the effervescence precursors in a glass mortar and 250 mg of the homogeneous powder is compressed in a manual hydraulic press. Each tablet contains ca. 7.5 mg of carbon nanotubes. The dispersion of carbon nanotubes was evaluated under four experimental conditions, in which the sorbent and effervescence precursors were added in different formats (simultaneously, sequentially, in powder and in a tablet). As it can be seen in Fig. (**3**), the best results were obtained when the ingredients were added in the form of effervescence tablet. Under these conditions, absolute recoveries in the range from 48% to 75% were achieved, which are very high for a microextraction technique.

Carbon nanotubes have been extensively used for the determination of organic, aromatic compounds, as they can be easily retained by means of π-π interactions. However, inorganic species (*e.g.* metal ions) can also be preconcentrated in the form of chelate. The final hydrophobic complex is adsorbed on the nanoparticle surface by means of Van der Waals forces and hydrophobic interactions. Both, multiwalled carbon nanotubes [26] and graphene [27] have been used in the dispersive format to determine metal ions at the trace and ultratrace levels by energy dispersive X-ray fluorescence spectrometry (EDXRF). In both applications, ammonium pyrrolidine dithiocarbamate (APDC) was selected as chelating reagent. Once retained, the nanoparticles enriched with the analytes were collected onto a filter and directly measured by EDXRF. Nickel, cobalt, copper and lead were extracted using this methodology. In view of the results, graphene is a better sorbent as lower limits of detection were obtained using even less amount of sorbent (0.4 mg of graphene *versus* 1 mg of MWCNTs) although it requires the presence of Triton X-100 for the extraction.

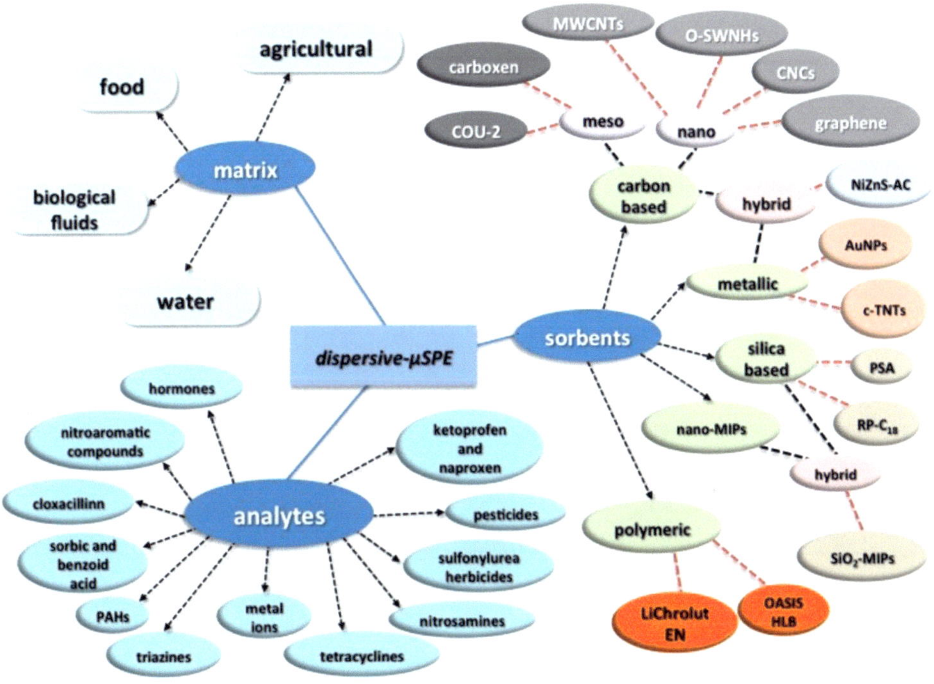

Fig. (1). Main application fields of dispersive μSPE and the most frequently used sorbents included in this chapter.

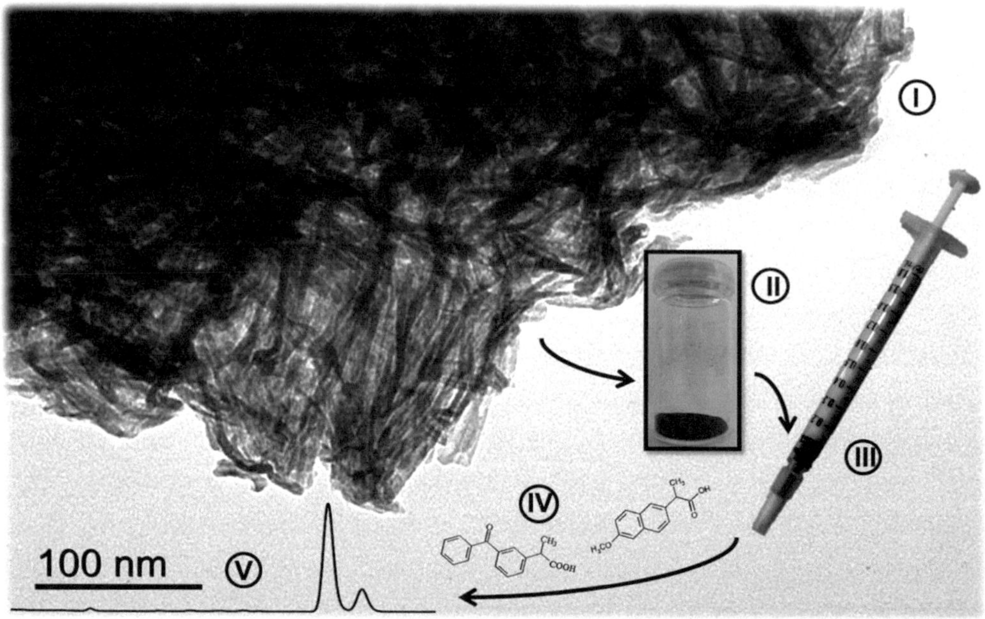

Fig. (2). Schematic representation of the dispersive-μSPE procedure using c-TNTs. (**I**) TEM micrograph of the c-TNTs; (**II**) solid obtained after treatment; (**III**), proposed extraction unit; (**IV**) target analytes; and (**V**) HPLC-UV chromatogram.

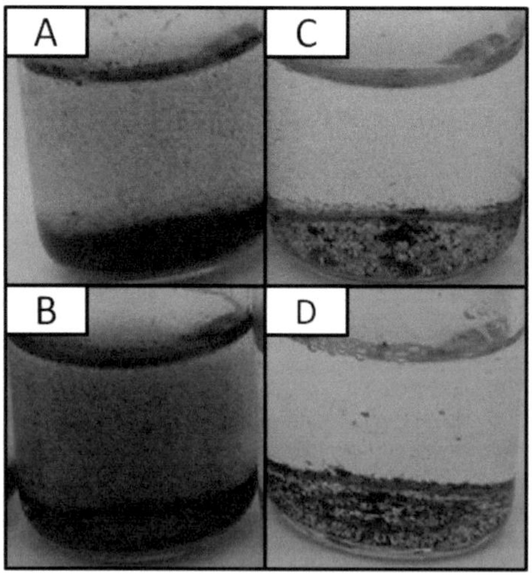

Fig. (3). Dispersions obtained after the addition of the effervescence precursors and the carbon nanotubes in powder (**A**), both in a tablet (**B**), carbon nanotubes without the effervescence precursors (**C**), and the tablet containing the effervescence precursors and the carbon nanotubes added after tablet dissolution (**D**).

Carbon molecular sieves such as Carboxen have also been proposed as sorbent in dispersive μ-SPE. N-nitrosamines were first extracted from meat by microwave-assisted extraction using 30 mL of 0.0025 M sodium hydroxide and further diluted up to 50 mL with deionised water prior to the dispersive procedure. In the second step, 100 mg of Carboxen were added and agitated in a vortex for 30 min at the maximum speed. After extraction, the solid with the analytes was separated in a filter and then eluted with 200 μL of dichloromethane and subjected to gas chromatographic-mass spectrometric analysis under chemical ionization mode. Two carbon molecular sieves, Carboxen 1000 and 1003 with different surface area and pore diameter were evaluated, being the former selected for the extraction. The authors conclude that the analytes were determined at the trace level with good precision, wide liner range and detection limits at the low ng/g, being a good alternative for the determination of nitrosamines in complex food matrices [28].

Finally, mesoporous carbon COU-2 has also been used for the determination of cloxacillin in waters [29]. The mesoporous materials are characterized by a high surface area, large pore volume, tunable mesoporous channels, well defined pore-size distribution and changeable surface properties [30]. COU-2 material presents a wormlike mesostructure with highly ordered mesopores [31]. For this application, 100 mg of COU-2 were added to 50 mL of sample and only agitation for 1 min was required for analyte extraction. Aliquots of 300 μL of methanol were used as eluent. The eluate was evaporated to dryness and reconstituted in water for HPLC analysis. The authors compared the performance of the mesoporous material with that of activated carbon. No peaks were obtained in this case owing to the strong interaction analyte-sorbent. The high surface area and the mesopore volume of COU-2 facilitates the diffusion of the analyte from the solid to the eluent during desorption. The analytical characteristics of the dispersive μ-SPE were comparable to those of conventional SPE by using less amount of sorbent.

8.3.5. Molecularly Imprinted Polymers

The use of selective sorbents is increasing in Analytical Chemistry, thanks to the simplification of the whole (bio)chemical measurement process. Molecularly imprinted polymers (MIPs) are built-in functionalized material, specifically designed for the isolation of a given molecule (template), around which the sorbent is fabricated (printed). The synthesized solid is highly selective as retention depends on shape, size and functionality criteria. Nanoscience and nanotechnology have played a key role in this particular area, as the extraction efficiency of MIPs is increased when they are prepared at the nanoscale level, *viz.* nanoMIPs, thanks to the morphology and accessibility of the recognition sites. Several synthetic processes have been proposed for the fabrication of nanoMIPs. Among them, the use of composite beads is highly popular as it permits to control the particle size as the polymeric material (silica or latex) is used as core shell, being the recognition element coated on its surface as an homogeneous layer. Following this method, molecularly imprinted polymer layer-coated silica nanoparticles have been designed towards sulfonylurea herbicides [32] in soil and crop samples, 17β-estradiol [33] and oestrogens in milk samples [34]. The imprinted nanoparticles exhibited a high selectivity, large adsorption capacity and fast kinetics. The sorption capacity of the SiO_2-MIP is higher than that of the non-imprinted polymers.

Molecularly imprinted microbeads have also been used for the determination of sulfamethazine by capillary electrophoresis. The microparticles were prepared *via* precipitation polymerization using the analyte as template, methacrylic acid as functional monomer and ethylene glycol dimethylacrylate as cross-linking monomer. The microparticles were highly dispersible in aqueous solution, so, they are very convenient for dispersive μ-SPE. The extraction was carried out at room temperature, stirring the sample plus the MIPs-microbeads for 5 min. The sorbent was separated by centrifugation and the analyte was eluted by 300 μL of methanol:acetic acid [35]. The dispersive method is highly selective with relative recoveries between 89-110% in different milk samples. The sensitivity was increased ca. 300 times compared to direct CE-UV analysis.

8.4. COMBINED DISPERSIVE MICROEXTRACTION TECHNIQUES

The analysis of complex samples usually requires extensive sample treatments. This is unavoidable if the analytes are presented at the trace levels when, in addition to sample clean-up, high preconcentration factors are mandatory. To fulfill these requirements, sequential extractions are usually implemented.

Taking into account the high efficiency of dispersive techniques, they can be used in the same analytical process combining both modalities, liquid and solid phase microextraction. In this way, the limitations of both techniques are overcome. In general, the dispersive solid phase extraction is used as clean up while the dispersive liquid phase microextraction is used for analyte preconcentration. The combined used of both dispersive methodologies at the microscale level is very scarce, however.

Zhou *et al.* have proposed a one step sample preparation technique which combines solid and liquid dispersive microextraction for the determination of organochlorine pesticides (OCPs) in tobacco [36]. The so-called hybrid field assisted solid-liquid-solid dispersive extraction (HF-SLSDE) consists of three phases: two solids (sample matrix and dispersive sorbent) and one liquid (extraction solvent). The clean-up is carried out on the sorbent while the analytes are isolated and preconcentrated in the solvent. The term hybrid field refers to the auxiliary energies used during the extraction: ultrasounds (US), microwave (MW) and stirring. Their application is aimed at accelerating the extraction process. An infrared sensor controls the temperature generated in the extraction vessel.

Several dispersive sorbents, namely Florisil, neutral alumina, underivatized silica, diatomaceous earth and activated carbon were considered and the clean-up ability together with the analytes adsorption were jointly evaluated. As expected, silica and diatomaceous earth showed a low efficiency in interference removal while the adsorption on activated carbon was non-selective and low extraction yields of the analytes were obtained. Therefore, Florisil was the best dispersive sorbent in terms of clean-up ability and OCPs extraction yields. As far as the solvent is concerned, hexane, ethyl acetate, acetone and hexane:ethyl acetate (9:1, v:v) mixture were tested. The most polar solvents were able to elute polar impurities from Florisil, which results in dirty chromatograms. Therefore, hexane was selected as optimum.

The authors also compared the effect of the hybrid field (MW-US) with that of using the auxiliary energies separately. The lowest recoveries and the poorest precision were obtained when ultrasound irradiation was employed. The microwave assisted SLSDE provided intermediate results. That means that the influence of the temperature owing to microwave energy is more important than the high pressure generated by cavitation. However, the hybrid field (MW-US) is more powerful than the single field.

Another interesting issue studied by the authors is the microwave absorptive property of Florisil. The heating rate of the hexane with Florisil was ca. 1.8 times higher than that in absence of the sorbent. That means that it is capable of absorbing the radiation and diffusing the heat to the solvent. This result in a fast temperature rise, thus improves the extraction speed and efficiency. The analytical figures of merit obtained are good enough for the given analytical problem.

In the green chemistry context, the effervescence can also be used to disperse a few microliters of an organic solvent, which is recovered by the aid of magnetic nanoparticles after the extraction [37]. In this case, the combination of dispersive techniques results in a reproducible collection on the liquid phase avoiding the need for centrifugation. Moreover, the efficient dispersion of the extractant by means of the CO_2 generated in the sample matrix does not require additional organic phase, which usually interferes in the extraction equilibrium by increasing the analyte solubility in the sample. The bare Fe_3O_4 magnetic nanoparticles can be easily synthesized in the laboratory. The extraction mixture is prepared as follows: 2.5 mL of glacial acetic acid (H^+ donor), 200 μL of 1-octanol (extractant) and 100 mg of bare Fe_3O_4 magnetic nanoparticles were sonicated for 5 min to achieve efficient homogenization. Aliquots of 250 μL of this mixture were used in each extraction. The procedure is schematically shown in Fig. (**4**). For analytes isolation and preconcentration, the aqueous standard or the water sample was enriched with sodium carbonate (CO_2 source) and then, the extractant mixture was slowly added by means of a syringe, the needle being completely immersed in the sample. In this way, the extractant and the nanoparticles were homogeneously distributed inside the sample while the effervescence takes place (ca. 5 s). By using an external magnet, the 1-octanol coated magnetic nanoparticles with the extracted analytes were separated from the sample, the target compounds being eluted using 100 μL of methanol and the magnet for nanoparticles isolation. This procedure was more efficient than the sequential addition of the effervescence precursors and the liquid-solid extractant phase. Also, the substitution of the effervescence by a conventional solvent-aided dispersion using acetonitrile provided lower extraction recoveries. The enrichment factors achieved for the selected triazines were in the range 21-185 with the limits of detection low enough to determine these herbicides in environmental water samples.

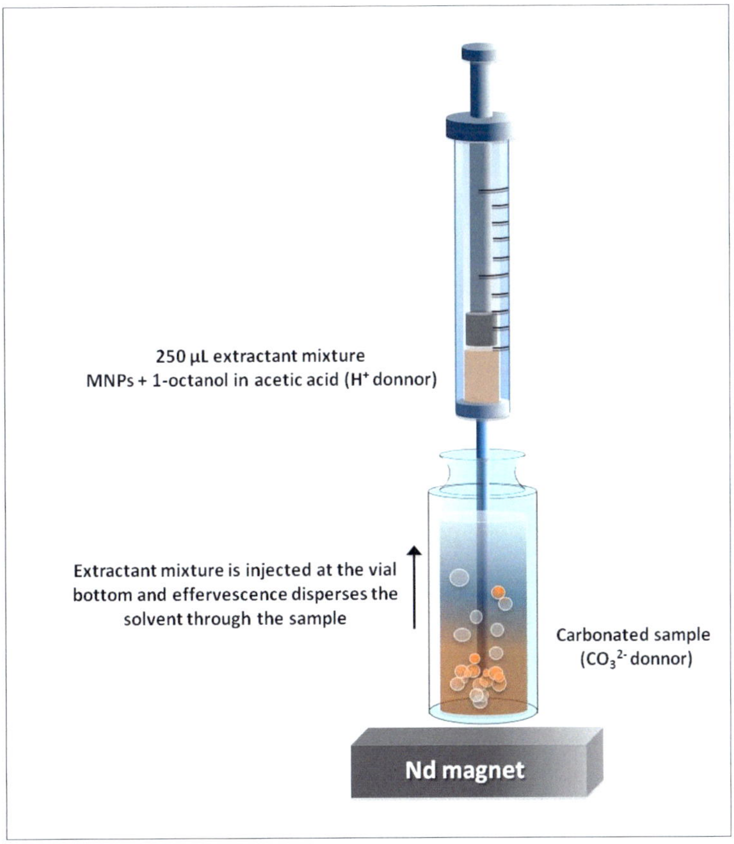

Fig. (4). Schematic representation of the procedure followed for the effervescence assisted liquid phase microextraction. (**A**) 250 μL MNPs + 1-octanol in acetic acid (H^+ donnor). (**B**) Carbonated sample (CO_3^{2-} donor). (**C**) Extractant mixture is injected at the vial bottom and effervescence disperses the solvent through the sample.

FUTURE TRENDS AND PERSPECTIVES

DSPE is a consolidated technique in food analysis taking into consideration its efficiency in selectivity improvement. The availability of commercial sets to implement this treatment technique testifies for its usefulness. The synthesis of novel sorbents has allowed the miniaturization of the technique and then it has moved the target to the analyte rather than the interferents. The resulted procedures are in coherence with the principles of Green Chemistry as far as the reduction of sample and reagents consumption is concerned. Dispersive μSPE is simpler than other microextraction techniques and also the apparatus required to be implemented are cheap and easy to use. The use of magnetic nanoparticles is also advantageous in this context as they provide the maximum simplicity for sorbent recovery in the analytical procedures.

Therefore, the applicability of dispersive micro solid phase extraction should be increased in the next years following two different research lines. First, the relevance of nanostructured solids in this context is unquestionable. Their outstanding sorbent properties yield high extraction recoveries using only a few milligrams of solid. The expected advances in Nanoscience and Nanotechnology will help to increase the applicability of dispersive micro solid phase extraction as the new nanostructured solids synthesized will be applied to novel analyte-sample binomials.

Next, the combination of dispersive micro solid phase extraction with other miniaturized extraction techniques is also a relevant research line. As it has been pointed out before, this strategy minimized or even overcomes the shortcomings of the individual dispersive technique. Up to now, the most successful combination uses a conventional DSPE with dispersive LPME. However, in order to miniaturize the whole analytical process, the sequential implementation of dispersive μSPE and LPME must be studied. Although the advantages of magnetic nanoparticles are unquestionable in this context, thanks to the simplification inherent, carbon-based, molecularly imprinted or polymeric nanostructured materials must be studied, as their potential is clearly demonstrated in the miniaturization and simplification of the analytical process [38].

CONFLICT OF INTEREST

The authors confirm that they have no conflict of interest to declare for this publication.

ACKNOWLEDGEMENTS

Financial support from the Spanish Ministry of Science and Innovation (grant CTQ2014-52939R) is gratefully acknowledged.

REFERENCES

[1] Spietelun, A.; Marcinkowski, Ł.; de la Guardia, M.; Namieśnik, J. Recent developments and future trends in solid phase microextraction techniques towards green analytical chemistry. *J. Chromatogr. A,* **2013**, *1321*, 1-13.
[http://dx.doi.org/10.1016/j.chroma.2013.10.030] [PMID: 24238710]

[2] Cruz-Vera, M.; Lucena, R.; Cárdenas, S.; Valcárcel, M. Sample treatments based on dispersive (micro)extraction. *Anal. Methods,* **2011**, *3*, 1719-1728.
[http://dx.doi.org/10.1039/c1ay05201b]

[3] Anastassiades, M.; Lehotay, S.J.; Stajnbaher, D.; Schenck, F.J. Fast and easy multiresidue method employing acetonitrile extraction/partitioning and "dispersive solid-phase extraction" for the determination of pesticide residues in produce. *J. AOAC Int.,* **2003**, *86*(2), 412-431.
[PMID: 12723926]

[4] Lehotay, S.J.; de Kok, A.; Hiemstra, M.; Van Bodegraven, P. Validation of a fast and easy method for the determination of residues from 229 pesticides in fruits and vegetables using gas and liquid chromatography and mass spectrometric detection. *J. AOAC Int.,* **2005**, *88*(2), 595-614.
[PMID: 15859089]

[5] Payá, P.; Anastassiades, M.; Mack, D.; Sigalova, I.; Tasdelen, B.; Oliva, J.; Barba, A. Analysis of pesticide residues using the Quick Easy Cheap Effective Rugged and Safe (QuEChERS) pesticide multiresidue method in combination with gas and liquid chromatography and tandem mass spectrometric detection. *Anal. Bioanal. Chem.,* **2007**, *389*(6), 1697-1714.
[http://dx.doi.org/10.1007/s00216-007-1610-7] [PMID: 17909760]

[6] Alcudia-León, M.C.; Lucena, R.; Cárdenas, S.; Valcárcel, M. Dispersive solid phase extraction for in-sorbent surface attenuated total reflection infrared detection. *Anal. Chem.,* **2009**, *81*(3), 1184-1190.
[http://dx.doi.org/10.1021/ac801507n] [PMID: 19117479]

[7] Garg, P.; Pardasani, D.; Mazumder, A.; Purohit, A.; Dubey, D.K. Dispersive solid-phase extraction for in-sorbent Fourier-transform infrared detection and identification of nerve agent simulants in analysis for verification of chemical weapon convention. *Anal. Bioanal. Chem.,* **2011**, *399*(2), 955-963.
[http://dx.doi.org/10.1007/s00216-010-4337-9] [PMID: 21058018]

[8] Lasarte-Aragonés, G.; Lucena, R.; Cárdenas, S.; Valcárcel, M. Effervescence-assisted dispersive micro-solid phase extraction. *J. Chromatogr. A,* **2011**, *1218*(51), 9128-9134.
[http://dx.doi.org/10.1016/j.chroma.2011.10.042] [PMID: 22071424]

[9] Galán-Cano, F.; Lucena, R.; Cárdenas, S.; Valcárcel, M. Direct coupling of dispersive micro-solid phase extraction and thermal desorption for sensitive gas chromatographic analysis. *Anal. Methods,* **2011**, *3*, 991-995.
[http://dx.doi.org/10.1039/c1ay05033h]

[10] Tsai, W.H.; Huang, T.C.; Huang, J.J.; Hsue, Y.H.; Chuang, H.Y. Dispersive solid-phase microextraction method for sample extraction in the analysis of four tetracyclines in water and milk samples by high-performance liquid chromatography with diode-array detection. *J. Chromatogr. A,* **2009**, *1216*(12), 2263-2269. [http://dx.doi.org/10.1016/j.chroma.2009.01.034] [PMID: 19185868]

[11] Wang, H.; Campiglia, A.D. Determination of polycyclic aromatic hydrocarbons in drinking water samples by solid-phase nanoextraction and high-performance liquid chromatography. *Anal. Chem.,* **2008**, *80*(21), 8202-8209. [http://dx.doi.org/10.1021/ac8014824] [PMID: 18834147]

[12] Wang, H.; Campiglia, A.D. Direct determination of benzo[a]pyrene in water samples by a gold nanoparticle-based solid phase extraction method and laser-excited time-resolved Shpol'skii spectrometry. *Talanta,* **2010**, *83*(1), 233-240. [http://dx.doi.org/10.1016/j.talanta.2010.09.013] [PMID: 21035669]

[13] Rendón-Ribera, A.; Toledo-Antonio, J.A. Cortés- Jacome, M.A.; Angeles-Chavez, C. Generation of highly reactive OH groups at the surface of TiO_2 nanotubes. *Catal. Today,* **2011**, *166*, 18-24. [http://dx.doi.org/10.1016/j.cattod.2010.03.045]

[14] Bavykin, D.; Walsh, F. *Titanate and titania nanotubes: synthesis, properties and Applications*; Royal Society Publishing: Cambridge, **2010**.

[15] Miyauchi, M.; Tokudome, H. *Inorganic and Metallic Nanotubular Materials*; Kijima, T., Ed.; Springer-Verlag Berlin: Heidelberg, **2010**, Vol. 117, pp. 45-57. [http://dx.doi.org/10.1007/978-3-642-03622-4_4]

[16] García-Valverde, M. T.; Lucena, R.; Cárdenas, S.; Valcárcel, M. Titanium dioxide nanotubes as sorbents in (micro)extraction techniques. *Trends Anal. Chem.,* 2014, 62, 37-45.

[17] García-Valverde, M.T.; Lucena, R.; Galán-Cano, F.; Cárdenas, S.; Valcárcel, M. Carbon coated titanium dioxide nanotubes: synthesis, characterization and potential application as sorbents in dispersive micro solid phase extraction. *J. Chromatogr. A,* **2014**, *1343*, 26-32. [http://dx.doi.org/10.1016/j.chroma.2014.03.062] [PMID: 24745843]

[18] Khodadoust, S.; Ghaedi, M.; Hadjmohammadi, M.R. Dispersive nano solid material-ultrasound assisted microextraction as a novel method for extraction and determination of bendiocarb and promecarb: response surface methodology. *Talanta,* **2013**, *116*, 637-646. [http://dx.doi.org/10.1016/j.talanta.2013.07.013] [PMID: 24148456]

[19] Valcárcel, M.; Cárdenas, S.; Simonet, B.M.; Moliner, Y.; Lucena, R. Carbon nanostructures as sorbent materials in analytical processes. *Trends Anal. Chem.,* **2008**, *27*, 34-43. [http://dx.doi.org/10.1016/j.trac.2007.10.012]

[20] Trojanowicz, M. Analytical applications of carbon nanotubes: a review. *Trends Anal. Chem.,* **2006**, *25*, 480-489. [http://dx.doi.org/10.1016/j.trac.2005.11.008]

[21] Socas-Rodríguez, B.; Herrera-Herrera, A.V.; Asensio-Ramos, M.; Hernández-Borges, J. Recent applications of carbon nanotube sorbents in analytical chemistry. *J. Chromatogr. A,* **2014**, *1357*, 110-146.

[http://dx.doi.org/10.1016/j.chroma.2014.05.035] [PMID: 24913369]

[22] Zhao, P.; Wang, L.; Zhou, L.; Zhang, F.; Kang, S.; Pan, C. Multi-walled carbon nanotubes as alternative reversed-dispersive solid phase extraction materials in pesticide multi-residue analysis with QuEChERS method. *J. Chromatogr. A,* **2012**, *1225*, 17-25. [http://dx.doi.org/10.1016/j.chroma.2011.12.070] [PMID: 22227363]

[23] Jiménez-Soto, J.M.; Cárdenas, S. Conical Carbon Nanoparticles in Analytical Chemistry. In: *Encyclopedia of Analytical Chemistry*; Meyers, R.A., Ed.; John Wiley: Chichester, **2012**. [http://dx.doi.org/10.1002/9780470027318.a9283]

[24] Jiménez-Soto, J.M.; Cárdenas, S.; Valcárcel, M. Evaluation of single-walled carbon nanohorns as sorbent in dispersive micro solid-phase extraction. *Anal. Chim. Acta,* **2012**, *714*, 76-81. [http://dx.doi.org/10.1016/j.aca.2011.11.055] [PMID: 22244139]

[25] Lasarte-Aragonés, G.; Lucena, R.; Cárdenas, S.; Valcárcel, M. Effervescence-assisted carbon nanotubes dispersion for the micro-solid-phase extraction of triazine herbicides from environmental waters. *Anal. Bioanal. Chem.,* **2013**, *405*(10), 3269-3277. [http://dx.doi.org/10.1007/s00216-013-6718-3] [PMID: 23354573]

[26] Skorek, R.; Zawisza, B.; Marguí, E.; Queralt, I.; Sitko, R. Dispersive micro solid-phase extraction using multiwalled carbon nanotubes for simultaneous determination of trace metal ions by energy-dispersive X-ray fluorescence spectrometry. *Appl. Spectrosc.,* **2013**, *67*(2), 204-209. [http://dx.doi.org/10.1366/12-06738] [PMID: 23622440]

[27] Kocot, K.; Sitko, R. Trace and ultratrace determination of heavy metal ions by energy-dispersive X-ray fluorescence spectrometry using graphene as solid sorbent in dispersive micro solid-phase extraction. *Spectrochim. Acta B At. Spectrosc.,* **2014**, *94-95*, 7-13. [http://dx.doi.org/10.1016/j.sab.2014.02.003]

[28] Huang, M.C.; Chen, H.C.; Fu, S.C.; Ding, W.H. Determination of volatile N-nitrosamines in meat products by microwave-assisted extraction coupled with dispersive micro solid-phase extraction and gas chromatography--chemical ionisation mass spectrometry. *Food Chem.,* **2013**, *138*(1), 227-233. [http://dx.doi.org/10.1016/j.foodchem.2012.09.119] [PMID: 23265481]

[29] Yahaya, N.; Mitome, T.; Nishiyama, N.; Sanagi, M.; Aini, W.; Ibrahim, W.; Nur, H. Rapid dispersive micro-solid phase extraction using mesoporous carbon COU-2 in the analysis of cloxacillin in water. *J. Pharm. Innov.,* **2013**, *8*, 240-246. [http://dx.doi.org/10.1007/s12247-013-9164-z]

[30] Zhao, L.; Qin, H.; Wu, R.; Zou, H. Recent advances of mesoporous materials in sample preparation. *J. Chromatogr. A,* **2012**, *1228*, 193-204. [http://dx.doi.org/10.1016/j.chroma.2011.09.051] [PMID: 21993515]

[31] Jin, J.; Nishiyama, N.; Egashira, Y.; Ueyama, K. Pore structure and pore size controls of ordered mesoporous carbons prepared from resorcinol/formaldehyde/triblock polymers. *Micropor. Mesopo. Mat.,* **2009**, *118*, 218-223. [http://dx.doi.org/10.1016/j.micromeso.2008.08.030]

[32] Peng, Y.; Xie, Y.; Luo, J.; Nie, L.; Chen, Y.; Chen, L.; Du, S.; Zhang, Z. Molecularly imprinted polymer layer-coated silica nanoparticles toward dispersive solid-phase extraction of trace sulfonylurea herbicides from soil and crop samples. *Anal. Chim. Acta,* **2010**, *674*(2), 190-200.

[http://dx.doi.org/10.1016/j.aca.2010.06.022] [PMID: 20678629]

[33] Ma, J.; Yuan, L.; Ding, M.; Wang, S.; Ren, F.; Zhang, J.; Du, S.; Li, F.; Zhou, X. The study of core-shell molecularly imprinted polymers of 17β-estradiol on the surface of silica nanoparticles. *Biosens. Bioelectron.,* **2011**, *26*(5), 2791-2795.
[http://dx.doi.org/10.1016/j.bios.2010.10.045] [PMID: 21094594]

[34] Yuan, L. MA, J.; Ding, M.; Wang, S.; Wu, X.; Li, Y.; Ma. K.; Zhou, X.; Li, F. Preparation of estriol-molecularly imprinted silica nanoparticles for determining oestrogens in milk samples. *Food Chem.,* **2012**, *131*, 1063-1068.
[http://dx.doi.org/10.1016/j.foodchem.2011.09.090]

[35] Chen, C.; Zhang, X.; Long, Z.; Zhang, J.; Zheng, C. Molecularly imprinted dispersive solid-phase microextraction for the determination of sulfamethazine by capillary electrophoresis. *Mikrochim. Acta,* **2012**, *178*, 293-299.
[http://dx.doi.org/10.1007/s00604-012-0833-2]

[36] Zhou, T.; Xiao, X.; Li, G. Hybrid field-assisted solid-liquid-solid dispersive extraction for the determination of organochlorine pesticides in tobacco with gas chromatography. *Anal. Chem.,* **2012**, *84*(1), 420-427.
[http://dx.doi.org/10.1021/ac202798w] [PMID: 22092268]

[37] Lasarte-Aragonés, G.; Lucena, R.; Cárdenas, S.; Valcárcel, M. Effervescence assisted dispersive liquid-liquid microextraction with extractant removal by magnetic nanoparticles. *Anal. Chim. Acta,* **2014**, *807*, 61-66.
[http://dx.doi.org/10.1016/j.aca.2013.11.029] [PMID: 24356221]

[38] Lucena, R.; Simonet, B.M.; Cárdenas, S.; Valcárcel, M. Potential of nanoparticles in sample preparation. *J. Chromatogr. A,* **2011**, *1218*(4), 620-637.
[http://dx.doi.org/10.1016/j.chroma.2010.10.069] [PMID: 21071033]

CHAPTER 9

Solid Phase Extraction Based on Magnetic Nanoparticles

Ángel Ríos[1,*] and **Mohammed Zougagh**[2,3]

[1] *Department of Analytical Chemistry and Food Technology, University of Castilla-La Mancha, E-13004, Ciudad Real, Spain*

[2] *Regional Institute for Applied Chemistry Research, IRICA, E-13004, Ciudad Real, Spain*

[3] *Albacete Science and Technology Park, E-02006 Albacete, Spain*

Abstract: Magnetic nanoparticles (MNPs) are attracting great interest for developing solid phase extraction (SPE). Nanoparticles involved in SPE processes present clear advantages with respect conventional SPE materials because of their high stability, simple preparation, and the short time involved in the sample preparation; avoiding time-consuming column passing and filtration, as it is in conventional SPE methods. In this e-book chapter, the progress involving MNPs used for SPE is presented and discussed. Taken into account the different types of MNPs used in this field, the different MNPs obtained by attaching inorganic components (*e.g.*, metal oxides, carbon, noble and semiconductor metals), or organic molecules (*e.g.*, surfactants and magnetic molecularly imprinted polymers, MMIP), are described through recent reported analytical applications of MNPs in SPE. The potential transfer of these analytical tools for daily work in routine laboratories is also pointed out.

Keywords: Analytical applications, Bioanalytical analysis, Clean-up, Environmental analysis, Food analysis, Functionalization, Magnetic nanoparticles, Magnetic solid phase extraction, Metal oxides, Modified surfactants, Molecular imprinted polymers, Preconcentration, Sample preparation, Silica, Synthesis schemes, Titanium oxide, Zirconium-coated.

* **Corresponding author Ángel Ríos:** Department of Analytical Chemistry and Food Technology, Faculty of Chemistry, University of Castilla – La Mancha. Av. Camilo José Cela, 10. E-13004 Ciudad Real, Spain; Tel: +34 926 295232; Fax: +34 926 295318; E-mail: angel.rios@uclm.es

M. Valcárcel, S. Cárdenas, R. Lucena (Eds.)

9.1. INTRODUCTION

Currently, analytical chemistry tends to exploit and improve the chemical properties of nanoparticles (NPs), especially the capacity of adsorption, for their use as a reversible sorbent in extraction and preconcentration processes, which were considered among the most attractive analytical applications. NPs with magnetic properties (magnetic nanoparticles, MNPs) can be easily handled by using external magnetic devices. Recently, SPE procedures involving MNPs (magnetic solid phase extraction, MSPE) have intensely been increased for the preconcentration of many analytes, as well as for elimination of interferences (clean-up), solving interesting analytical problems. Thus: (i) high extraction efficiency; (ii) increase of the extraction rates; (iii) easy recovery of the sorbent after the extraction by using a magnetic device, involving clear advantages with respect to centrifugation and filtration procedures.

The use of MNPs in analytical field has recently been reported in different applications. Thus, for the preconcentration of pollutants in water [1, 2], soil [3] and biological samples [4]; and the selective sorption of biological compounds as proteins [5] and DNA [6]. Aspects related to the magnetization are also discussed [7] from a practical point of view. From an applied side, two main issues can be considered: (i) the sample analyses involving reduced sizes of samples; and (ii) the use of very small amount of samples.

9.2. PRACTICAL ASPECTS

MSPE is a physical process in which they magnetic material (solid) and a liquid (sample or analytes) are involved. The magnetic solid phase presents a more attraction to the analytes or interferences than the solvents in which they are dissolved. The use of MNPs for performing MSPE has introduced the advantages that the nanoscale and nanomaterials present. In fact, sorptive microextraction processes can be easily carried out involving MNPs. The general batch procedures based on the use of MNPs and used in SPE are shown in Fig. (**1**). In the first step, and after conditioning, the MNPs are introduced in the sample solution containing the sample (analytes and interferences). Sorption of the analytes and/or the interferences takes place. According to the main purpose of the sample preparation, two procedures can be distinguished in the second step. Thus, if the analytes must be preconcentrated to increase the sensibility, MNPs are associated with the analytes and this combination is concentrated by using and external magnetic device. After separation of the remaining solution, analytes are eluted from the MNPs using an appropriate solvent. Then, the determination step takes place. On the other hand, if the interferences must be removed for sample clean-up purposes, they are retained in the MNPs and, thus, concentrated in the magnetic field. The remaining solution is easily separated and analytes determined in it.

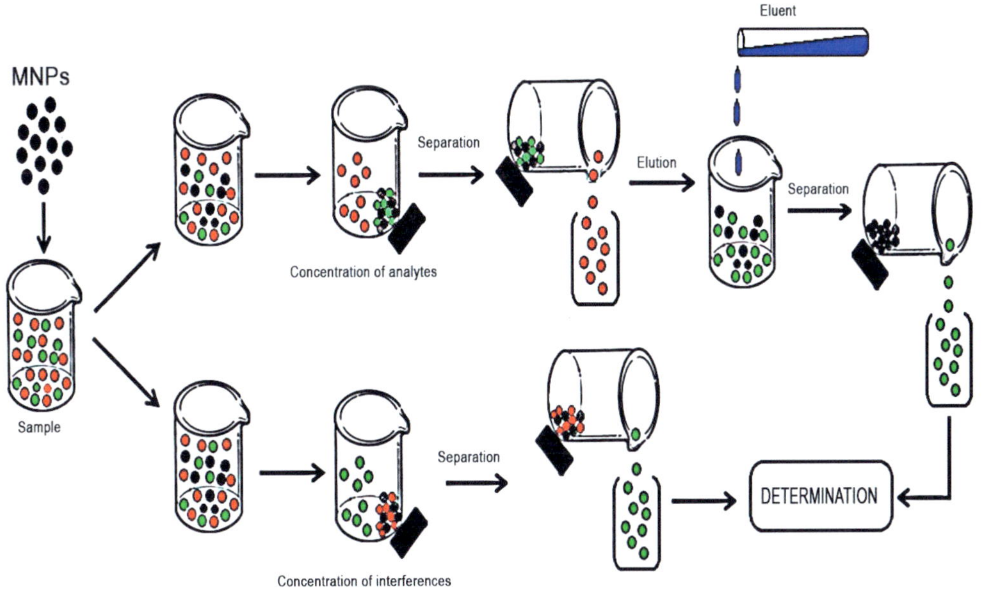

Fig. (1). Extraction experimental alternatives to carry out SPE with MNPs.

9.3. MAIN TYPES OF HYBRID MAGNETIC NANOPARTICLES USED IN SPE

The procedures employed for the synthesis of a hybrid MNPs used in SPE includes the synthesis of the magnetic particle and their surface modification for producing the final coating hybrid nanoparticles. The main type of hybrid nanomaterials used for this purpose is shown in Fig. (**2**), the following sections briefly describe the involved procedures.

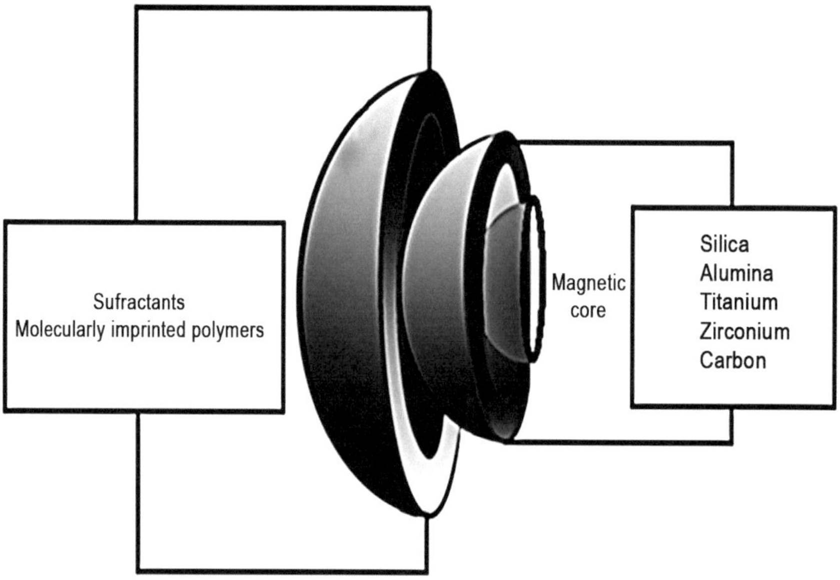

Fig. (2). Main types of hybrid magnetic nanoparticles used for SPE.

9.3.1. Synthesis of MNPs

MNPs have been synthetized using different iron oxides with (Fe_2O_3 and Fe_3O_4), presenting high magnetic moments and showing compatibility with biological systems, as well as involving a simple preparation process. Some related ferrite materials were also used (*e.g.*, $MnFe_2O_4$, $MgFe_2O_4$, $ZnFe_2O_4$, $CuFe_2O_4$, $NiFe_2O_4$ and $CoFe_2O_4$). Hug *et al.* [8] prepared different MNPs using these ferrites for Cr(VI) removal taking into account different experimental variables, comparing The adsorption capabilities were in the following order: $MnFe_2O_4 > MgFe_2O_4 > ZnFe_2O_4 > CuFe_2O_4 > NiFe_2O_4 > CoFe_2O_4$. Pure metals and alloys, such as Fe, Co, Ni, Mn_3O_4, FePt and FePd can be also used in the synthesis of MNPs, but with clear disadvantages with respect the previous ones.

Several methods were described for the synthesis of MNPs, including co-precipitation, hydrothermal reactions, sol–gel method, microemulsions techniques, flow injection synthesis method, aerosol assisted chemical vapour deposition (AACVD) method, sonochemical reactions, and electrospraying methods. All these methods have been used to synthesize particles with small size distribution and homogeneous composition. For further information about these synthesis procedures, readers can consult more specialized review publications [9, 10]. The usual method for the generation of MNPs is by chemical co-precipitation. This method is easy way to synthetize iron oxides (FeOOH, Fe_3O_4 or γ-Fe_2O_3) from ferric and ferrous salts under the presence of inert atmosphere. Ferric and ferrous chlorides ($FeCl_3$ and $FeCl_2$) are dissolved into deoxygenated distilled water, and stirred for long time. Chemical precipitation can be achieved at low temperature by stirring in the presence of NaOH under presence of inert atmosphere. After cooling the precipitated was separated and washed. Finally, the obtained MNPs are washed and dried in oven. The overall chemical reaction is:

$$Fe^{2+} + 2Fe^{3+} + 8OH^- \rightarrow Fe_3O_4 + 4\,H_2O$$

The problem of stability can be avoided by converting these MNPs into maghemite.

9.3.2. Surface Modification of MNPs

As bare magnetic cores do not present specific bindings and they are only stable under specific experimental conditions, they are generally coated using different selected compounds. In many cases these processes include the introduction of functionalized groups in the NMPs surface. Physical and chemical interactions are involved. For instance, van der Walls forces or the introduction of carboxylic and phosphate groups. It is also well known that hydroxyl groups (Fe-OH) are formed on the iron oxide surface in the co-precipitation process. According to the experimental conditions (pH and ionic strength, basically), iron oxides can be positively (Fe-OH_2^+) or negatively (Fe-O^-) charged. Thus, it has been reported pK_a values of 4.4 (pK_1) and 9.0 (pK_2) for magnetite [11]. Nevertheless, this procedure did not allow achieving appropriated stability for practical uses. In order to obtain a higher stability and specific interactions or binding with target analytes, these nanomaterials are coated with inorganic materials (metal oxides, noble and semiconductor metals and carbon), and organic materials such as surfactants and polymers.

9.3.2.1. Metal Oxides-Coated MNPs

9.3.2.1.1. Silica

Silanization is one of the most common alternatives to modify the surface of many materials. This is also the case of surface modification of the magnetic cores of MNPs in order to achieve the functionalization of these nanoparticles. In this way, silica gel is an excellent support for SPE procedures because its particular properties: (i) it is stable under acidic conditions; (ii) it does not enlarged by swelling; (iii) it presents a high mass exchange; and (iv) it is characterized by a high thermal stability. In addition, silica-functionalized magnetic nanoparticles incorporate different interesting capabilities such as they are their stability when used in aqueous solutions, the facility to modify the surface and to produce the binding of different functional groups by silane centers, as well as the possibility of a simple control of the interactions between nanoparticles. This is the reason why silica coated magnetic nanoparticles have been used for the extraction of many polar and non-polar analytes from various samples in a variety of cases [12]. Polar compounds can be adsorbed directly in free silane groups. Chemical interactions can be also manipulated by the strategic modification of the silica surface. For instance, the replacement of the free silane groups with amino groups ($-NH_2$) produces the reduction of the interaction of polar compounds, whereas the coupling of silane groups with C18 chains in silica material results in a strong non-polar material. In this case, reversible interactions are produced between the basically non-polar compounds and the modified surface of the sorbent material. A sol-gel process is involved in the formation of the silica metal oxides-coated MNPs. By using the procedure proposed by Stöber, the size, porosity, and morphology of the MNPs are achieved. Following this procedure, spherical and uniformly dispersed materials are obtained. This procedure is simple, effective, and it allows the synthesis of sized MNPs. This process is based on the use of a fixed mixture of ethanol, deionized water, and ammonia under a controlled temperature. Afterwards, a mixture of tetraethoxysilane (TEOS, $Si(C_2H_5O)_4$) and ethanol is added while vigorously stirring. Ultrasound radiation can be also used for the effective mixing. Under an inert gas atmosphere, the hydrolysis of Si-O groups in the presence of TEOS takes place, resulting in hydroxyl groups and then, by condensation, the final production of –Si-O-Si- sequences, remaining the magnetic characteristics of Fe_3O_4. In particular, inner magnetic Fe_3O_4 core endues the NPs with magnetic properties, while the outer silica shell shows high chemical stability and biocompatibility.

A variety of works reported the formation of magnetite-silica composite NPs (*e.g.*, [13 - 16]). Thus, a new method based on the use of ionic liquids (ILs) supported on magnetic nanoparticles (ILs-MNPs) of sulfonylurea herbicides (SUHs) was developed by our research group [2]. The ILs-Fe_3O_4/SiO_2 MNPs used in the preconcentration step were synthesised, by incorporating N-methylimidazolium ionic liquid and Fe_3O_4 MNPs in a silica matrix, following the sheme presented in Fig. (**3**). The Fe_3O_4 MNPs were firstly prepared by using the coprecipitation method, and dispersed in the 2-propanol/water (1/3) solution. After which, a NH_3 solution was added, under a room temperature conditions in the presence of inert atmosphere, followed by the addition of tetraethyl orthosilicate with stirring for 24 h, thus allowing the silica shell to grow on the surface of the MNPs. The Fe_3O_4/SiO_2 MNPs were washed with water and dried under vacuum. In the end step, Fe_3O_4/SiO_2 MNPs were mixed with a silane-coupling agent attached with N-methylimidazolium ionic liquid. The mixture was refluxed under stirring for 72 h. After reaction, the solid product was isolated by a magnet, washed and dried under vacuum. The authors demonstrated that ILs immobilized on Fe_3O_4/SiO_2 MNPs are suitable for extraction and preconcentration of SUHs, as a consequence of the high surface area and the high adsorption capability of these magnetic adsorbents.

9.3.2.1.2. Alumina

Alumina-coated magnetite nanoparticles (Fe_3O_4/Al_2O_3 MNPs) can be synthetized by alkoxide hydrolysis precipitation method [17, 18]. The Fe_3O_4 MNPs were firstly prepared and dispersed into an ethanol solution of aluminium isopropoxide. After which, an ethanol-water mixture was added drop wise to the previous suspension with vigorous stirring. Thus, aluminium isopropoxide can be easily hydrolysed into aluminium hydroxide deposited on the surface of Fe_3O_4 MNPs through electrostatic attraction, because the hydroxide ions on the surface of Fe_3O_4 MNPs can catalyze the hydrolysis reaction. After various cycles of separation/washing/redispersion with ethanol, the Fe_3O_4/Al_2O_3 MNPs obtained were oven dried and calcinated.

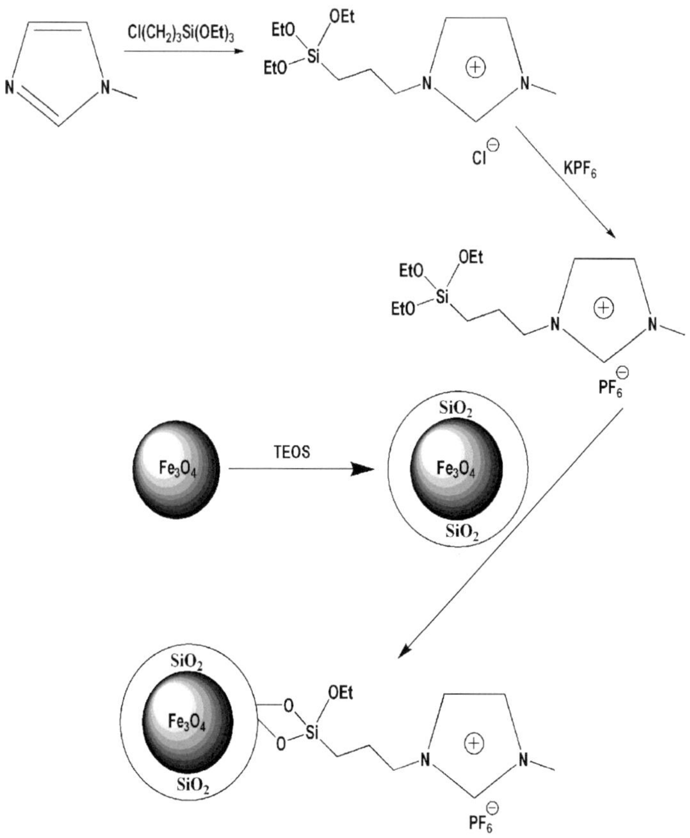

Fig. (3). Schematic synthesis of ionic liquid-functionalised silica-coated MNPs. (reproduced by permission of reference 2).

Alumina is a good shell for MNPs that can be easily used in SPE, because it presents high values of surface area, sorption capacity and chemical activity. Many molecules can be adsorbed by alumina thanks to the interactions with: (i) the aluminum metal; (ii) the hydroxyl groups by hydrogen bonding; and (iii) through an ion exchange process if the nanoparticle surfaces are charged. Thus, it allowed interactions with electronegative atoms, such as N, O, P or S, and with aromatic compounds. Other cationic species can be retained by conditioning the surface with a basic solution because in this way a net negative charge resulted. On the other hand, the surface conditioning by an acidic solution resulted in a surface with scarce capacity to retain basic compounds. Fe_3O_4/Al_2O_3 MNPs can be modified by ion surfactants that adsorb on metal oxides-coated MNPs, such as Fe_3O_4/Al_2O_3 MNPs, by the formation of aggregates as hemimicelles and admicelles [19], which present a high potential to complete adsorption of some analytes. In this way, Sun *et al.* [17] developed a Fe_3O_4/Al_2O_3 MNPs, modified by sodium dodecyl sulphate (SDS), for the selective extraction of trimethoprim (TMP) from environmental water samples, based on mixed hemomicelles SPE. The use of surfactants, such as SDS, introduces interesting possibilities. In fact, SDS can be adsorbed on the Fe_3O_4/Al_2O_3 MNPs surface and thus to complete the adsorption of TMP. The method showed good precision and recovery values. In comparison to the use of non-magnetic micro-adsorbent materials, the use of alumina-coated magnetite nanoparticles allowed a rapid and efficient extraction.

9.3.2.1.3. Other Materials

Titanium dioxide is another very important material which has also been used for the coating and functionalization of MNPs [20]. Titanium-coated magnetite nanoparticles (Fe_3O_4/TiO_2 MNPs) can be synthetized by the hydrolysis of tetrabutyl titanate (TBT) in an ethanol/water mixture in the presence of the dispersed Fe_3O_4 MNPs [21 - 23]. Zirconium-coated magnetite nanoparticles (Fe_3O_4/ZrO_2 MNPs) can be prepared by a sol-gel method as reported by Wu *et al.* [24]. Zirconyl chloride, dissolved in an ethanol-water mixture, was added to Fe_3O_4 NPs, and then ultrasonicated. After a complete reaction, the mixture was heated in a muffle furnace to obtain Fe_3O_4/ZrO_2 MNPs. These MNPs were used for the preconcentration of trace amounts of Cr, Cu, Cd and Pb. Then, the adsorbed metals were on-line eluted for their determination by inductively coupled plasma mass spectrometry (ICP-MS) [25]. Wei *et al.* [26] developed a novel strategy based on the use of MNPs coated with zirconium phosphonate for the enrichment of phosphopeptide from complex samples. The preparation of zirconium phosphonate MNPs is depicted in Fig. (**4**).

9.3.2.2. Carbon Nanotubes Magnetic Nanoparticles

Magnetic CNTs (MCNTs) composites are hybrids of magnetite (Fe_3O_4) and/or maghemite (γ-Fe_2O_3) with single-walled (SWCNTs) or multiwall CNTs

(MWCNTs) [27]. It is demonstrated that more attention has been paid to carbon-protected MNPs because their intrinsic advantages with respect polymers or silica, summarized by their higher thermal and chemical stabilities, in addition to the biocompatibility. Composites such as maghemite-CNT nanohybrid and magnetite-CNT nanohybrid materials (this last one more used) combine the unique optical, electrical, and mechanical properties of CNTs with the paramagnetic or ferromagnetic properties of MNPs at room temperature. Based on these features, they have been used as tips for magnetic force microscopes, in the wastewater treatment, in biosensors, drug delivery systems, and biomanipulators. These hybrid MNPs can be prepared following different procedures, as it is reported in the literature [28 - 31].

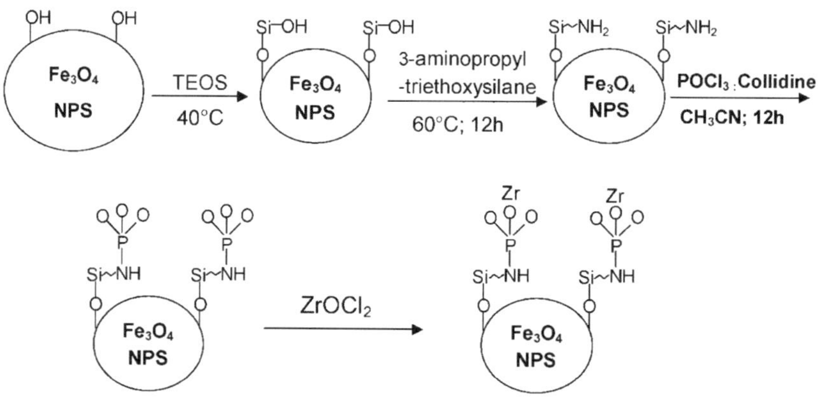

Fig. (4). Scheme for the preparation of zirconium phosphonate MNPs (reproduced by permission of reference 26).

It is very promising the synthesis of MNPs onto the CNTs surface [32]. In this way, Sun *et al.* [33] synthetized MWCNTs with maghemite from the pyrolysis of ferrocene, using different temperatures. Youn *et al.* coated SWCNTs with iron oxide NPs along the nanotube *via* a magnet or evaporation method [34]. In this work, indium tin oxide surfaces were used for the specific alienation of the nanotubes. Tan *et al.* reported the preparation of magnetic CNTs based on a thermolysis and oxidation process from a previous solution of dispersed nanotubes in iron pentacarbonyl [35]. The *in situ* decoration of carbon nanotubes can also be obtained by chemical precipitation. This procedure was used by Cao *et al.* for decorating CNTs with iron oxide particles [36]. Following this method, a wide variety of hybrid nanoparticle size can be obtained. Other possibility for producing MNPs onto CNTs surfaces is by thermal decomposition of precursors such as mixtures of iron chloride/iron acetate and MWCNTs in ethylene glycol. This synthesis procedure is very efficient and it can be easily controlled. Cid-Morales *et al.* have been used this procedure of synthesis [32]. The scanning electro micrograph images (SEMs), as it is shown in Fig. (**5**), revealed that the composite is formed by deposition of the Fe_2O_3 on the external walls of the MWCNTs.

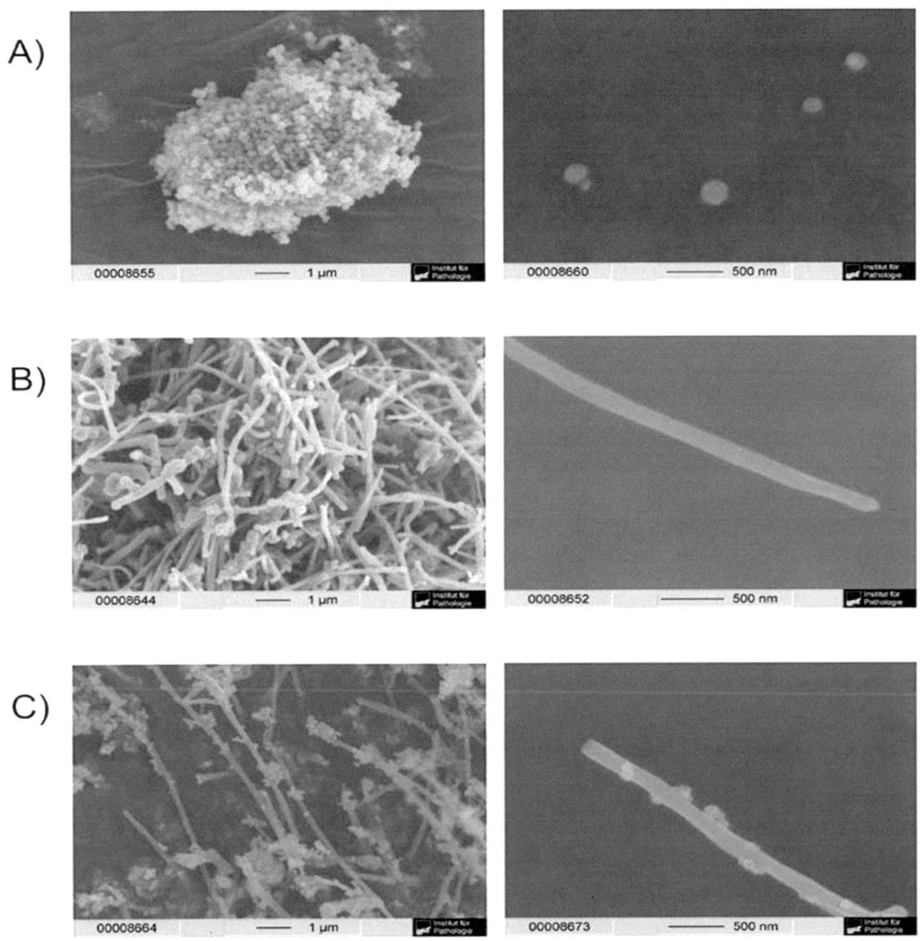

Fig. (5). Scanning electro micrograph images of the magnetic manoparticles (MNPs) **(a)**, multiwall carbon nanotubes (MWCNTs) **(b)** and MNPs decorated by MWCNTs **(c)**. (Reproduced by permission of reference 32).

Magnetic metals can be encapsulated inside carbon nanotubes (SWCNTs or MWCNTs), showing interesting perspectives [37, 38]. Thus, cobalt nanoparticles were encapsulated in SWCNTs, preserving very well the magnetic properties of this material, and presenting an unusual enhancement of the surface magnetic anisotropy [37]. Iron nanoparticles were also encapsulated into different types of carbon nanotubes, or even the synthesis of graphene-based Fe_3O_4 MNPs (G-Fe_3O_4 MNPs), which was used as selective sorbent material (MSPE) for triazine herbicides (atrazine, prometon, propazine and prometryn) in environmental water samples with excellent results [38].

9.3.2.3. Surfactant-modified MNPs

MNPs can also be modified with ionic surfactants. Mineral oxides, such as alumina, silica, titanium dioxide and ferric oxyhydroxides, can be used as sorbent materials for ionic surfactants. The formed aducts are hemimicelles and admicelles, which show a good capacity as sorbent materials in SPE [39, 40]. Hemimicelles are produced by the interaction of the head groups with MNPs of an opposite charge, and the hydrocarbon tail chain remains into the solution. Afterwards, admicelles were formed by the saturation of the MNPs surface with the tails of surfactants hydrocarbons chains and their corresponding hydrophobic interactions. Both, hemimicilles and admicelles, can be simultaneously formed on the MNPs surface, producing the so-called mixed hemimicelles, as (Fig. **6**) illustrated. In this case, the adsorption is controlled by both hydrophobic interactions and electrostatic attraction. Based on this approach, a wide variety of organic pollutants have been preconcentrated from different complex environmental samples using mixed hemimicelles (MHSPE) [41, 42]. In this case, the sorbent materials used were produced by adsorbing ionic surfactants onto the surface of metal oxides, such as sodium dodecyl sulphate (SDS)-coated alumina or CTAB-coated silica. The main advantages of using these types of sorbent materials are the versatility, the wide field of application, the high extraction efficiency, and the low cost. An additional feature needs to be remarked for this material: whereas the outer surface of hemimicelle is hydrophobic, the admicelles are ionic. This fact allows introducing different mechanisms for the retention of organic compounds. The combined advantages of the mixed hemimicelles and the magnetic-silica nanoparticles allow the preparation of nanosized SPE sorbent materials presenting interesting advantages such as high surface area, high chemical stability and good magnetic properties.

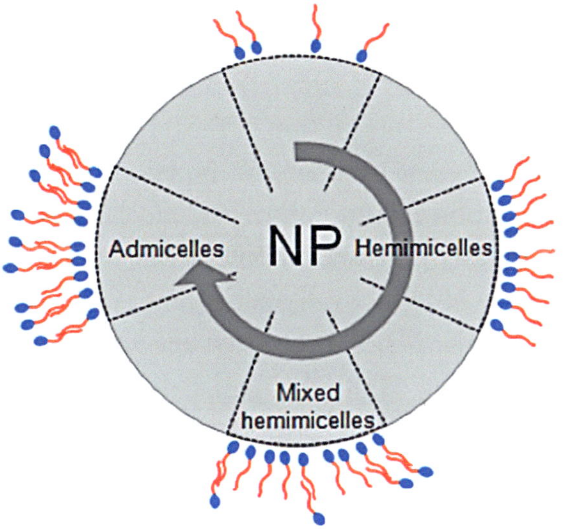

Fig. (6). Adsorption of ionic surfactants on the surface of Fe_3O_4 NPs.

9.3.2.4. Molecularly Imprinted Polymer-Magnetic Nanoparticles

Magnetic molecularly imprinted polymers (MMIPs) are synthesized by copolymerizing the suitable monomer(s), cross-linking agent and magnetic NPs in the presence of the molecule to which the sorbent is intended to be selective for. This molecule is known as the template molecule. The high degree of selectivity arises because, during the synthesis of the MMIP, the molecule that the MMIP is intended to be selective is involved in the polymerization process, leaving a finger-print for further recognition. The synthesis of MMIPs starts in the same way as with any other kind of polymerization: all the monomers involved in the polymerization process are mixed together and then left to polymerize. However, selection of the monomers involved in the polymerization process is critical for the MMIP to have the desired selectivity. In the synthesis of MMIPs, all the monomers used in the polymerization process have a very well-defined role and are known as functional monomers and cross-linking agents.

The general procedure for obtaining MMIPs is to mix the template molecule with the functional monomer. The functional monomer must always have complimentary functionalities to the template molecule because, once the MMIP is obtained, the functional monomer is the part responsible for the selective retention of the analyte onto the MMIP. Due to this complementarity, a complex template molecule-functional monomer is formed by self-assembly of those molecules when the functional monomer and the template molecule are mixed together. This complex is essential for a proper recognition of the target molecule by the MMIP. Afterwards, a cross-linking agent is added. Its function is to link the polymeric chains growing in the solution and to deliver the mechanical properties for the intended application. As in any polymerization process, an initiator agent must be introduced into the solution to induce polymerization after all of the compounds involved in the synthesis of the MMIP have been added. This agent normally has an unstable structure that, once exposed to either UV radiation or heat, easily decomposes into radicals, which are responsible for starting the polymerization process. Once the polymer is obtained, the template molecule is extracted, leaving behind different cavities that are not only complimentary in size and shape to the template molecule. Additionally, the functionalities inside this cavity reinforce the selective retention of the target molecule on to the MMIP. This synergy established between these two factors is the added value that MMIPs have to offer over other sorbents used in SPE. The used magnetic NPs, such as Fe_3O_4, is an important aspect because the surface of NPs is modified or functionalized, for a compatible polymerization reaction and providing magnetic properties.

As it is reported by Zhang *et al.* [43], the advantage of MMIP used for SPE is clear: "the participation of a magnetic component in the imprinted polymer can build a controllable rebinding process and allow magnetic separation to replace the centrifugation and filtration step in a convenient and economical way". When MIP particles contain magnetic components, the adsorption can be achieved by dispersing them in solution, and they are then easily separated from the matrix by applying an external magnet device. It means that SPE based on MIP with magnetic separations allow a very simple and efficient way for preconcentration, avoiding the need to pack the SPE in cartridges and the time-consuming process of loading a large-volume sample. Therefore, MMIP combine the magnetically susceptible characteristic with the selectivity for the guest molecule. Some interesting examples can be found in the literature (*e.g.*, [44, 45]).

9.3.2.5. Quantum Dots Magnetic Nanoparticles

One type of nanomaterials that has increased attention in last decades are quantum dots (QDs). They are basically inorganic semiconductor materials made from several elements from the periodic groups II-VI, II-V or IV-VI. Besides, there are different compositions of quantum dots based in their core, core/shell, core/shell/shell, as CdSe, CdTe, CdS, ZnSe, CdSe/ZnS, CdSeS/ZnS, CdTe/CdS, among others. The nanoscale size of QDs is the responsible for the exceptional optoelectronic properties, due to the "quantum confinement effect", including strong size-dependent properties result in a tunable emission profiles, narrow spectral bands and high emission quantum yields. Therefore, great efforts have been paid to synthetize fluorescent magnetic nanoparticles (FMNPs) with both fluorescence and magnetism functionalities for potential application [7]. These materials can be used for extracting analytes from complex samples and, then, measuring the fluorescence provides by the FMNPs, related to the presence of these particular analytes in the sample.

Although the interest of this approach, some problems related to the fluorescence quenching and the low fluorescence quantum yield of the FMNPs can be appeared [46]. One common procedure to prepare water soluble FMNPs is the layer-b-layer approach. This procedure was used for the synthesis of Fe_3O_4(core)-QD(shell) [7]. In this case, quenching effect was avoided by controlling the distance between the core and the shell. In other case, quenching is avoided by the silanization of QDs. Zhang *et al.* used this procedure by QDs silica coating, with very efficient results [46].

9.4. ANALYTICAL APPLICATIONS OF MNPS IN SPE

In this section some representative contributions with different MNPs, classified as described in section 3, are reported. These contributions are summarized in Table **1**, at which MNPs are used as sorbent material in environmental, bioanalytical, pharmaceutical or food applications. In this way, additional interesting information dealing with metal/metalloid preconcentration and/or separation, can be found in the review by Giakisikli *et al.* [71].

Table 1. Application of magnetic nanoparticles in the solid phase extraction.

MNPs coating	Amount of sorbent	Analyte	Sample and sample volume	Elution conditions	Separation/ detection	LOD (ng mL^{-1})	Ref.
SiO_2/γ-MPTMS	50 mg	Cd^{2+}, Cu^{2+}, Hg^{2+}, Pb^{2+}	Biological, environmental (50-100 mg)	HCl/ thioura	ICP-MS	24-107	[47]
SiO_2/ Bismuthiol	100 mg	Cr^{3+}, Cu^{2+}, Pb^{2+}	Waters (200 mL)	HNO_3	ICP-OES	0.043-0.085	[48]
SiO_2/C_{18}	10 mg	Ergosterol	Cigarettes (0.7 g)	n-hexane	GC-MS	10	[49]
SiO_2/C_{18}	0.2 mg	Methylprednisolone	Rat plasma (0.5 mL)	n-hexane	HPLC-DAD	10	[50]
SiO_2	Used in capillary column	Antibiotics	Not applied (200 μL)	Methanol/ H_2O	CLC-DAD	----	[51]
SiO_2/ILs	60 mg	Sulfonylurea herbicides	Waters (50 mL)	Acetonitrile	CLC-DAD	1.1-2.9	[52]
Al_2O_3	10 mg	Phosphopeptides	Tryptic digest products (200 μL)	Acetonitrile/ TFA	MALDI–TOF-MS	---	[53]
Ti_2O_3	0.02 mg	Phosphopeptides	Bovine serum albumin, β-casein (200 μL)	Nitric acid	MALDI–TOF-MS	---	[23]
ZrO_2	50 mg	Cr,$^{3+}$ Cr^{6+}	Water, biological (75 mL)	Nitric acid	FAAS	0.69	[24]
MWCNTs	1.28 mg	(Fluoro)quinolones	Human Plasma (300 μL)	Methanol/NH_4OH	UPLC-DAD	19.2-48.2	[32]
MWCNTs	4 mg	Phthalate acid esters	Beverages, Water, Perfume (10 mL)	Acetone	GC-MS	0.0049-0.038	[54]
CNTs	30 mg	Cu^{2+}, Pb^{2+}	Water (50 mL)	Not eluted	AAS	----	[55]
MWCNTs/ polydimethylsiloxane	10 mg	(Fluoro)quinolones	Water, honey (10 mL, 1g)	HCl/Methanol	CLC-DAD	2.2-3.7	[56]
MWCNTs	10 mg	*Escherichia coli*	Pharmaceutical (1 mL)	Not eluted: Incubated	TEM	-----	[57]
CTAB and CPC	100 mg	Phenolic compounds	Water (5 mL)	Acetonitrile	HPLC-FLD	0.012-0.034	[58]
CTAB	10 mg	Chlorophenols	Water (700 mL)	Methanol	HPLC-MS	0.11-0.15	[59]
OTMAB	110 mg	Sulfonamides	Water (500 mL)	Methanol	HPLC-DAD	0.024-0.033	[60]
SDS	40 mg	Hg^{2+}	Water (500 mL)	1-propanol	ICP-OES	0.04	[61]
CTAB or CPC/SiO_2	100 mg	Phenolic compounds	Water (500 mL)	Acetic acid	HPLC-FLD	0.007-0.020	[62]

(Table 1) contd.....

MNPs coating	Amount of sorbent	Analyte	Sample and sample volume	Elution conditions	Separation/ detection	LOD (ng mL^{-1})	Ref.
SDS/Al_2O_3	100 mg	Trimethoprim	Water (500 mL)	Methanol (3x1mL)	HPLC-DAD	0.09-0.24	[63]
MIP	100 mg	Fluoroquinolone antibiotics	Water (500 mL)	Methanol (3x1mL)	HPLC-MS	0.0032-0.0062	[64]
MIP	50 mg	Bisphenol A	Water (200 mL) and milk (50 mL)	Methanol/Acetic acid	HPLC-DAD	0.08	[65]
MIP	100 mg	Atrazine	Soybean, millet, lettuce and soil (1g)	Methanol	HPLC-MS	0.7	[66]
MIP	50 mg	Tetracycline antibiotics	Egg and tissues (2 g)	Methanol/ acetic acid	HPLC-MS	0.07-0.15	[67]
MIP	50 mg	Catecholamines	Urine (2 mL)	Acetic acid	CE-DAD	0.04-0.06	[68]
MIP	500 mg	Sulfonylurea herbicides	Water (500 mL)	Methanol/ acetic acid	CLC-DAD	0.02-0.03	[69]
QDs	2 mL (0.1 μM)	Anticycline E antibodies	Serum (200 μL)	Not eluted	FLD		[70]

LOD: Limit of detection, γ-MPTMS: γ-mercaptopropyltrimethoxysilane, ICP-MS: Inductively coupled plasma mass spectrometry, CTAB: cetyltrimethylammonium bromide, CPC: Cetylpyridinium chloride, OTMAB: octadecyltrimethylammo-nium bromide.

9.4.1. Analytical Application Involving Metal Oxides-coated MNPs

Metal oxides coated magnetic materials have been widely used for the extraction of different polar and non-polar analytes from various samples. Thus, Huang and Hu synthesized silica-coated Fe_3O_4 NPs, afterwards modified with 3-mercaptopropyltrimethoxysilane (MPTMS) [47]. These nanoparticles contain thiol groups, which allowed the selective extraction of Cd^{2+}, Cu^{2+}, Hg^{2+}, and Pb^{2+} by the formation of the corresponding metal-thiol complexes. The posterior elution of the absorbed metal ions was carried out by using a 1 M HCl – 2% thiourea mixing solution. The efficiency of the extraction-elution process is very high, allowing the determination of these metallic ions at the pg L^{-1} level. Other thiol modified ligands have also been used for the extraction of trace amount of metallic ions. For instance, Suleiman *et al.* used magnetic silica sorbent of bismuthiol-II-immobilized MNPs for the selective extraction of Cr, Cu and Pb from polluted water samples [48]. The sorbent material is prepared by the sol-gel procedure. Therefore, the modification of MNPs *via* silica nanoparticles introduces interesting possibilities, as the silica surface can be easily functionalized accordingly. The main disadvantage of these sorbent materials is the instability under alkaline media. Hence, some strategies should be developed to avoid this limitation. Magnetic silica NPs functionalized with C18 phases *via* a microwave-assisted derivatization process was also reported for the determination of ergosterol in cigarettes [49]. The preconcentration of this compound is very efficient because the high surface area and the strong magnetism of the sorbent material. These MNPs were also proposed for the preconcentration and the determination of methylprednisolone in rat plasma by high performance liquid chromatography (HPLC) [50].

A magnetic in-tube solid phase microextraction process was proposed by Moliner *et al.* as a new approach to implement an in-tube SPME [51]. By using this methodology, the authors reached high extraction efficiencies for acetylsalicylic acid, atenolol, acetaminophen, diclofenac, and ibuprofen. On the other hand, silica-coated MNPs incorporating N-methylimidazolium ionic liquid (ILs) are used for extracting and preconcentrate sulfonylurea herbicides (SUHs) from polluted water samples [52]. In this case, the ionic liquid introduces hydrophobic interactions for the selective extraction of SUHs, allowing the clean-up and the preconcentration of the residues of these pesticides. Phosphopeptide compounds were extracted by Li *et al.* thanks to their strong interactions with Fe_3O_4/Al_2O_3 MNPs in standard phosphoproteins β-casein and bovine serum albumin (BSA) [53]. The capacity of the Fe_3O_4/Al_2O_3 MNPs for retaining the phosphopeptides, and the magnetic properties for an easy isolation of the extracted compounds are combined. Other authors used titanium coated MNPs (Fe_3O_4/TiO_2 MNPs) for the selective preconcentration of phosphopeptides from complex samples [23].

Zirconium was also used for the synthesis of zirconium-coated MNPs, which were employed as sorbent material to carry out the speciation of chromium species in water and biological samples by FAAS [24]. The pH plays a key role for the selective adsorption of chromium species on the Fe_3O_4/ZrO_2 MNPs surface. Thus, Cr(III) was selectively adsorbed at pH 8.0 in the presence of Cr(VI). The analytical method is, in addition, very sensitive and precise for the determination of trace amounts of Cr(III).

9.4.2. Analytical Applications Involving Carbon-protected Magnetic Nanoparticles

The combination of CNTs with MNPs represents an interesting alternative which allows their use as sorbent materials in SPE, since an easier manipulation of CNTs can be achieved with an external magnetic field provided by a permanent magnet with a high energy product. The extraction and the preconcentration of different analytes were carried out by using this type of nanomaterials; for instance, as previous step for the determination of fluoroquinolones [63], and phthalate acid esters [54]. Additionally, metallic ions, such as Pb(II) and Cu(II), were extracted from water samples using CNTs-iron oxide magnetic composites [55]. Again, a simple magnetic device allowed the separation of the adsorbent material containing the analytes from the bulk medium. Recovery values around 98% were obtained. Magnetic MWCNTs were also prepared as MWCNT-MNPs by Morales-Cid for the clean-up of fluoroquinolones from human plasma samples [32]. Afterwards, these compounds were separated and determined by Ultrahigh Pressure Liquid Chromatography. Average recoveries about 89% were found using this sorbent material. Xu *et al.* [56] prepared surface functionalized magnetic particles with polydimethylsiloxane and MWCNT in a two steps reaction. The MMWCNTs were applied to MSPE of the fluoroquinolones ofloxacin, norfloxacin, ciprofloxacin, enrofloxacin. Then, these compounds were separated and determined by capillary liquid chromatography. MMWCNTs showed also an antimicrobial activities to bacterial cells (*Escherichia coli*), while the magnetic nanotubes are biocompatible even with a higher concentration than that of MWCNTs [57].

9.4.3. Analytical Applications Involving Surfactants

Ionic surfactants can be also used, as previously stated, to support hemimicelles and admicelles on the MNPs surface through the establishment of hydrophobic/electrostatic interactions. Thus, cetyltrimethylammonium bromide (CTAB), octadecyltrimethylammonium (OTAB) and sodium dodecyl sulfate (SDS) were used for this purpose. These surfactant compounds introduce hydrophobic and electrostatic groups involved in the extraction process, and hence their use as SPE materials. For instance, CTAB was used as coated surfactant of Fe_3O_4 to form a magnetic MHSPE material for the extraction of phenolic compounds from polluted water samples [58, 59]. In this case, the surfactant introduced the selectivity to the phenolic compounds. OTAB was also used as a coated agent onto magnetite NPs. The MMHSPE procedure then developed showed very good results for the preconcentration of sulfonamides from polluted water samples [60]. In the case of using SDS, a sorbent material based on MNPs was developed for the preconcentration of mercury(II) from water samples [61]. Hemimicelles were formed on the Fe_3O_4 surface, which establish the specific interaction with mercury ion. The extraction is very efficient and mercury can be determined with high sensitivity. Nevertheless, it was observed magnetism was lost when pH was below 4.0. To avoid this problem, magnetite NP was firstly coated with silica or alumina, and modified by the surfactants to form mixed hemimicelles [62, 63].

9.4.4. Analytical Applications Involving MIPs

When magnetic components are encapsulated into MIPs, the resulting MMIP composite polymer presents magnetic properties and a high selectivity for the guest molecules. Because of a high selectivity of MMIPs, presenting in some cases similar or higher recognition levels compared with natural systems such as monoclonal antibodies. They have been classified as a suitable material for clean-up, selective separation and extraction of analytes from complex samples. Thus, the most important applications of MMIPs have been their use as selective adsorbents in analytical processes for clean-up and/or preconcentration of samples [64 - 69]. Different synthesis protocols can be followed to achieve the appropriate selectivity and sensitivity. Thus, using ciprofloxacin as template molecule, an MMIP was developed involving the magnetic separation and the selective extraction of fluoroquinolones from environmental water samples [64]. In addition to the template molecule, methacrylic acid as functional monomer, ethylene-glycol dimethacrylate as cross-linking agent and Fe_3O_4 magnetite as magnetic component were needed to the synthesis of the MMIP. The developed analytical method allowed the determination of fluoroquinolones at the ng L^{-1} levels. Other interesting examples are reported in the literature. Ji *et al.* synthesized magnetic MIP microspheres with a porous structure and the desired level of magnetic susceptibility by miniemulsion polymerization [65]. Zhang *et al.*

prepared MMIP beads of atrazine (template molecule) by suspension polymerization applying microwave heating [66]. In other example, MMIP were used for the extraction of tetracycline antibiotics [67], catecholamines [68] and sulphonylurea herbicides [69]. In these examples hydrophobic Fe_3O_4 magnetite was used as the magnetically susceptible component, with template molecules such as oxytetracycline, dopamine or methylsulfuron-methyl, and methacrylic acid as functional monomer. The polymeric matrix components commonly used were trimethylolpropane trimethacrylate, styrene and divinylbenzene.

9.4.5. Analytical Applications Involving QDs Magnetic Nanoparticles

Some few references have been reported combining the MNPs with QDs, although the potential of the merging in the same nanomaterial of magnetic properties with fluorescence response is very attractive from an analytical point of view. One of these references comes from Wang *et al.* [70]. These authors prepared water soluble nanocomposite particles consisting of a magnetic core (Fe_2O_3) and a shell of luminescent QDs (CdSe–ZnS). The solubility in water is very important for analytical purposes, but in addition, in this case, the material presented a high emission quantum yield, and it can be easily separated from the bulk solution by a magnetic device. This material was very useful for the immobilization of anticyclin E antibodies on their surface, allowing the extraction of MCF-7 breast cancer cells from serum solutions. The separated breast cells were easily observed by fluorescence imaging microscopy thanks to the strong luminescence of the luminescent-magnetic nanocomposite particles. In another application, He *et al.* [72] synthesized hydrophilic high-luminescent magnetic nanocomposites composed of both fluorescent clusters (QDs) and MNPs. Biotin–streptavidin was used as molecular recognition system. The developed system demonstrated its good response for streptavidin concentrations between 23 and 92 nM.

CONCLUSION

Nanotechnology has emerged as a clear trend, with an increasing potential in analytical science, particularly by the unique characteristics of nanoparticles. In particular, the use of MNPs in sample preparation is one important trend topic in analytical science, as the application of magnetic separation procedures simplifies the sample treatment. As shown in Table **2**, the preparation of sorbent does not need to be packed into the cartridge (as in the conventional SPE), and hence the separation of the extracted phase can be carried out in an easy way by applying an external magnetic field resulting in short time in the sample preparation, and the MNPs showed a great stability, with possibilities of reusability for several times. The usefulness of MNPs, involved in the sample preparation step, has been very well demonstrated in many clean-up and preconcentration procedures, as it has been reported in different sections of this chapter. The improvements of the physicochemical properties of MNPs will introduce interesting advantages and new possibilities. Therefore, this exciting field will open interesting expectative, as well as new challenges. Without any doubt new procedures of synthesis should be developed, including new functionalization methods for MNPs. Problems related to stability should be solved, as well as to improve life-span and selectivity. From an analytical point of view, the robustness of the extraction, preconcentration and/or clean-up processes based on the use of MNPs as SPE sorbent materials will be a critical point to assure the transference to routine laboratories, thus demonstrating their real usefulness.

Table 2. Comparison of magnetic solid phase extraction (MSPE) and conventional SPE.

Factor	**MSPE**	**Conventional SPE**
Preparation of sorbent	Simple/ used without packing	Labour-intesive/ need packing in cartridge or column
Method development	Simple/ separation by magnet	Labour-intesive/ separation by filtration
Sorbent consumption	Low (<100 mg)	High (500-1000 mg)
Sample volume	Medium (<500 mL)	High (up to 1 L)
Eluent volume	Low (<2 mL)	Medium (<5 mL)
Process time	Short (< 15 min)	Long (up to 1 hour)

CONFLICT OF INTEREST

The authors confirm that they have no conflict of interest to declare for this publication.

ACKNOWLEDGEMENTS

Financial support from the Spanish Ministry of Economy and Competitiveness (CTQ2013-48411-P), and Project PEIC-2014-001-P (JCCM) are gratefully acknowledged.

REFERENCES

[1] Karatapanis, A.E.; Fiamegos, Y.; Stalikas, C.D. Silica-modified magnetic nanoparticles functionalized with cetylpyridinium bromide for the preconcentration of metals after complexation with 8-hydroxyquinoline. *Talanta,* **2011**, *84*(3), 834-839. [http://dx.doi.org/10.1016/j.talanta.2011.02.013] [PMID: 21482290]

[2] Bouri, M.; Gurau, M.; Salghi, R.; Cretescu, I.; Zougagh, M.; Ríos, Á. Ionic liquids supported on magnetic nanoparticles as a sorbent preconcentration material for sulfonylurea herbicides prior to their determination by capillary liquid chromatography. *Anal. Bioanal. Chem.,* **2012**, *404*(5), 1529-1538. [http://dx.doi.org/10.1007/s00216-012-6221-2] [PMID: 22832671]

[3] Sun, L.; Sun, X.; Du, X.; Yue, Y.; Chen, L.; Xu, H.; Zeng, Q.; Wang, H.; Ding, L. Determination of sulfonamides in soil samples based on alumina-coated magnetite nanoparticles as adsorbents. *Anal. Chim. Acta,* **2010**, *665*(2), 185-192. [http://dx.doi.org/10.1016/j.aca.2010.03.044] [PMID: 20417329]

[4] Beiraghi, A.; Pourghazi, K.; Amoli-Diva, M.; Razmara, A. Magnetic solid phase extraction of mefenamic acid from biological samples based on the formation of mixed hemimicelle aggregates on Fe_3O_4 nanoparticles prior to its HPLC-UV detection. *J. Chromatogr. B Analyt. Technol. Biomed. Life Sci.,* **2014**, *945-946*, 46-52. [http://dx.doi.org/10.1016/j.jchromb.2013.11.039] [PMID: 24321760]

[5] Pochet, L.; Heus, F.; Jonker, N.; Lingeman, H.; Smit, A.B.; Niessen, W.M.; Kool, J. Online magnetic bead based dynamic protein affinity selection coupled to LC-MS for the screening of acetylcholine binding protein ligands. *J. Chromatogr. B Analyt. Technol. Biomed. Life Sci.,* **2011**, *879*(20), 1781-1788. [http://dx.doi.org/10.1016/j.jchromb.2011.04.023] [PMID: 21565564]

[6] Shi, R.; Wang, Y.; Hu, Y.; Chen, L.; Wan, Q.H. Preparation of magnetite-loaded silica microspheres for solid-phase extraction of genomic DNA from soy-based foodstuffs. *J. Chromatogr. A,* **2009**, *1216*(36), 6382-6386. [http://dx.doi.org/10.1016/j.chroma.2009.07.016] [PMID: 19632684]

[7] Rios, A.; Zougagh, M.; Bouri, M. Magnetic (nano)materials as an useful tool for sample preparation in analytical methods. A review. *Anal. Methods,* **2013**, *5*, 4558-4573. [http://dx.doi.org/10.1039/c3ay40306h]

[8] Hug, J.; Lo, I.M.; Chen, G. Comparative study of various magnetic nanoparticles for Cr(VI) removal. *Separ. Purif. Tech.,* **2007**, *56*, 249-256. [http://dx.doi.org/10.1016/j.seppur.2007.02.009]

[9] Lu, A.H.; Salabas, E.L.; Schüth, F. Magnetic nanoparticles: synthesis, protection, functionalization,

and application. *Angew. Chem. Int. Ed. Engl.,* **2007**, *46*(8), 1222-1244.
[http://dx.doi.org/10.1002/anie.200602866] [PMID: 17278160]

[10] Ramimoghadam, D.; Bagheri, S.; Abd Hamid, B.S. Progress in electrochemical synthesis of magnetic iron oxide nanoparticles. *J. Magn. Magn. Mater.,* **2014**, *368*, 207-229.
[http://dx.doi.org/10.1016/j.jmmm.2014.05.015]

[11] Borlido, L.; Azevedo, A.M.; Roque, A.C.; Aires-Barros, M.R. Magnetic separations in biotechnology. *Biotechnol. Adv.,* **2013**, *31*(8), 1374-1385.
[http://dx.doi.org/10.1016/j.biotechadv.2013.05.009] [PMID: 23747736]

[12] Wierucka, M.; Biziuk, M. Application of magnetic nanoparticles for magnetic solid-phase extraction in preparing biological, environmental and food samples. *Trends Analyt. Chem.,* **2014**, *59*, 50-58.
[http://dx.doi.org/10.1016/j.trac.2014.04.007]

[13] Lu, Z.; Dai, J.; Song, X.; Wang, G.; Yang, W. Facile synthesis of Fe_3O_4/SiO_2 composite nanoparticles from primary silica particles. *Colloids Surf. A Physicochem. Eng. Asp.,* **2008**, *317*, 450-456.
[http://dx.doi.org/10.1016/j.colsurfa.2007.11.020]

[14] Wu, B.P.; Zhu, J.H.; Xu, Z.H. Template-assisted synthesis of mesoporous magnetic nanocomposite particles. *Adv. Funct. Mater.,* **2004**, *14*, 345-351.
[http://dx.doi.org/10.1002/adfm.200305455]

[15] Zhao, W.; Gu, J.; Zhang, L.; Chen, H.; Shi, J. Fabrication of uniform magnetic nanocomposite spheres with a magnetic core/mesoporous silica shell structure. *J. Am. Chem. Soc.,* **2005**, *127*(25), 8916-8917.
[http://dx.doi.org/10.1021/ja051113r] [PMID: 15969545]

[16] Barnakov, Y.A.; Yu, M.H.; Rosenzweig, Z. Manipulation of the magnetic properties of magnetite-silica nanocomposite materials by controlled Stober synthesis. *Langmuir,* **2005**, *21*(16), 7524-7527.
[http://dx.doi.org/10.1021/la0508893] [PMID: 16042489]

[17] Sun, L.; Zhang, C.; Chen, L.; Liu, J.; Jin, H.; Xu, H.; Ding, L. Preparation of alumina-coated magnetite nanoparticle for extraction of trimethoprim from environmental water samples based on mixed hemimicelles solid-phase extraction. *Anal. Chim. Acta,* **2009**, *638*(2), 162-168.
[http://dx.doi.org/10.1016/j.aca.2009.02.039] [PMID: 19327455]

[18] Sun, L.; Sun, X.; Du, X.; Yue, Y.; Chen, L.; Xu, H.; Zeng, Q.; Wang, H.; Ding, L. Determination of sulfonamides in soil samples based on alumina-coated magnetite nanoparticles as adsorbents. *Anal. Chim. Acta,* **2010**, *665*(2), 185-192.
[http://dx.doi.org/10.1016/j.aca.2010.03.044] [PMID: 20417329]

[19] Merino, F.; Rubio, S.; Pérez-Bendito, D. Solid-phase extraction of amphiphiles based on mixed hemimicelle/admicelle formation: application to the concentration of benzalkonium surfactants in sewage and river water. *Anal. Chem.,* **2003**, *75*(24), 6799-6806.
[http://dx.doi.org/10.1021/ac030224a] [PMID: 14670038]

[20] Greene, D.; Serrano, R.; Govan, J. Synthesis, characterization and photocatalytic studies of cobalt ferrite-silica-titania nanocomposites. *Nanomaterials (Basel),* **2014**, *4*, 331-343.
[http://dx.doi.org/10.3390/nano4020331]

[21] Ma, W.F.; Zhang, Y.; Li, L.L.; You, L.J.; Zhang, P.; Zhang, Y.T.; Li, J.M.; Yu, M.; Guo, J.; Lu, H.J.; Wang, C.C. Tailor-made magnetic Fe_3O_4@mTiO_2 microspheres with a tunable mesoporous anatase

shell for highly selective and effective enrichment of phosphopeptides. *ACS Nano,* **2012**, *6*(4), 3179-3188.
[http://dx.doi.org/10.1021/nn3009646] [PMID: 22452444]

[22] Li, Y.; Wu, J.; Qi, D.; Xu, X.; Deng, C.; Yang, P.; Zhang, X. Novel approach for the synthesis of Fe_3O_4@TiO_2 core-shell microspheres and their application to the highly specific capture of phosphopeptides for MALDI-TOF MS analysis. *Chem. Commun. (Camb.),* **2008**, (5), 564-566.
[http://dx.doi.org/10.1039/B716055K] [PMID: 18209790]

[23] Ma, W.F.; Zhang, Y.; Li, L.L.; You, L.J.; Zhang, P.; Zhang, Y.T.; Li, J.M.; Yu, M.; Guo, J.; Lu, H.J.; Wang, C.C. Tailor-made magnetic Fe_3O_4@mTiO_2 microspheres with a tunable mesoporous anatase shell for highly selective and effective enrichment of phosphopeptides. *ACS Nano,* **2012**, *6*(4), 3179-3188.
[http://dx.doi.org/10.1021/nn3009646] [PMID: 22452444]

[24] Wu, Y.W.; Zhang, J.; Liu, J.F. Fe_3O_4@ZrO_2 nanoparticles magnetic solid phase extraction coupled with flame atomic absorption spectrometry for chromium (III) speciation in environmental and biological samples. *Appl. Surf. Sci.,* **2012**, *258*, 67-72.

[25] Wu, Y.; Hu, B.; Jiang, Z.; Feng, Y.; Lu, P.; Li, B. Sol-gel zirconia coating capillary microextraction on-line hyphenated with inductively coupled plasma mass spectrometry for the determination of Cr, Cu, Cd and Pb in biological samples. *Rapid Commun. Mass Spectrom.,* **2006**, *20*(23), 3527-3534.
[http://dx.doi.org/10.1002/rcm.2763] [PMID: 17078104]

[26] Wei, J.; Zhang, Y.; Wang, J.; Tan, F.; Liu, J.; Cai, Y.; Qian, X. Highly efficient enrichment of phosphopeptides by magnetic nanoparticles coated with zirconium phosphonate for phosphoproteome analysis. *Rapid Commun. Mass Spectrom.,* **2008**, *22*(7), 1069-1080.
[http://dx.doi.org/10.1002/rcm.3485] [PMID: 18327884]

[27] Gao, C.; Li, W.; Morimoto, H.; Nagaoka, Y.; Maekawa, T. Magnetic carbon nanotubes: synthesis by electrostatic self-assembly approach and application in biomanipulations. *J. Phys. Chem. B,* **2006**, *110*(14), 7213-7220.
[http://dx.doi.org/10.1021/jp0602474] [PMID: 16599489]

[28] Korneva, G.; Ye, H.; Gogotsi, Y.; Halverson, D.; Friedman, G.; Bradley, J.C.; Kornev, K.G. Carbon nanotubes loaded with magnetic particles. *Nano Lett.,* **2005**, *5*(5), 879-884.
[http://dx.doi.org/10.1021/nl0502928] [PMID: 15884887]

[29] Deng, Y.; Deng, C.; Yang, D.; Wang, C.; Fu, S.; Zhang, X. Preparation, characterization and application of magnetic silica nanoparticle functionalized multi-walled carbon nanotubes. *Chem. Commun. (Camb.),* **2005**, (44), 5548-5550.
[http://dx.doi.org/10.1039/b511683j] [PMID: 16358059]

[30] Li, W.W.; Gao, C.; Qian, H.F. Multiamino-functionalized carbon nanotubes and their applications in loading quantum dots and magnetic nanoparticles. *J. Mater. Chem.,* **2006**, *16*, 1852-1859.
[http://dx.doi.org/10.1039/b600190d]

[31] Jia, B.; Gao, L.; Sun, J. Self-assembly of magnetite beads along nanotubes *via* a simple hydrothermal process. *Carbon,* **2007**, *45*, 1476-1481.
[http://dx.doi.org/10.1016/j.carbon.2007.03.025]

[32] Morales-Cid, G.; Fekete, A.; Simonet, B.M.; Lehmann, R.; Cárdenas, S.; Zhang, X.; Valcárcel, M.;

Schmitt-Kopplin, P. *In situ* synthesis of magnetic multiwalled carbon nanotube composites for the clean-up of (fluoro)quinolones from human plasma prior to ultrahigh pressure liquid chromatography analysis. *Anal. Chem.,* **2010**, *82*(7), 2743-2752.
[http://dx.doi.org/10.1021/ac902631h] [PMID: 20201520]

[33] Sun, Z.; Liu, Z.; Wang, Y. Fabrication and characterization of magnetic carbon nanotube composites. *J. Mater. Chem.,* **2005**, *15*, 4497-4501.
[http://dx.doi.org/10.1039/b509968d]

[34] Youn, S.C.; Jung, D.H.; Ko, Y.K.; Jin, Y.W.; Kim, J.M.; Jung, H.T. Vertical alignment of carbon nanotubes using the magneto-evaporation method. *J. Am. Chem. Soc.,* **2009**, *131*(2), 742-748.
[http://dx.doi.org/10.1021/ja8073209] [PMID: 19072279]

[35] Tan, F.Y.; Fan, X.B.; Zhang, G.L. Coating and filling of carbon nanotubes with homogeneous magnetic nanoparticles. *Mater. Lett.,* **2007**, *61*, 1805-1808.
[http://dx.doi.org/10.1016/j.matlet.2006.07.163]

[36] Cao, H.Q.; Zhu, M.F.; Li, Y.G. Novel carbon nanotube iron oxide magnetic nanocomposites. *J. Magn. Magn. Mater.,* **2006**, *305*, 321-324.
[http://dx.doi.org/10.1016/j.jmmm.2006.01.021]

[37] Cleuziou, J.P.; Wernsdorfer, W.; Ondarçuhu, T.; Monthioux, M. Electrical detection of individual magnetic nanoparticles encapsulated in carbon nanotubes. *ACS Nano,* **2011**, *5*(3), 2348-2355.
[http://dx.doi.org/10.1021/nn2000349] [PMID: 21344889]

[38] Cao, C.; Ma, Z.; Ma, C. Synthesis and characterization of Fe/C core–shell nanoparticles. *Mater. Lett.,* **2012**, *88*, 61-64.
[http://dx.doi.org/10.1016/j.matlet.2012.08.041]

[39] Saitoh, T.; Nakayama, Y.; Hiraide, M. Concentration of chlorophenols in water with sodium dodecylsulfate-γ-alumina admicelles for high-performance liquid chromatographic analysis. *J. Chromatogr. A,* **2002**, *972*(2), 205-209.
[http://dx.doi.org/10.1016/S0021-9673(02)01118-4] [PMID: 12416878]

[40] Merino, F.; Rubio, S.; Pérez-Bendito, D. Evaluation and optimization of an on-line admicelle-based extraction-liquid chromatography approach for the analysis of ionic organic compounds. *Anal. Chem.,* **2004**, *76*(14), 3878-3886.
[http://dx.doi.org/10.1021/ac049736v] [PMID: 15253620]

[41] Merino, F.; Rubio, S.; Perez-Bendito, D. Solid-phase extraction of amphiphilic based on mixed hemimicelles/admicelles formation: application to the benzalkonium surfactants in sewage and river. *Anal. Chem.,* **2003**, *75*, 6799-6806.
[http://dx.doi.org/10.1021/ac030224a] [PMID: 14670038]

[42] Rubio, S.; Perez-Bendito, D. Supramolecular assemblies for extracting organic compounds. *Trends Analyt. Chem.,* **2003**, *22*, 470-485.
[http://dx.doi.org/10.1016/S0165-9936(03)00706-4]

[43] Zhang, Y.; Liu, R.; Hu, Y.; Li, G. Microwave heating in preparation of magnetic molecularly imprinted polymer beads for trace triazines analysis in complicated samples. *Anal. Chem.,* **2009**, *81*(3), 967-976.
[http://dx.doi.org/10.1021/ac8018262] [PMID: 19178336]

[44] Li, Y.; Yin, X.F.; Chen, F.R. Synthesis of magnetic molecularly imprinted polymer nanowires using a nanoporous alumina template. *Macromolecules,* **2006**, *39*, 4497-4499. [http://dx.doi.org/10.1021/ma0526185]

[45] Lu, S.; Cheng, G.; Zhang, H. Preparation and characteristics of Tryptophan-imprinted Fe_3O_4/P(TRIM) composite microspheres with magnetic susceptibility by inverse emulsion–suspension polymerization. *J. Appl. Polym. Sci.,* **2006**, *99*, 3241-3250. [http://dx.doi.org/10.1002/app.22997]

[46] Zhang, B.; Chen, B.; Wang, Y.; Guo, F.; Li, Z.; Shi, D. Preparation of highly fluorescent magnetic nanoparticles for analytes-enrichment and subsequent biodetection. *J. Colloid Interface Sci.,* **2011**, *353*(2), 426-432. [http://dx.doi.org/10.1016/j.jcis.2010.09.084] [PMID: 20974471]

[47] Huang, C.; Hu, B. Silica-coated magnetic nanoparticles modified with γ-mercaptopropyltrimethoxysilane for fast and selective solid phase extraction of trace amounts of Cd, Cu, Hg, and Pb in environmental and biological samples prior to their determination by inductively coupled plasma mass spectrometry. *Spectrochim. Acta B At. Spectrosc.,* **2008**, *63*, 437-444. [http://dx.doi.org/10.1016/j.sab.2007.12.010]

[48] Suleiman, J.S.; Hu, B.; Peng, H.; Huang, C. Separation/preconcentration of trace amounts of Cr, Cu and Pb in environmental samples by magnetic solid-phase extraction with Bismuthiol-II-immobilized magnetic nanoparticles and their determination by ICP-OES. *Talanta,* **2009**, *77*(5), 1579-1583. [http://dx.doi.org/10.1016/j.talanta.2008.09.049] [PMID: 19159767]

[49] Sha, Y.; Deng, C.; Liu, B. Development of C18-functionalized magnetic silica nanoparticles as sample preparation technique for the determination of ergosterol in cigarettes by microwave-assisted derivatization and gas chromatography/mass spectrometry. *J. Chromatogr. A,* **2008**, *1198-1199*, 27-33. [http://dx.doi.org/10.1016/j.chroma.2008.05.049] [PMID: 18533171]

[50] Yu, P.; Wang, Q.; Zhang, X.; Zhang, X.; Shen, S.; Wang, Y. Development of superparamagnetic high-magnetization C18-functionalized magnetic silica nanoparticles as sorbents for enrichment and determination of methylprednisolone in rat plasma by high performance liquid chromatography. *Anal. Chim. Acta,* **2010**, *678*(1), 50-55. [http://dx.doi.org/10.1016/j.aca.2010.08.012] [PMID: 20869504]

[51] Moliner-Martínez, Y.; Prima-Garcia, H.; Ribera, A.; Coronado, E.; Campíns-Falcó, P. Magnetic in-tube solid phase microextraction. *Anal. Chem.,* **2012**, *84*(16), 7233-7240. [http://dx.doi.org/10.1021/ac301660k] [PMID: 22861152]

[52] Bouri, M.; Gurau, M.; Salghi, R.; Cretescu, I.; Zougagh, M.; Rios, Á. Ionic liquids supported on magnetic nanoparticles as a sorbent preconcentration material for sulfonylurea herbicides prior to their determination by capillary liquid chromatography. *Anal. Bioanal. Chem.,* **2012**, *404*(5), 1529-1538. [http://dx.doi.org/10.1007/s00216-012-6221-2] [PMID: 22832671]

[53] Li, Y.; Liu, Y.; Tang, J.; Lin, H.; Yao, N.; Shen, X.; Deng, C.; Yang, P.; Zhang, X. Fe_3O_4@Al_2O_3 magnetic core-shell microspheres for rapid and highly specific capture of phosphopeptides with mass spectrometry analysis. *J. Chromatogr. A,* **2007**, *1172*(1), 57-71. [http://dx.doi.org/10.1016/j.chroma.2007.09.062] [PMID: 17936290]

[54] Luo, Y.B.; Yu, Q.W.; Yuan, B.F.; Feng, Y.Q. Fast microextraction of phthalate acid esters from beverage, environmental water and perfume samples by magnetic multi-walled carbon nanotubes. *Talanta,* **2012**, *90*, 123-131.
[http://dx.doi.org/10.1016/j.talanta.2012.01.015] [PMID: 22340126]

[55] Peng, X.; Luan, Z.; Di, Z. Carbon nanotubes-iron oxides magnetic composites as adsorbent for removal of Pb(II) and Cu(II) from water. *Carbon,* **2005**, *4*, 880-883.
[http://dx.doi.org/10.1016/j.carbon.2004.11.009]

[56] Xu, S.; Jiang, C.; Lin, Y.; Jia, L. Magnetic nanoparticles modified with polydimethylsiloxane and multi-walled carbon nanotubes for solid-phase extraction of fluoroquinolones. *Mikrochim. Acta,* **2012**, *179*, 257-264.
[http://dx.doi.org/10.1007/s00604-012-0894-2]

[57] Shan, Y.; Chen, K.; Yu, X. Preparation and characterization of biocompatible magnetic carbon nanotubes. *Appl. Surf. Sci.,* **2010**, *257*, 362-366.
[http://dx.doi.org/10.1016/j.apsusc.2010.06.078]

[58] Zhao, X.; Shi, Y.; Cai, Y.; Mou, S. Cetyltrimethylammonium Bromide-coated magnetic nanoparticles for the preconcentration of phenolic compounds from environmental water samples. *Environ. Sci. Technol.,* **2008**, *42*(4), 1201-1206.
[http://dx.doi.org/10.1021/es071817w] [PMID: 18351093]

[59] Li, J.; Zhao, X.; Shi, Y.; Cai, Y.; Mou, S.; Jiang, G. Mixed hemimicelles solid-phase extraction based on cetyltrimethylammonium bromide-coated nano-magnets Fe_3O_4 for the determination of chlorophenols in environmental water samples coupled with liquid chromatography/spectrophotometry detection. *J. Chromatogr. A,* **2008**, *1180*(1-2), 24-31.
[http://dx.doi.org/10.1016/j.chroma.2007.12.028] [PMID: 18179801]

[60] Sun, L.; Chen, L.; Sun, X.; Du, X.; Yue, Y.; He, D.; Xu, H.; Zeng, Q.; Wang, H.; Ding, L. Analysis of sulfonamides in environmental water samples based on magnetic mixed hemimicelles solid-phase extraction coupled with HPLC-UV detection. *Chemosphere,* **2009**, *77*(10), 1306-1312.
[http://dx.doi.org/10.1016/j.chemosphere.2009.09.049] [PMID: 19836824]

[61] Faraji, M.; Yamini, Y.; Rezaee, M. Extraction of trace amounts of mercury with sodium dodecyle sulphate-coated magnetite nanoparticles and its determination by flow injection inductively coupled plasma-optical emission spectrometry. *Talanta,* **2010**, *81*(3), 831-836.
[http://dx.doi.org/10.1016/j.talanta.2010.01.023] [PMID: 20298861]

[62] Zhao, X.; Shi, Y.; Wang, T.; Cai, Y.; Jiang, G. Preparation of silica-magnetite nanoparticle mixed hemimicelle sorbents for extraction of several typical phenolic compounds from environmental water samples. *J. Chromatogr. A,* **2008**, *1188*(2), 140-147.
[http://dx.doi.org/10.1016/j.chroma.2008.02.069] [PMID: 18329033]

[63] Sun, L.; Zhang, C.; Chen, L.; Liu, J.; Jin, H.; Xu, H.; Ding, L. Preparation of alumina-coated magnetite nanoparticle for extraction of trimethoprim from environmental water samples based on mixed hemimicelles solid-phase extraction. *Anal. Chim. Acta,* **2009**, *638*(2), 162-168.
[http://dx.doi.org/10.1016/j.aca.2009.02.039] [PMID: 19327455]

[64] Chen, L.; Zhang, X.; Xu, Y.; Du, X.; Sun, X.; Sun, L.; Wang, H.; Zhao, Q.; Yu, A.; Zhang, H.; Ding, L. Determination of fluoroquinolone antibiotics in environmental water samples based on magnetic

molecularly imprinted polymer extraction followed by liquid chromatography-tandem mass spectrometry. *Anal. Chim. Acta,* **2010**, *662*(1), 31-38.
[http://dx.doi.org/10.1016/j.aca.2010.01.001] [PMID: 20152262]

[65] Ji, Y.; Yin, J.; Xu, Z.; Zhao, C.; Huang, H.; Zhang, H.; Wang, C. Preparation of magnetic molecularly imprinted polymer for rapid determination of bisphenol A in environmental water and milk samples. *Anal. Bioanal. Chem.,* **2009**, *395*(4), 1125-1133.
[http://dx.doi.org/10.1007/s00216-009-3020-5] [PMID: 19690840]

[66] Zhang, Y.; Liu, R.; Hu, Y.; Li, G. Microwave heating in preparation of magnetic molecularly imprinted polymer beads for trace triazines analysis in complicated samples. *Anal. Chem.,* **2009**, *81*(3), 967-976.
[http://dx.doi.org/10.1021/ac8018262] [PMID: 19178336]

[67] Chen, L.; Liu, J.; Zeng, Q.; Wang, H.; Yu, A.; Zhang, H.; Ding, L. Preparation of magnetic molecularly imprinted polymer for the separation of tetracycline antibiotics from egg and tissue samples. *J. Chromatogr. A,* **2009**, *1216*(18), 3710-3719.
[http://dx.doi.org/10.1016/j.chroma.2009.02.044] [PMID: 19268956]

[68] Bouri, M.; Lerma-García, M.J.; Salghi, R.; Zougagh, M.; Ríos, A. Selective extraction and determination of catecholamines in urine samples by using a dopamine magnetic molecularly imprinted polymer and capillary electrophoresis. *Talanta,* **2012**, *99*, 897-903.
[http://dx.doi.org/10.1016/j.talanta.2012.07.053] [PMID: 22967640]

[69] Lerma-García, M.J.; Zougagh, M.; Ríos, A. Magnetic molecular imprint-based extraction of sulfonylurea herbicides and their determination by capillary liquid chromatography. *Mikrochim. Acta,* **2013**, *180*, 363-370.
[http://dx.doi.org/10.1007/s00604-013-0942-6]

[70] Wang, D.; He, J.; Rosenzweig, N.; Rosenzweig, Z. Superparamagnetic Fe_2O_3 beads–CdSe/ ZnS quantum dots core–shell nanocomposite particles for cell separation. *Nano Lett.,* **2004**, *4*, 409-413.
[http://dx.doi.org/10.1021/nl035010n]

[71] Giakisikli, G.; Anthemidis, A.N. Magnetic materials as sorbents for metal/metalloid preconcentration and/or separation. A review. *Anal. Chim. Acta,* **2013**, *789*, 1-16.
[http://dx.doi.org/10.1016/j.aca.2013.04.021] [PMID: 23856225]

[72] You, X.; He, R.; Gao, F.; Shao, J.; Pan, B.; Cui, D. Hydrophilic high-luminescent magnetic nanocomposites. *Nanotechnology,* **2007**, *18*(3), 035701-035706.
[http://dx.doi.org/10.1088/0957-4484/18/3/035701] [PMID: 19636132]

CHAPTER 10

Solid-Phase Microextraction Under the Thin Film Format

M. Roldán-Pijuán, R. Lucena. S. Cárdenas* and **M. Valcárcel**

Department of Analytical Chemistry, Institute of Fine Chemistry and Nanochemistry, Marie Curie Building. Campus de Rabanales. University of Córdoba.14071 Córdoba, Spain

Abstract: Thin film microextraction (TFME) is considered as a type of solid-phase microextraction (SPME). There has been a growing interest in TFME as a novel sample preparation technique, which was originally introduced to address the limiting uptake rate and capacity sometimes observed with fiber microextraction. The inherent properties of TFME technique such as the excellent sample clean-up and the larger surface to volume ratio, enhance the sensitivity and the extraction rates. This e-book article is mainly focused on the fundamental principles behind and in the diverse existing TFME configurations, paying particular attention to cotter pin supported format, copper mesh holder and 96-blade format.

Keywords: : Autosampler, Field sampling, *In-vivo* sampling, Membrane-based TFME, Microextraction, On-site extraction, Sample preparation, Solid phase microextraction (SPME), Sorptive tape extraction, Stirring integrated TFME, Thin film microextraction (TFME), 96-blade (thin film) SPME.

10.1. INTRODUCTION

Solid-phase microextraction (SPME) is a sample preparation technique developed by Pawliszyn in the early 1990s which integrates sample clean-up, analyte isolation, and pre-concentration in a single step [1]. This technique appeared

* **Address correspondence to S. Cárdenas:** Department of Analytical Chemistry, Institute of Fine Chemistry and Nanochemistry, Marie Curie Building. Campus de Rabanales. University of Córdoba.14071 Córdoba, Spain; Tel: 0034 957 218616; E-mail:qa1caarm@uco.es

M. Valcárcel, S. Cárdenas, R. Lucena (Eds.)

to address the need for a rapid and solvent-free extraction technique. SPME is based on the partitioning of the analytes between the sample and the extraction phase, which is typically immobilized on a fiber support. For direct immersion, the amount of analyte extracted when equilibrium conditions are reached (n_e) by SPME can be described as [2]:

$$n_e = \frac{K_{es} V_e V_s C_0}{K_{es} V_e + V_s} \quad \textbf{(1)}$$

Where K_{es} is the distribution constant between the extraction phase and the sample matrix, V_e represents the sorbent phase volume, V_s is the sample volume and C_0 is the initial concentration of a analyte in the sample matrix. Since in the majority of the extractions the sample volume is higher than the extraction phase, this equation can be simplified to:

$$n_e = K_{es} V_e C_0 \quad \textbf{(2)}$$

According to **Equation 2**, the mass of analyte extracted depends on the sorbent phase volume (V_e). There are two possible strategies to increase the extraction phase volume:(i) make greater in size the thickness of the coating or (ii) expand the surface area. However, when thicker coatings are employed larger times are needed to reach the equilibrium (t_e), which gives rise low sample throughput because the overall rate of extraction is controlled by the diffusion from the bulk solution through the boundary layer to the extraction coating (**Equation 3**):

$$t_e = t_{95\%} = 3\frac{\delta K_{es}(b-a)}{D} \quad \textbf{(3)}$$

where *(b-a)* is the thickness of the coating, *D* represents the diffusion coefficient and δ is the thickness of the boundary layer. Based on **Equation 3** [3],the thinner the thickness of the sorbent phase is, the shorter the time required to reach the extraction equilibrium it would be.

On the other hand, large surface area accelerates the initial rate of extraction as it is shown in **Equation 4** [3]. As a consequence, a thin film with a large surface area-to-volume ratio ensue in an improvement of the extraction efficiency without sacrifice the extraction time assuming the same conditions.

$$\frac{dn}{dt} = \left(\frac{DA}{\delta}\right) C_o \qquad \textbf{(4)}$$

Where (n) is the amount of analyte extracted over the sampling time (t) and (A) is the surface area of the extraction phase. On the basis of the previous discussion, the superficial area of a SPME phase directly affects to the extraction kinetics and this theoretical fact has led to the development of thin film microextraction (TFME) by Prof. Pawliszyn group in 2003 [3]. According to the inventors group, TFME can be divided in two different modes depending on the use of the membranes or brushes as extracting phase [4].

10.2. THIN FILM MICROEXTRACTION UNDER THE MEMBRANE CONFIGURATION

Membrane-based TFME makes use of a thin membrane of polydimethylsiloxane (PDMS), the classical SPME coating, as extraction phase. In the classical configuration the membrane is attached to a deactivated stainless steel rod and it is deployed, in a flag shape, to exploit all its potential area during extraction (see Fig. **1**). After the extraction, which can be accomplished both in the direct immersion and headspace mode, the membrane is coiled around the rod which is finally introduced in a glass liner for gas chromatography/mass spectrometry (GC/MS) analysis. The first approach of TFME, which was directly towards the analysis of several polycyclic aromatic hydrocarbons (PAHs) from water samples, demonstrated the superior features of TFME over classical SPME [3]. In summary, TFME presents a higher extraction rate and capacity owing to the larger area and higher volume of extractant phase (at least 4 times) in the membrane format over the fiber shape. In fact, while a typical 100 µm PDMS fiber presents an area of 10 mm^2, a 1 cm x 1 cm membrane sheet presents a surface area of 200 mm^2 (two faces of 100 mm^2 each one) due to its planar structure.

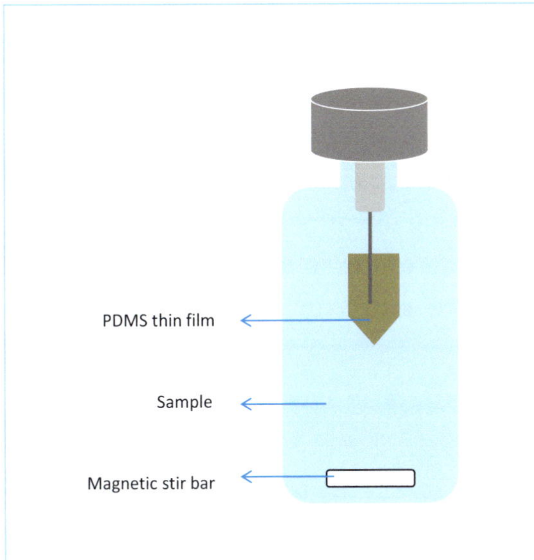

Fig. (1). Cotter pin supported format of TFME.

The stirring of the sample during extraction is critical to achieve an acceptable extraction rate. In fact, as it can be deduced form **Equation 3**, the time required to reach the extraction equilibrium depends on the thickness of the boundary layer which may be controlled with the efficient stirring of the sample solution. This aspect, applicable to all the microextraction techniques, plays a key role on those extraction/stirring integrated techniques. Stir bar sorptive extraction, developed by Prof. Sandra group, in 1999 [5], is the clear paradigm in this context and presents a clear similarity with TFME since a PDMS phase coated on a glass jacketed iron bar is used as extractant. Qin *et al.* compared TFME and SBSE in terms of kinetics and efficiency using some PAHs as model analytes [6]. However, the different stirring performance of both techniques (SBSE integrates extraction and stirring in the same device whereas, as it is shown in Fig. (**1**), classical TFME is performed with an external magnetic bar) limits their direct comparison.

To normalize both performances, the researchers modified the TFME procedure integrating the thin film in an electric drill, which produces the stirring of the extraction phase rather than the solution. Moreover, the researchers developed a new holder to attach the thin film, avoiding their bending during the stirring. In the light of the extraction rate profiles obtained for different PAHs, TFME resulted to be faster than SBSE. This observed behavior can be explained taking into account **Equations 3** and **4**. On the one hand, t_e increases with the thickness of the extraction phase which is higher in SBSE than in TFME. On the other hand, the initial extraction rate increases with the extractant area which is larger in TFME. Additionally, stirring integrated TFME opens a door to on-site extractions (Fig. **2**) which is quite interesting in environmental applications since the target analytes usually are more stable in the extractant phase than in the original sample.

Fig. (2). On-site sampling. Reproduced with permission of Elsevier from Reference [7].

This aspect has been studied in depth by the same research group of the first development using some PAHs as model analyte. In this work they developed a new configuration based on a copper mesh holder (Fig. **3**) in order to protect the thin film during the extraction and to make easier the transference of the film to the glass liner for chromatographic analysis [7].

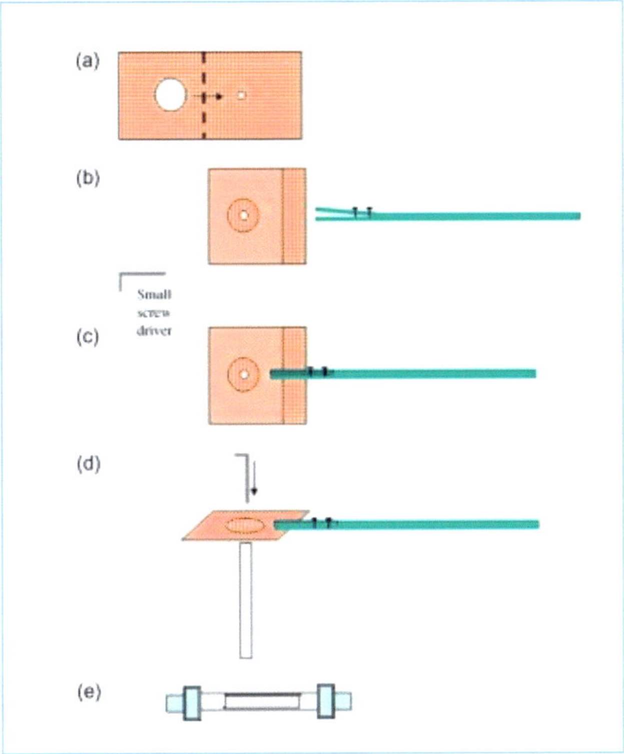

Fig. (3). Thin film copper mesh.(**a**) Fold (**b**, **c**) fix on a rod (**d**) push into a liner (**e**) seal the liner for transportation. Reproduced with permission of Elsevier from Reference [7].

TFME can be also applied in field sampling in order to measure the time-weighted average concentration of target pollutants in environmental specimens. This approach has been evaluated for the monitoring of PAHs in river water [8] using a thin film of PDMS (127 µm in thickness) as sorptive phase. The thin film (see Fig. **4**) is cut into a house-shape format that consists of a square of 2cmx2cm with an additional triangular section of 1 cm in height. The thin film is attached to a stainless steel wire that plays two roles: (i) it serves as support of the thin film and (ii) allows its introduction on a GC liner (see Fig. **4**) for final analysis. For field sampling, the thin film is introduced in a copper cage to protect the sorptive phase from degradation and contamination of algae. The sampling is performed for a defined time and after sampling the thin film is recovered and stored in a sealed vial until analysis. Before analysis, the thin film is washed with water, rolled up and introduced in a glass inlet for GC/MS analysis. The inherent irreproducibility of field sampling, due to the different variables that may affect the extraction, is overcome using an internal standard (deuterated analytes) impregnating the membrane. The internal standard losses during the extraction provided information about the sampling conditions (time, volume of sample), normalizing their effect.

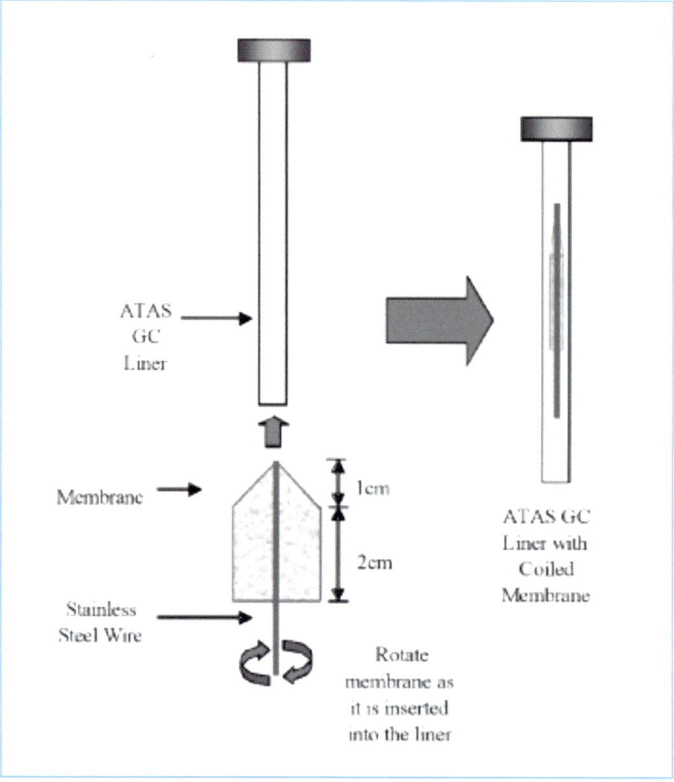

Fig. (4). PDMS thin-film configuration and its coupling into the ATAS GC liners for analysis. Reproduced with permission of Oxford Journals from Reference [8].

Although the first developments of TFME were focused on the extraction of liquid samples, TFME can be applied for the extraction of gaseous samples which is quite interesting in environmental analysis. In this sense, Eom *et al.* employed the technique for the monitoring of ambient air for the detection of biomarkers of the presence of Cimex lectularius [9]. In the extraction procedure, the thin film is located in the back-side of a small electric fan which circulates air in order to increase the sampled volume. As in any gas-solid partitioning process, the extraction temperature plays a key role on the sorption of the analytes. The reduction of the sorptive phase temperature should increase the extraction efficiency. In view of this, Jiang and Pawliszyn have recently proposed the use of cooled membranes in gas sampling [10] using a special extraction device (Fig. **5**). The device allows the continuous flow of sample gas through the sorptive phase and the membrane cooling by means of an electric fan and thermoelectric cooler, respectively. In addition, the device permits the direct introduction of the membrane after the extraction in a thermal desorption tube for GC/MS analysis avoiding losses of the target analytes.

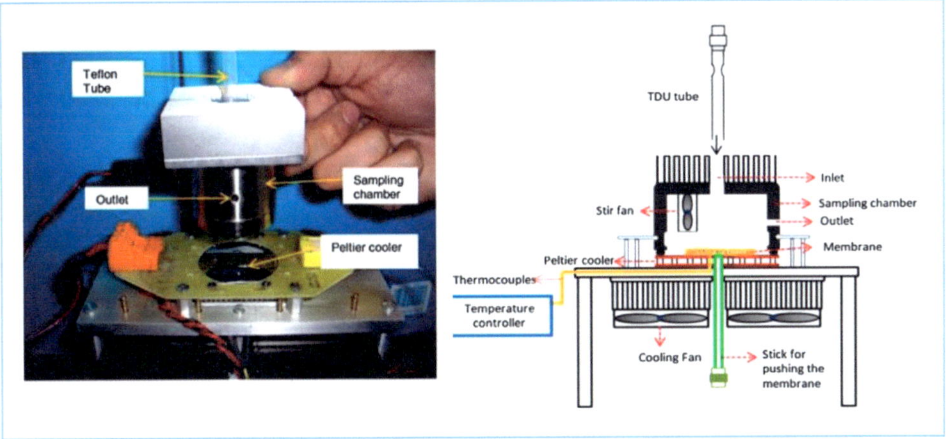

Fig. (5). Configuration of cooled membrane device. Reproduced with permission of Elsevier from Reference [10].

The great versatility of the technique allows its application to process solid samples. In this context, sorptive tape extraction (STE) was proposed by Prof. Sandra´s group in 2006 as a microextraction technique adapted to the sampling of the surface of a given system [11]. As TFME, STE is based on the use of a flexible thin film of PDMS phase which is attached to the sample acting as chemical trap which retains the analytes on the basis of its sample/PDMS partition coefficient. In its first proposal, STE was applied for the qualitative and quantitative characterization of skin sebum. A thin film of PDMS is attached to a defined area of the skin with a plaster and the sampling/extraction is developed in a defined time, usually 15 min. The extracted analytes can be thermally desorbed or chemical eluted with acetone prior to their final determination by GC/MS. Although the thickness (0.5 mm) and length (15 mm) of the PDMS are fixed, the width of the film is different depending on the final procedure. In this sense, the width is 4 mm and 12 mm for thermal desorption and chemical elution, respectively, involving a different sampled area in each application. STE provided comparable results than classical techniques employed in sebum analysis. It is simpler, more precise and present a wide application field that may comprise not only the sebum characterization (different compounds and concentrations) but also the *in-vivo* analysis of the efficiency of skin treatment that is of outstanding importance in the clinical context.

The same research group has extended the application of STE to the analysis of biological solid matrices which can be directly or indirectly sampled, depending whether the thin tape is located on the surface of the material in direct contact (DC)-STE, or in close contact with its headspace (HS-STE) [12]. Two different extraction vessels can be employed depending on the type of matrix to be analyzed using in both cases a STE holder, a stainless steel piece which acts as support of the thin tape. For *in-vivo* sampling of plant leaves, a vial with four inlets (as it can be observed in Fig. (**6A**)) is employed. One of the inlets is pierced allowing the introduction of the leaves in the extraction vessel while the STE holder is introduced through another inlet. The effect of stress on the plant can be easily studied with this device. For skin analysis, an extraction vessel that houses the STE holder, as it is depicted in Fig. (**6B**), is employed. This approach isolates the skin from the environment thanks to its bell shape allowing the determination of natural volatiles produced by the skin or synthetic volatiles generated after the application of cosmetic products such as perfumes.

Bichi *et al.* also employed STE to determine potential allergens in cosmetic creams applied in the skin of voluntaries [13]. This approach can be employed to understand the skin absorption of these compounds and also their decay after application. For this purpose, cosmetic creams spiked with the target allergens were applied on the back of the hand of the voluntaries and six PDMS tapes were applied covering a surface of 32 cm^2. After the extraction, the tapes were thermally desorbed, the analytes being analyzed by GC. The results confirmed again the potential of STE in this type of studies. Special attention should be given to the recovery obtained since almost all of them fulfilled the 70-130% criterion.

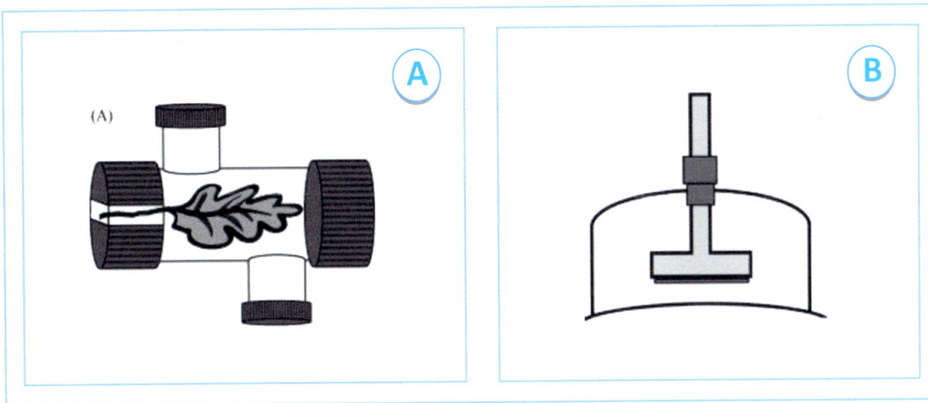

Fig. (6). Tools for *in-vivo* sampling: (**A**) HS-STE of solid samples and (**B**) DC-STE of biological matrices. Reproduced with permission of Elsevier from Reference [12].

STE has been adapted to the analysis of lipids composition in cosmetic lipsticks [14]. The cosmetics are applied to a polymeric substrate which is put in close contact with a silica TLC plate that acts a STE material. The compounds in the STE are finally analyzed laser desorption/ionization mass spectrometry imaging.

TFME was employed by Jiang *et al.* for the monitoring of volatile compounds released from human skin, which can be employed for metabolic studies [15]. The extraction can be performed either in DI or HS modes, although the latter approach provides a higher selectivity since it allows the extraction of only compounds with high and medium volatility. In this approach, a circular PDMS film was employed as extractant phase and it is sandwiched between two stainless steel meshes to avoid its direct contact with the skin surface. The bottom mesh is used as separation between the membrane and the skin whereas the top mesh is employed to fix the whole device into the skin. Finally, the system is covered with aluminum foil to avoid external contamination from the environment. The proposed configuration is evaluated in both in-vial and *in-vivo* manifolds showing acceptable reproducibility even when different films are employed. In this communication several interesting variables are evaluated, the appropriate storage conditions as well as the comparison between direct contact and headspace are remarkable. The approach was finally applied to two different metabolic studies. In the first one, some markers of the consumption of garlic are analyzed in the skin emission and on the other case ethanol is determined.

Jahnke *et al.* evaluated the possibilities and limitations of using TFME for the extraction of several polychlorinated biphenyls (PCBs) directly from the fish tissue [16]. For this purpose, disks of PDMS of variable thickness are inserted through slots, previously done with a scalpel, in the fish tissue. For the extraction, the fish is wrapped into aluminum foil and stored during the extraction process in the fridge. After the extraction, the thin films are chemically eluted and the final extract is analyzed by GC/MS. As the lipid composition is not homogeneous in all the fish tissues, several films are placed in different locations and finally pooled before analysis. This technique presents a great potential but the large times required for extraction is the main shortcoming.

10.3. 96-BLADE BRUSHES (THIN FILM) SPME

The increasing research interest to improve current sample preparation technique is directed towards the development of fast and high-throughput analysis [17 - 21]. Sample extraction rate is enhanced when multiple samples are processed simultaneously. The performance of SPME in a 96-well-plate enables parallel analysis of 96 samples [22]. The first coupling of TFME to this automatic 96-well-plate was developed by the research group of Prof. Pawliszyn [23]. This system uses 96 small solid phase extraction (SPE) disk membranes fixed to the stainless steel wires for the determination of benzodiazepines in biological samples. With the aim to enhance both the stability and the precision in the development of the coating in the concept of 96-thin film SPME, the same authors developed a flat stainless steel thin film coated with a layer of octadecyl (C_{18}) [24]. An automated in-house 96-fiber SPME system for high-throughput was developed. By increasing the contact surface area between the SPME coating and the sample, the recovery of the extraction is noticeably increased. However, in-house multi-fiber device presents some shortcomings such as the instability of the fiber during the agitation or the complications in making the thin film flatten. To overcome these limitations, in 2011 was reported sprayed C_{18}-polyacrylonitrile

(PAN) coating making use of commercially available automated 96-blade (thin film) technology [25] designed by Professional Analytical System (PAS) (Magdala, Germany). Briefly, the concept 96-blade SPME device is made of a 96-blade SPME device that fits in the 96-well-plate and works as a part of an autosampler. The high throughput analysis of the samples enhance the work rate and simplifies the tedious sample preparation steps. Furthermore, automation and precision process control is achieved since each step is timing regulate. Fig. (**7**) illustrates a photograph of the current configuration of Concept 96-autosampler [26]. The standard instrument consists of four working positions (each equipped with an orbital shaker) allowing the performance of the following steps: (i) preconditioning, (ii) extraction, (iii) washing and (iv) solvent desorption and (optional) evaporation step.

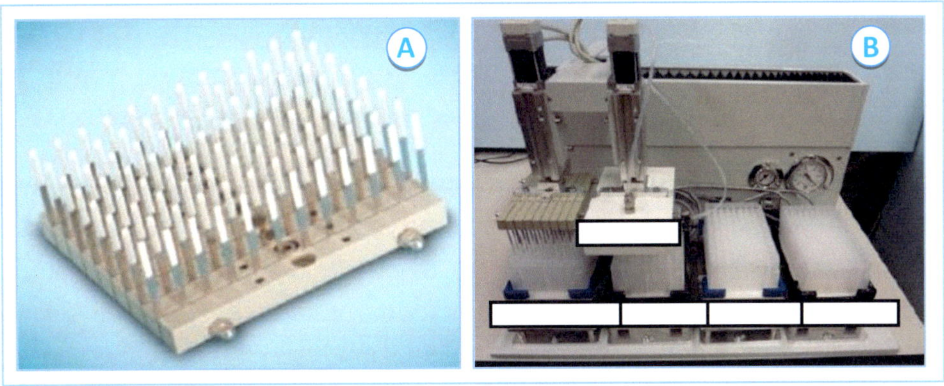

Fig. (7). (**A**) 96-blade (thin film) SPME (**B**) Concept 96-autosampler system. (**A**) Reproduced with permission of ACS Publications from Reference [25] and (**B**) adapted from Reference [26] with permission of Springer.

Prior to the extraction, 96-blade (thin film) SPME is preconditioned to activate in an efficient way the coatings. Typically, a mixture of methanol and water (1:1, v/v) and 15-30 min are employed. As a general rule, taking into account that high stirring speed are generally aployed to perform the extraction, the sample volume should be avoid the liquid to spill over and possible contaminations of other samples. On the other hand, the selected volume should be large enough to ensure that the totality of the coating is dipped in the sample. After the extraction and before the desorption, a fast wash step in a weak solvent is needed. This step is mandatory when dealing with biological samples so as to wash properly the coating and avoid the transference of biomolecules to the elution solvent. Usually, a wash with nanopure water for 10-20 second is enough. Concerning the chemical elution step, two strategies can be employed: (i) direct solvent desorption and (ii) solvent desorption followed by evaporation/reconstitution. Direct solvent desorption should be performed in a large volume to guarantee that the entire length of the extraction phase is eluted. Typically, acetonitrile:water (1:1, v/v) is employed with average desorption time in the range of 40-120 min. The second approach is based on thc direct solvent desorption followed by solvent evaporation to dryness and reconstitution into an small volume of solvent suitable for LC. The instrument has an special nitrogen unit so as to carry out both the evaporation of the solvent and analyte enrichment. This strategy is suitable for non-volatile compounds hence analyte loss during evaporation is not an issue. However, it can be also employed with semivolatile analytes if an internal standard is added to quantify analyte losses. A second desorption step in fresh solvent is usually performed in order to avoid undesirable carry-over effects.

Trends in 96-blade (thin film) SPME systems coating has been in a continuous evolution in the last decade. A biocompatible coating PAN-over C_{18}-PAN was applied to the direct analysis of diazepam from whole blood [27]. Polystyrene-divinylbenzene (PS-DVB) PAN and phenylboronic acid (PBA) PAN were employed as coating allowing the simultaneous analysis of different compounds with a wide range of polarity (log P= -3.7 to 2.8) from biological samples [28]. Also C_{18}-PAN has been reported to provide high recovery with compounds in wide range of polarity (log P=0.14 to 4.98) [29]. C_{18}-strong cation exchange (SCX) is another coating chemistry for the TFME system. This coating has been employed for the determination of carbamazepine and triclosan in wastewater [30]. Later, ionic liquid (IL)-based coatings have been also reported. The synthesis of octadecylimidazolium-modified silica ($SiImC_{18}$) PAN 96-blade SPME was employed as an extraction coating for aminoacids (highly polar nature) from complex matrices coupled to LC-MS/MS analysis [31]. Thin film C_{18}-silica glass coating by sol-gel was employed for the determination of benzodiazepines [32]. 96-blade SPME based on the use of carbon nanomaterials was reported for the first time in 2013 [33]. In this proposal, carboxilated multiwalled carbon nanotubes/polydimethylsiloxane (MWCNTs-COOH/PDMS) combines the high adsorption capacity of MWCNTs with the high chemical stability of PDMS. The developed coating is reusable up to 110 extractions and it has been employed for the analysis of phenolic compounds in waters samples. Later, the extraction of these phenolic compounds in complex matrices (wine, berry and grape) has been accomplished making use of PS-DVB-PAN SPME coating [34]. Hydrophilic lipophilic balance (HLB) coated-blades were suitable for the simultaneous determination of nine benzylic and aliphatic quaternary ammonium compounds coupled to LC-MS/MS analysis [35].

10.4. OTHER MICROEXTRACTION TECHNIQUES BASED ON THIN FILMS OR MEMBRANES AS EXTRACTION PHASES

As it was previously indicated, the stirring of the sample during the extraction increases the analyte transference to the extractant phase accelerating the process. Integrated extraction/stirring devices offers additional advantages. On the one hand, the agitation is more intense near to the boundary layer. On the other hand, the potential retention of excessively non-polar compounds on the polymeric stirring bar used is avoided. The use of membrane and thin films as extraction phases in stirring/extraction integrated techniques was proposed almost simultaneously in 2009 by the research groups of Prof. Valcárcel and Prof. Richter. The first research group proposed the stir membrane extraction (SME) [36] technique, which employs a thin polymeric membrane adapted to a stirring device as extraction phase, while the latter proposed the rotating disk sorptive extraction (RDSE) [37] where a thin film of PDMS integrated in a rotating disk is employed to extract the target analytes. The typical extraction units employed in both techniques are presented in Fig. (**8**).

The typical SME configuration uses polytetrafluorethylene (PTFE) membranes to isolate and preconcentrate hydrophobic analytes. The extraction is performed introducing the extraction unit into the sample and stirring it for a fix time. Later on, the unit is withdrawn from the sample and the membrane can be chemically eluted [36] or directly monitored by a spectroscopic technique [38]. Following both procedures, SME has been applied to the extraction of PAHs and to determine the global hydrocarbon index in water samples, respectively. The comparison of SME with SBSE indicates superior extraction rates that can be ascribed to the planar format of the extractant phase.

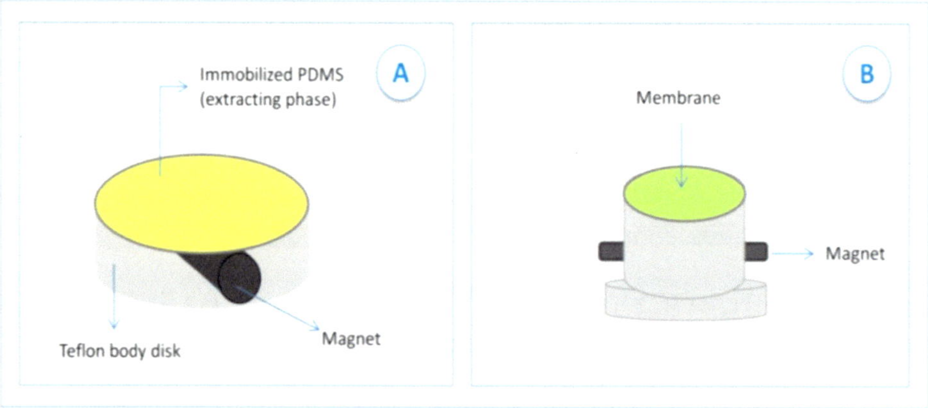

Fig. (8). (**A**) Rotating disk sorptive extraction and (**B**) stir membrane extraction devices.

RDSE was proposed as an alternative to classic SBSE in order to avoid the contact between the sorptive phase and the vessel walls that may produce the cracking of this phase. This modification let use high-rate stirring in comparison witg SBSE, increasing the mass transference from the sample to the sorptive extraction media. The typical extraction procedure involves the introduction of the unit into the sample where it is stirred during a defined time in order to isolate the target compounds. Once the extraction is performed, the unit is taken out from the sample and the PDMS film is eluted [39, 40] or spectroscopically analyzed [41].

10.5. NEW MATERIALS FOR THIN FILM CONSTRUCTION

Although PDMS is the classical phase in TFME, several attempts have been focused on the development of novel sorptive phases. Bagheri and Aghakhani proposed in 2012 the use of polymeric composite membranes of polyaniline (PANI) and Nylon-6 (N6) as sorptive phase under the HS mode for the extraction of several chlorobenzenes from water samples [42]. The PANI-N6 membrane was synthesized by electrospinning over a collector made of aluminum foil providing a polymeric mat with nanofibers diameters in the range of 100–300 nm. For the extraction, the membrane is located in gauze metallic holder which is attached to the upper part of the vial, where the sample is located, to avoid the direct immersion of the membrane. The composite membrane, where the aromatic rings presented in the PANI structure, are responsible for the extraction of the analytes by π-π interactions, provided an excellent extraction performance allowing the determination of the analytes in the ng/L using GC/MS as instrumental technique.

Saraji *et al.* proposed in 2013, the use of cellulose membranes in TFME [43]. As cellulose is a hydrophilic polymer which cannot interact strongly enough with organic compounds in an aqueous environment, the authors proposed its derivatization to include functional groups able to interact with the target analytes, three estrogenic compounds. Phenyl isocyanate (PHIC), cyclohexyl isocyanate (CHIC), diphenyldichlorosilane (DPDS), and octadecyltrichlorosilane (ODTS) were selected as modifiers of the cellulose membrane and the extraction performance of obtained materials were evaluated. The hydrophobic/hydrophilic balance of the materials was crucial on their extraction performance. On the one hand, the membrane should present a high interaction, usually by non polar and polar forces, with the targets. On the other hand, the membrane should be hydrophilic enough to allow the penetration of the aqueous sample. Considering this, PHIC-cellulose membrane provided the best extraction recoveries.

Kabir *et al.*, who have reviewed the potential of sol-gel phases in sorptive microextraction [44], have recently proposed a novel microextraction technique called fabric phase sorptive extraction (FPSE) [45]. This technique uses a high porosity cellulose membrane as support of a sol-gel organic coating which is uniformly distributed on the network. The high porosity makes easier the penetration of the aqueous phase enhancing the partitioning of the analytes. In addition, the high density of the sol-gel coating, which is 400 times higher than in a conventional SPME fiber, provides FP with a high sorptive capacity. In the first application, FPSE has been applied for the determination of 17α-ethynylestradiol, β-estradiol and bisphenol A (BPA) in urine and water samples by HPLC-FD [46].

CONCLUSION AND TRENDS

TFME has been proved to be an excellent alternative in sample preparation. TFME delivers remarkable advantages such as: (i) high degree of clean-up because of the selectivity towards the target analytes over the matrix interferents (ii) high sensibility since it is a flat film with a high surface area-to-volume ratio of extracting phase and in consequence (iii) short analysis time.

Current research of this technique is focused on two lines of investigation. On the one hand, efforts are mainly toward the development of new coating chemistries. In this context, the reproducibility of the reported coatings can be further enhanced with the development of automatic preparation coating procedures. Furthermore, the synthesis of biocompatible coating with high chemical stability would provide long-term reusability since to date, most of thin-film for bioanalytical purpose are disposable. On the other hand, the development of new thin film configurations would its implementation in environmental, toxicology, forensic and clinic fields. In addition to these two main objectives, research must also focus on the search of novel development for thermal desorption and direct combination with analytical instrumentation.

CONFLICT OF INTEREST

The authors confirm that they have no conflict of interest to declare for this publication.

ACKNOWLEDGEMENTS

The author would like thank the Spanish Ministry of Science and Innovation for project CTQ2014-52939R. M. Roldán-Pijuán expresses her gratitude for the predoctoral grant (refs AP2009-3499) from the Spanish Ministry of Education.

REFERENCES

[1] Arthur, C.L.; Pawliszyn, J. Solid Phase Microextraction with Thermal Desorption Using Fused Silica Optical Fibers. *Anal. Chem.,* **1990**, *62*, 2145-2148. [http://dx.doi.org/10.1021/ac00218a019]

[2] Pawliszyn, J., Ed. *Solid-Phase Microextraction: Theory and Practice*; Wiley-VCH: New York, **1997**.

[3] Bruheim, I.; Liu, X.; Pawliszyn, J. Thin-film microextraction. *Anal. Chem.,* **2003**, *75*(4), 1002-1010. [http://dx.doi.org/10.1021/ac026162q] [PMID: 12622398]

[4] Webpage of Prof. Pawliszyn research group. https://uwaterloo.ca/pawliszyn-group/research/thin-film. Accessed on August 2014.

[5] Baltusen, E.; Sandra, P.; David, F.; Cramers, C. Stir bar sorptive extraction (SBSE). A novel extraction

technique for aqueous samples: theory and principles. *J. Microcolumn,* **1999**, *11*, 737-747. [http://dx.doi.org/10.1002/(SICI)1520-667X(1999)11:10<737::AID-MCS7>3.0.CO;2-4]

[6] Qin, Z.; Bragg, L.; Ouyang, G.; Pawliszyn, J. Comparison of thin-film microextraction and stir bar sorptive extraction for the analysis of polycyclic aromatic hydrocarbons in aqueous samples with controlled agitation conditions. *J. Chromatogr. A,* **2008**, *1196-1197*, 89-95. [http://dx.doi.org/10.1016/j.chroma.2008.03.063] [PMID: 18423652]

[7] Qin, Z.; Bragg, L.; Ouyang, G.; Niri, V.H.; Pawliszyn, J. Solid-phase microextraction under controlled agitation conditions for rapid on-site sampling of organic pollutants in water. *J. Chromatogr. A,* **2009**, *1216*(42), 6979-6985. [http://dx.doi.org/10.1016/j.chroma.2009.08.052] [PMID: 19762034]

[8] Bragg, L.; Qin, Z.; Alaee, M.; Pawliszyn, J. Field sampling with a polydimethylsiloxane thin-film. *J. Chromatogr. Sci.,* **2006**, *44*(6), 317-323. [http://dx.doi.org/10.1093/chromsci/44.6.317] [PMID: 16884586]

[9] Eom, I-Y.; Risticevic, S.; Pawliszyn, J. Simultaneous sampling and analysis of indoor air infested with Cimex lectularius L. (Hemiptera: Cimicidae) by solid phase microextraction, thin film microextraction and needle trap device. *Anal. Chim. Acta,* **2012**, *716*, 2-10. [http://dx.doi.org/10.1016/j.aca.2011.06.010] [PMID: 22284870]

[10] Jiang, R.; Pawliszyn, J. Cooled membrane for high sensitivity gas sampling. *J. Chromatogr. A,* **2014**, *1338*, 17-23. [http://dx.doi.org/10.1016/j.chroma.2014.02.070] [PMID: 24647099]

[11] Sisalli, S.; Adao, A.; Lebel, M.; Le Fur, I.; Sandra, P. Sorptive Tape Extraction — A Novel Sampling Method for the *in vivo* Study of Skin. *LC GC Eur.,* **2006**, *19*, 33.

[12] Bicchi, C.; Cordero, C.; Liberto, E.; Rubiolo, P.; Sgorbini, B.; Sandra, P. Sorptive tape extraction in the analysis of the volatile fraction emitted from biological solid matrices. *J. Chromatogr. A,* **2007**, *1148*(2), 137-144. [http://dx.doi.org/10.1016/j.chroma.2007.03.007] [PMID: 17395189]

[13] Sgorbini, B.; Ruosi, M.R.; Cordero, C.; Liberto, E.; Rubiolo, P.; Bicchi, C. Quantitative determination of some volatile suspected allergens in cosmetic creams spread on skin by direct contact sorptive tape extraction-gas chromatography-mass spectrometry. *J. Chromatogr. A,* **2010**, *1217*(16), 2599-2605. [http://dx.doi.org/10.1016/j.chroma.2009.12.052] [PMID: 20074740]

[14] de Oliveira, D.N.; Siqueira, M.; Sartor, S.; Catharino, R. Direct analysis of lipsticks by Sorptive tape-like extraction laser desorption/ionization mass spectrometry imaging. *Int. J. Cosmet. Sci.,* **2013**, *35*(5), 467-471. [http://dx.doi.org/10.1111/ics.12066] [PMID: 23651468]

[15] Jiang, R.; Cudjoe, E.; Bojko, B.; Abaffy, T.; Pawliszyn, J. A non-invasive method for *in vivo* skin volatile compounds sampling. *Anal. Chim. Acta,* **2013**, *804*, 111-119. [http://dx.doi.org/10.1016/j.aca.2013.09.056] [PMID: 24267071]

[16] Jahnke, A.; Mayer, P.; Broman, D.; McLachlan, M.S. Possibilities and limitations of equilibrium sampling using polydimethylsiloxane in fish tissue. *Chemosphere,* **2009**, *77*(6), 764-770. [http://dx.doi.org/10.1016/j.chemosphere.2009.08.025] [PMID: 19765800]

[17] Peng, S.X.; Henson, C.; Strojnowski, M.J.; Golebiowski, A.; Klopfenstein, S.R. Automated high-throughput liquid-liquid extraction for initial purification of combinatorial libraries. *Anal. Chem.*, **2000**, *72*(2), 261-266.
[http://dx.doi.org/10.1021/ac990946v] [PMID: 10658318]

[18] Kataoka, H.; Lord, H.L.; Yamamoto, S.; Narimatsu, S.; Pawliszyn, J. Development of automated in-tube SPME/LC/MS method for drug analysis. *J. Microcolumn Sep.*, **2000**, *12*, 493-500.
[http://dx.doi.org/10.1002/1520-667X(2000)12:9<493::AID-MCS3>3.0.CO;2-C]

[19] Dotsikas, Y.; Apostolou, C.; Kousoulos, C.; Tsatsou, G.; Loukas, Y.L. An improved high-throughput liquid chromatographic/tandem mass spectrometric method for terbinafine quantification in human plasma, using automated liquid-liquid extraction based on 96-well format plates. *Biomed. Chromatogr.*, **2007**, *21*(2), 201-208.
[http://dx.doi.org/10.1002/bmc.738] [PMID: 17221909]

[20] Peng, S.X.; Branch, T.M.; King, S.L. Fully automated 96-well liquid-liquid extraction for analysis of biological samples by liquid chromatography with tandem mass spectrometry. *Anal. Chem.*, **2001**, *73*(3), 708-714.
[http://dx.doi.org/10.1021/ac001036c] [PMID: 11217789]

[21] Rogacs, A.; Santiago, J.G. Temperature effects on electrophoresis. *Anal. Chem.*, **2013**, *85*(10), 5103-5113.
[http://dx.doi.org/10.1021/ac400447k] [PMID: 23627294]

[22] Vuckovic, D.; Cudjoe, E.; Hein, D.; Pawliszyn, J. Automation of solid-phase microextraction in high-throughput format and applications to drug analysis. *Anal. Chem.*, **2008**, *80*(18), 6870-6880.
[http://dx.doi.org/10.1021/ac800936r] [PMID: 18712934]

[23] Cudjoe, E.; Pawliszyn, J. A new approach to the application of solid phase extraction disks with LC-MS/MS for the analysis of drugs on a 96-well plate format. *J. Pharm. Biomed. Anal.*, **2009**, *50*(4), 556-562.
[http://dx.doi.org/10.1016/j.jpba.2008.07.014] [PMID: 18771872]

[24] Cudjoe, E.; Vuckovic, D.; Hein, D.; Pawliszyn, J. Investigation of the effect of the extraction phase geometry on the performance of automated solid-phase microextraction. *Anal. Chem.*, **2009**, *81*(11), 4226-4232.
[http://dx.doi.org/10.1021/ac802524w] [PMID: 19413335]

[25] Mirnaghi, F.S.; Chen, Y.; Sidisky, L.M.; Pawliszyn, J. Optimization of the coating procedure for a high-throughput 96-blade solid phase microextraction system coupled with LC-MS/MS for analysis of complex samples. *Anal. Chem.*, **2011**, *83*(15), 6018-6025.
[http://dx.doi.org/10.1021/ac2010185] [PMID: 21711040]

[26] Mirnaghi, F.S.; Pawliszyn, J.; Hein, D. Thin-Film Microextraction Coupled with Mass Spectrometryand Liquid Chromatography–Mass Spectrometry. *Chromatographia*, **2013**, *76*, 1215-1223.
[http://dx.doi.org/10.1007/s10337-013-2443-5]

[27] Mirnaghi, F.S.; Pawliszyn, J. Reusable solid-phase microextraction coating for direct immersion whole-blood analysis and extracted blood spot sampling coupled with liquid chromatography-tandem mass spectrometry and direct analysis in real time-tandem mass spectrometry. *Anal. Chem.*, **2012**,

84(19), 8301-8309.
[http://dx.doi.org/10.1021/ac3018229] [PMID: 22928515]

[28] Mirnaghi, F.S.; Pawliszyn, J. Development of coatings for automated 96-blade solid phase microextraction-liquid chromatography-tandem mass spectrometry system, capable of extracting a wide polarity range of analytes from biological fluids. *J. Chromatogr. A,* **2012**, *1261*, 91-98.
[http://dx.doi.org/10.1016/j.chroma.2012.07.012] [PMID: 22824218]

[29] Mirnaghi, F.S.; Goryński, K.; Rodriguez-Lafuente, A.; Boyacı, E.; Bojko, B.; Pawliszyn, J. Microextraction *versus* exhaustive extraction approaches for simultaneous analysis of compounds in wide range of polarity. *J. Chromatogr. A,* **2013**, *1316*, 37-43.
[http://dx.doi.org/10.1016/j.chroma.2013.09.084] [PMID: 24119750]

[30] Strittmatter, N.; Düring, R.A.; Takáts, Z. Analysis of wastewater samples by direct combination of thin-film microextraction and desorption electrospray ionization mass spectrometry. *Analyst (Lond.),* **2012**, *137*(17), 4037-4044.
[http://dx.doi.org/10.1039/c2an35411j] [PMID: 22814299]

[31] Mousavi, F.; Pawliszyn, J. Silica-based ionic liquid coating for 96-blade system for extraction of aminoacids from complex matrixes. *Anal. Chim. Acta,* **2013**, *803*, 66-74.
[http://dx.doi.org/10.1016/j.aca.2013.07.005] [PMID: 24216198]

[32] Mirnaghi, F.S.; Monton, M.R.; Pawliszyn, J. Thin-film octadecyl-silica glass coating for automated 96-blade solid-phase microextraction coupled with liquid chromatography-tandem mass spectrometry for analysis of benzodiazepines. *J. Chromatogr. A,* **2012**, *1246*, 2-8.
[http://dx.doi.org/10.1016/j.chroma.2011.11.030] [PMID: 22197254]

[33] Kueseng, P.; Pawliszyn, J. Carboxylated multiwalled carbon nanotubes/polydimethylsiloxane, a new coating for 96-blade solid-phase microextraction for determination of phenolic compounds in water. *J. Chromatogr. A,* **2013**, *1317*, 199-202.
[http://dx.doi.org/10.1016/j.chroma.2013.08.038] [PMID: 23972463]

[34] Mirnaghi, F.S.; Mousavi, F.; Rocha, S.M.; Pawliszyn, J. Automated determination of phenolic compounds in wine, berry, and grape samples using 96-blade solid phase microextraction system coupled with liquid chromatography-tandem mass spectrometry. *J. Chromatogr. A,* **2013**, *1276*, 12-19.
[http://dx.doi.org/10.1016/j.chroma.2012.12.043] [PMID: 23332787]

[35] Boyacı, E.; Sparham, C.; Pawliszyn, J. Thin-film microextraction coupled to LC-ESI-MS/MS for determination of quaternary ammonium compounds in water samples. *Anal. Bioanal. Chem.,* **2014**, *406*(2), 409-420.
[http://dx.doi.org/10.1007/s00216-013-7507-8] [PMID: 24337184]

[36] Alcudia-León, M.C.; Lucena, R.; Cárdenas, S.; Valcárcel, M. Stir membrane extraction: a useful approach for liquid sample pretreatment. *Anal. Chem.,* **2009**, *81*(21), 8957-8961.
[http://dx.doi.org/10.1021/ac9016192] [PMID: 19874057]

[37] Richter, P.; Leiva, C.; Choque, C.; Giordano, A.; Sepúlveda, B. Rotating-disk sorptive extraction of nonylphenol from water samples. *J. Chromatogr. A,* **2009**, *1216*(49), 8598-8602.
[http://dx.doi.org/10.1016/j.chroma.2009.10.044] [PMID: 19879588]

[38] Alcudia-León, M.C.; Lendl, B.; Lucena, R.; Cárdenas, S.; Valcárcel, M. Sensitive in-surface infrared

monitoring coupled to stir membrane extraction for the selective determination of total hydrocarbon index in waters. *Anal. Bioanal. Chem.,* **2010**, *398*(3), 1427-1433. [http://dx.doi.org/10.1007/s00216-010-4040-x] [PMID: 20669007]

[39] Giordano, A.; Richter, P.; Ahumada, I. Determination of pesticides in river water using rotating disk sorptive extraction and gas chromatography-mass spectrometry. *Talanta,* **2011**, *85*(5), 2425-2429. [http://dx.doi.org/10.1016/j.talanta.2011.07.087] [PMID: 21962663]

[40] Jachero, L.; Sepúlveda, B.; Ahumada, I.; Fuentes, E.; Richter, P. Rotating disk sorptive extraction of triclosan and methyl-triclosan from water samples. *Anal. Bioanal. Chem.,* **2013**, *405*(24), 7711-7716. [http://dx.doi.org/10.1007/s00216-012-6679-y] [PMID: 23322351]

[41] Richter, P.; Cañas, A.; Muñoz, C.; Leiva, C.; Ahumada, I. Rotating disk sorbent extraction for pre-concentration of chromogenic organic compounds and direct determination by solid phase spectrophotometry. *Anal. Chim. Acta,* **2011**, *695*(1-2), 73-76. [http://dx.doi.org/10.1016/j.aca.2011.04.009] [PMID: 21601032]

[42] Bagheri, H.; Aghakhani, A. Polyaniline-nylon-6 electrospun nanofibers for headspace adsorptive microextraction. *Anal. Chim. Acta,* **2012**, *713*, 63-69. [http://dx.doi.org/10.1016/j.aca.2011.11.027] [PMID: 22200309]

[43] Saraji, M.; Farajmand, B. Chemically modified cellulose paper as a thin film microextraction phase. *J. Chromatogr. A,* **2013**, *1314*, 24-30. [http://dx.doi.org/10.1016/j.chroma.2013.09.018] [PMID: 24075457]

[44] Kabir, A.; Furton, K.G.; Malik, A. Innovations in sol-gel microextraction phases for solvent-free sample preparation in analytical chemistry. *Trends Analyt. Chem.,* **2013**, *45*, 197-218. [http://dx.doi.org/10.1016/j.trac.2012.11.014]

[45] Kabir, A.; Furton, K.G. Fabric phase sorptive extractors (FPSE) *US Patent Application,* **2014**, *14*(216), 121.

[46] Kumar, R.; Gaurav, ; Heena, ; Malik, A.K.; Kabir, A.; Furton, K.G. Efficient analysis of selected estrogens using fabric phase sorptive extraction and high performance liquid chromatography-fluorescence detection. *J. Chromatogr. A,* **2014**, *1359*, 16-25. [http://dx.doi.org/10.1016/j.chroma.2014.07.013] [PMID: 25082524]

CHAPTER 11

Single-drop Microextraction and Related Techniques

Francisco Pena-Pereira, Isela Lavilla and **Carlos Bendicho**[*]

Departamento de Química Analítica y Alimentaria, Área de Química Analítica, Facultad de Química, Universidad de Vigo, Campus As Lagoas-Marcosende s/n, 36310 Vigo, Spain

Abstract: The number of applications of liquid phase microextraction (LPME) techniques has undergone a dramatic increase during the last years. Miniaturisation of sample preparation encompasses several advantages, *i.e.*, low consumption of extractants, integrated operation, ease of clean-up, large preconcentration factors, apart from an increased greenness, as compared to classical solvent extraction. Among LPME, single-drop microextration (SDME) approaches have deserved much interest. Versatile and adaptable procedures to every each analyte and matrix have been reported. Extractants are not only limited to organic solvents, but ionic liquids (ILs) and even aqueous solvents can be also employed. Whereas, analytical techniques based on chromatographic/electrophoretic separations and some modes of atomic spectrometry were applied in early applications, SDME has also been exploited in combination with other detection approaches such as UV-*vis* spectrophotometry, fluorospectrometry, chemiluminescence, *etc.* This has spread the use of SDME to almost every application area. In this chapter, the state of art of SDME and its main modes is reviewed.

Keywords: Anions, Atomic-spectrometry, Continuous-flow, Direct-immersion, Directly-suspended, Electrophoresis, Elements, Gas-chromatography, Headspace, Ionic-liquids, Liquid-chromatography, Mass-spectrometry, Microextraction, Organic-compounds, Organometals, Single-drop, Speciation, Three-phases, Trace-analysis, Two-phases.

[*] **Corresponding author Carlos Bendicho:** Departamento de Química Analítica y Alimentaria, Área de Química Analítica, Facultad de Química, Universidad de Vigo, Campus As Lagoas-Marcosende s/n, 36310 Vigo, Spain; Tel: +34-986-812281; Fax: +34-986-812556; E-mail: bendicho@uvigo.es

M. Valcárcel, S. Cárdenas, R. Lucena (Eds.)

11.1. INTRODUCTION

In an attempt to overcome some of the drawbacks inherent to classical solvent extraction techniques (*i.e.*, liquid-liquid extraction), a plethora of miniaturised techniques have been proposed during last years.

Since first works by Dasgupta' group, using drops as collectors of gaseous substances such as ammonia and sulphur dioxide from air [1], several techniques have been developed that have in common the disposal of a microvolume of extractant phase (*i.e.*, acceptor phase) using mostly a chromatographic syringe as support. The implementation of a syringe has provided a smart and wise way to accomplish solvent drop microextraction.

Initial prototypes addressing solvent drop microextraction were developed by Liu and Dasgupta in 1996 [2], using a chloroform microdrop suspended in a larger sample drop to extract sodium dodecyl sulphate (SDS). First report making use of a microsyringe for performing drop microextraction was published by Jeannot and Cantwell [3], who also suggested for the first time the name of 'single-drop microextration' (SDME).

The SDME technique encompasses a variety of approaches depending on how the solvent microvolume is configured (*i.e.*, at the tip of a microsyringe or directly suspended), the number of phases involved in the transfer of the analyte from the donor to the acceptor phase (*i.e.*, two- of three-phase systems) or the confinement of the sample solution in a vial (static approach) or its displacement over the acceptor phase (dynamic approaches). Mass transport is generally limited by diffusion, so agitation of the sample solution is needed to improve the convective transport in order to shorten the diffusion distance.

Classification of SDME techniques usually relies on the number of phases involved in the process. Among two-phases SDME, the following modes have been reported: direct immersion single-drop microextraction (DI-SDME), headspace single-drop microextraction (HS-SDME), continuous flow microextraction (CFME), directly suspended drop microextraction (DSDME), solidification floating organic drop microextraction (SFODME) [4 - 7]. Some variants of any of the above modes, such as drop-to-drop microextraction

(DDME) can be considered as miniaturised versions of DI-SDME.

The liquid-liquid-liquid microextraction (LLLME) mode involves a three-phase mass transfer and can be considered a solvent microextraction with simultaneous back-extraction. This mode is well suited for hydrophilic compounds in combination with those techniques requiring an aqueous acceptor solution such as reverse-phase high performance liquid chromatography (HPLC) and capillary electrophoresis (CE). All these approaches will be discussed in detail in next section.

11.2. SDME: MICROEXTRACTION MODES

As has been previously mentioned, a variety of SDME modes and related techniques has been reported in the literature with the aim of achieving efficient extraction and preconcentration of target compounds, while solving the limitations identified in preliminary SDME modes. The evolution of the number of publications per year concerning SDME modes and related techniques, as well as the frequency of use of these miniaturised sample preparation approaches, is shown in Fig. (**1**).

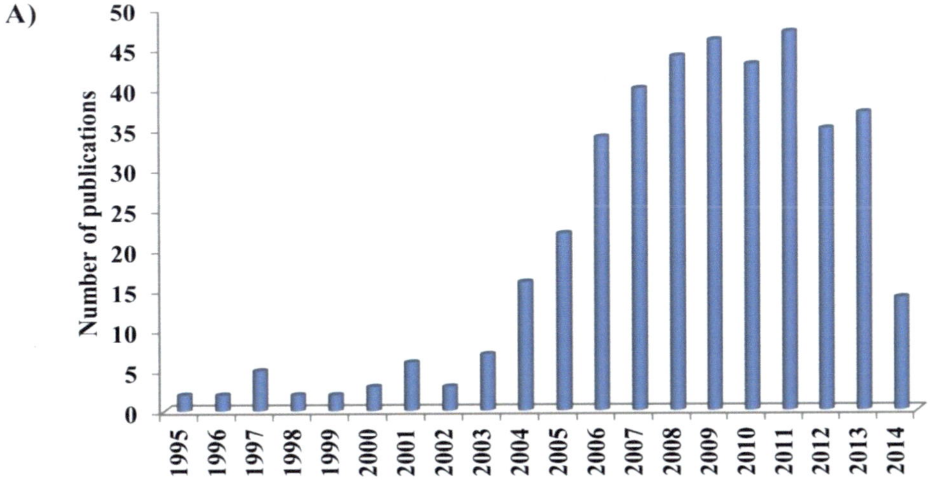

Fig. 1 contd.....

B)

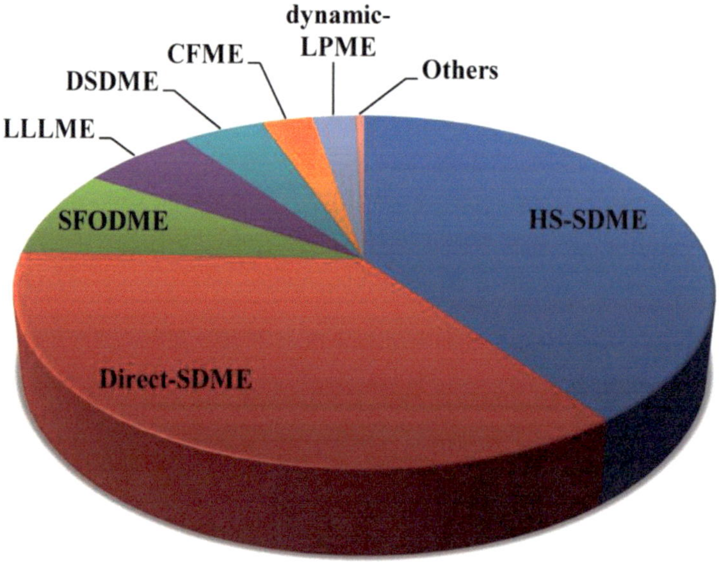

Fig. (1). (**A**) Evolution of the number of publications per year devoted to SDME and related techniques; (**B**) Frequency of application of SDME and related techniques.

In this section, the most popular SDME modes and related techniques are classified bearing in mind the phases involved in the extraction process, namely biphasic (liquid-liquid) and triphasic (liquid-gas-liquid and liquid-liquid-liquid) systems. A schematic representation of the SDME modes and related techniques described below is shown in Fig. (**2**).

11.2.1. Byphasic Systems

The extraction process in byphasic systems is based on the direct contact of a reduced volume of an immiscible extractant phase with the sample solution. A diversity of modes has been reported by exploiting specific physicochemical properties of the extractant phases and the way in which they are exposed to the sample. As miniaturised approaches derived from conventional liquid-liquid extraction, byphasic systems can be used to extract target compounds of very different nature (volatile and non-volatile, ionisable and non-ionisable compounds). DI-SDME is the first reported and most commonly used byphasic SDME mode [2, 3]. In DI-SDME, a microdrop of extractant phase is directly exposed from the tip of a syringe needle to a continuously stirred sample solution (Fig. **2A**). The main shortcoming of DI-SDME is the limited stability of the drop under relatively strong agitation conditions, high temperatures, and extended extraction times. More importantly, the extraction kinetics are highly influenced by these parameters, so the conditions used in DI-SDME usually hinder the achievement of high enrichment factors. Aqueous samples are by far the most common type of sample analysed after DI-SDME, although a reversed DI-SDME was recently reported for the extraction and preconcentration of polar compounds from water immiscible samples [8].

A specific case of DI-SDME mode applied with microliter volume samples was introduced by Liu and Dasgupta in 1996 under the denomination of drop-in-drop system [2]. This concept was more recently extended to the analysis of scarce samples by using a microsyringe as described for DI-SDME [9]. In this case, a drop of immiscible solvent is directly immersed into a drop of sample placed in a microvial, as shown in Fig. (**2B**). Compared with conventional DI-SDME, it involves a static extraction process, where agitation is not employed due to its negative effect on the stability of the microdrop. The application of a surface acoustic wave device to improve the extraction kinetics of drop-to-drop systems was recently reported [10].

Another system analogous to DI-SDME was introduced by Liu and Lee under the denomination of continuous flow microextraction [11]. In CFME, the sample solution is not magnetically stirred but pumped into a continuous flow extraction chamber, where it interacts with the drop (Fig. **2C**). Due to the almost negligible decrease on the analytes concentration during the extraction process, recycling of the sample by introducing the waste outlet of tubing into the sample reservoir has been considered as a convenient and effective way to perform CFME with reduced sample consumption [12].

With the aim of avoiding the risk of microdrop depletion from the needle tip experienced in SDME modes, a related technique called directly suspended droplet microextraction was introduced [13]. In DSDME, a microdrop of an immiscible extractant phase with a density lower than water is injected in the centre of the vortex formed at the top of a highly stirred sample solution

(Fig. **2D**). After extracting the relevant compounds, an aliquot of the enriched drop is taken for analysis by means of a microsyringe. DSDME enables the achievement of higher extraction efficiencies than those obtained with SDME modes due to the improved convective mass transfer, which favourably affects the extraction kinetics. It is worth mentioning that evaporation and solubilisation of the drop can occur in DSDME during the extraction process. Therefore, extractant phases with low vapour pressure, high boiling point, reduced water solubility and lower density than water, are necessary requirements so that they can be applicable in DSDME.

The microextraction mode known as SFODME was introduced by Khalili Zanjani in 2007 [14]. In this sample pretreatment technique, the extractant phase must be immiscible with water, have a lower density than water and a melting point near room temperature (in the range 10-30 °C). In SFODME, the extraction process is carried out in an analogous way to that described for DSDME, but temperature control plays a major role in SFODME (Fig. **2E**). Thus, the extraction system is heated above the melting point of the extractant phase during the preconcentration step, and cooled down after completion of the process. The enriched organic solvent is therefore solidified and easily transferred into a small conical vial by means of a spatula, where the solvent melts, then being ready for analysis. Unlike DI-SDME or CFME, the immiscible extractant phase is not directly exposed to the sample at the needle tip of a microsyringe, so the problems related with drop stability reported under strong agitation, high temperatures or extended extraction times are not experienced in SFODME, thus allowing the improvement of extraction kinetics by exploiting the positive effect of these experimental parameters. Nevertheless, the specific requirements established for the application of this technique limit the set of solvents that can be potentially used in SFODME. Furthermore, SFODME involves a significant number of consecutive steps that makes the procedure more tedious and time-consuming.

Recently, a microextraction method based on a dynamic single-drop in a narrow-bore tube was introduced by Farajzadeh *et al.* [15]. This miniaturised sample preparation method consists on the repeated movement of a single drop of immiscible solvent with lower density than water through a narrow-bore tube (120 cm x 5 mm) that contains the sample solution. The movement of the droplet is favoured by an air bubble intentionally introduced together with the solvent microdrop. The repeated movement of the drop through the tube improves the mass transfer when compared with static systems. By reversing upside down the tube, the drop is allowed to repeatedly contact with the sample for extraction of the target analytes. The enriched drop is finally collected after the required number of extraction cycles for analysis.

In dynamic liquid-phase microextraction (dynamic-LPME), a microsyringe is used as both microseparatory funnel and syringe for injection of the enriched extract in the corresponding analytical instrument [16]. This technique is based on exposing a microliter volume solvent to an aqueous sample by the repeated up and down movement of the syringe plunger (Fig. **2F**). Thus, a renewable microfilm is formed on the inner surface of the syringe. The film-like geometry of the extractant phase provides an increased surface area when compared with the spherical configuration of SDME modes. Thus, dynamic-LPME provides higher enrichment factors than DI-SDME and related modes. However, the intensive manual operation yields worse repeatability. Nevertheless, it can be dramatically improved by automating the dynamic-LPME technique.

11.2.2. Triphasic Systems

Liquid-gas-liquid techniques are advantageous options for the extraction of volatile and semivolatile compounds (or analyte derivatives when derivatisation reactions are performed). In these systems, volatiles are transferred from the sample solution to the headspace and, subsequently, to the extractant phase. On the contrary, non-volatile matrix compounds are obviously not extracted, and the interference from high molecular weight compounds can also be minimised or even eliminated. Depending on the configuration of the extractant phase (drop or film), two modes can be described, namely, HS-SDME and dynamic headspace liquid-phase microextraction (dynamic HS-LPME).

In HS-SDME (Fig. **2G**), a drop of extractant phase is exposed to the headspace above the sample for extraction [1, 17, 18]. Mass transfer of the analytes from the sample to the headspace is the rate-limiting step of the microextraction process as a result of the much lower diffusion in the liquid as compared to the gas phase.

Efficient mass transfer to the headspace can be achieved by means of a high stirring rate (which can be employed in this case with minimal risk of drop dislodgement from the needle tip), salting out effect and high temperatures. Microwave and ultrasound irradiation, as well as hydrodistillation, have also been employed for accelerating the analyte transport from condensed phase to the gaseous phase. Unlike DI-SDME, both water miscible and immiscible solvents can be used as extractant phases in HS-SDME. Furthermore, high stirring rates can be effectively employed to improve the extraction process, since the drop is not in direct contact with the stirred sample solution. Evaporation of the microdrop during the extraction process, especially under extended extraction times and high temperatures, is the main limitation of this SDME mode, and therefore extractant phases with very low volatility are recommended.

In dynamic HS-LPME (Fig. **2H**), extraction of volatiles is performed within the syringe barrel by repeatedly contacting a reduced volume of organic solvent with the headspace above the sample [19]. A fresh aliquot of the gaseous phase is thus exposed to the extractant phase in each extraction cycle. Remarkably, the selection of the extractant phase is more flexible in dynamic HS-LPME than in HS-SDME, since evaporation losses are reduced.

Liquid-liquid-liquid systems are recommended approaches for the extraction of ionisable compounds. In comparison with the above mentioned biphasic modes, liquid-liquid-liquid systems are advantageous in terms of selectivity, since non-ionisable matrix compounds are not transferred to the acceptor phase. In addition, the aqueous nature of the acceptor solution makes them directly compatible with the running buffer and the mobile phase in CE and HPLC. To achieve an efficient mass transfer, the neutral form of the ionisable analytes must show high partition coefficients in the organic phase employed as membrane, and complete ionisation at the pH selected for the acceptor phase to be effectively back-extracted.

Based on the preliminary work by Ma and Cantwell [20], two analogous liquid-liquid-liquid systems have been described in the literature, namely, LLLME [21] and DSDME with single drop back-extraction [22].

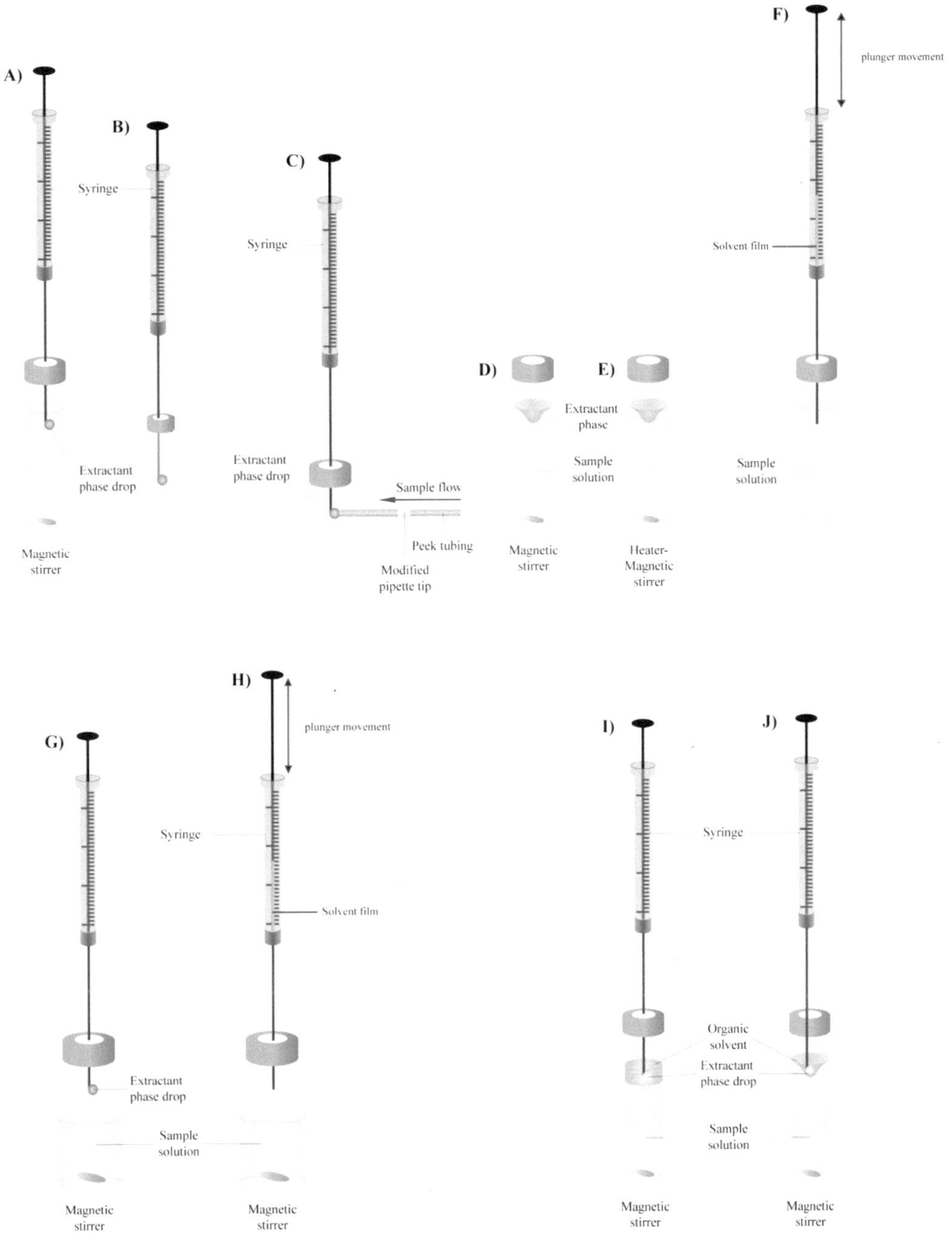

Fig. (2). Schematic representation of SDME and related techniques. (**A**) DI-SDME, (**B**) drop-in-drop SDME; (**C**) CFME; (**D**) DSDME; (**E**) SFODME; (**F**) dynamic-LPME; (**G**) HS-SDME; (**H**) dynamic HS-LPME; (**I**) LLLME; (**J**) DSDME with single drop back-extraction.

In LLLME, analytes are extracted from the sample solution into a liquid membrane of immiscible organic solvent with lower density than water, and subsequently back-extracted from this intermediate phase into a single drop of acidic or alkaline aqueous solution (Fig. **2I**). After extraction for a prescribed time, the drop is retracted back into the syringe and injected in the corresponding analytical technique for analysis. In LLLME, the microdrop of acceptor solution is kept immersed into the organic phase during the whole microextraction process by means of a syringe, so extraction and back-extraction processes are simultaneously performed. On the contrary, the extraction process is carried out following two consecutive steps in DSDME with single drop back-extraction. The first step involves the extraction of the neutral form of the analyte by the procedure described above for DSDME and, subsequently, the aqueous drop is exposed from the needle tip of a syringe to the enriched organic phase for back-extraction (Fig. **2J**).

The use of electrically induced transfer of charged compounds across a free liquid membrane has been recently reported [23, 24]. In brief, chemically inert tubing is sequentially filled with aqueous acceptor solution, organic solvent and donor solution, and connected to a d.c. power supply [23]. Alternatively, an aqueous acceptor droplet hanging from a conductive pipette tip could be immersed into an immiscible organic phase for the three-phase electroextraction of target analytes from the aqueous donor phase [24]. The application of an electric field enables the migration of the charged species across the free liquid membrane into the aqueous acceptor phase. Even though the enrichment factors achieved are not especially high, three-phase electroextraction systems provide an interesting alternative to miniaturised sample preparation systems used for the analysis of scarce samples, where mass transfer is commonly governed by diffusion.

11.3. EXTRACTANT PHASES IN SDME

A variety of extractant phases has been employed in SDME, including organic solvents, ionic liquids (ILs), surfactants, supramolecular systems, and even aqueous drops. Some examples of extractant phases employed in SDME and related techniques are shown in Table **1**. The selection of the most appropriate extractant phase depends to a large extent on the nature of target analytes, the

SDME mode and the analytical instrumentation employed.

Table 1. Extractant phase requirements in SDME and related techniques.

Microextraction mode	Basic extractant phase requirements	Extractant phases
Two phases		
DI-SDME	Immiscible with the sample High boiling point Low vapour pressure Low water solubility	cyclohexane, hexane, octane, iso-octane, toluene, xylene, 1-octanol, n-butyl acetate, chloroform, 1,2-dichloroethane, benzonitrile, $[C_6MIM][PF_6]$, coacervatives (decanoic acid, tetrabutyl ammonium hydroxide and water)
CFME	Immiscible with the sample High boiling point Low vapour pressure Low water solubility	cyclohexane, iso-octane, benzene, toluene, xylene,, n-butyl acetate, chloroform, carbon tetrachloride
DSDME	Immiscible with the sample High boiling point Low vapour pressure Low water solubility Lower density than the sample	hexane, heptane, iso-octane, toluene, xylene, 1-butanol, 1-octanol, undecanol, 2-octanone, n-butyl acetate, hexyl acetate
SFODME	Immiscible with the sample High boiling point Low vapour pressure Low water solubility Lower density than the sample Melting point near to room temperature	n-hexadecane, 1-undecanol, 1-dodecanol, 2-dodecanol, 1-chlorooctadecane, 1,10-dichlorodecane, 1-bromohexadecane, coacervatives (decanoic acid, tetrabutyl ammonium hydroxide and water)
Dynamic-LPME	Immiscible with the sample High boiling point Low vapour pressure Low water solubility	hexane, octane, benzene, toluene, xylene, methylene chloride, chloroform
Three phases		
Liquid-gas-liquid		
HS-SDME	High boiling point Low vapour pressure	water, octane, decane, hexadecane, toluene, xylene, octanol, ethylene glycol, benzyl alcohol, butyl acetate, tetrachloroethylene, N,N'-dimethylformamide, $[C_6MIM][PF_6]$

(Table 1) contd.....

Microextraction mode	**Basic extractant phase requirements**	**Extractant phases**
Dynamic HS-LPME	High boiling point Low vapour pressure	hexane, iso-octane, cyclohexane, benzene, toluene
Liquid-liquid-liquid		
LLLME	*Intermediate phase:* Immiscible with the sample High boiling point Low vapour pressure Low water solubility Lower density than the sample *Acceptor phase:* Immiscible with the intermediate phase	*Intermediate phase:* hexane, heptane, iso-octane, toluene, xylene, 1-butanol, 1-octanol, undecanol, 2-octanone, n-butyl acetate, hexyl acetate *Acceptor phase:* Acidic or basic aqueous drop
DSDME with single drop back-extraction	*Intermediate phase:* Immiscible with the sample High boiling point Low vapour pressure Low water solubility Lower density than the sample *Acceptor phase:* Immiscible with the intermediate phase	*Intermediate phase:* hexane, iso-octane, benzene, toluene, xylene, 1-octanol, 2-octanone, hexyl acetate *Acceptor phase:* Acidic or basic aqueous drop

Several physicochemical properties are of paramount importance when choosing extractant phases for SDME. The formation and stability of the drops, the extractability of target analytes, as well as the compatibility of the extractant phase with analytical instrumentation are highly influenced by a set of physicochemical properties, including boiling point, vapour pressure, water solubility, density, viscosity, *etc*. The knowledge of these properties is therefore crucial for an appropriate method development involving SDME as sample preparation technique.

Organic solvents are by far the most common extractant phases employed in SDME. In fact, organic solvents have been employed as extractant phases in each of the SDME modes described in the previous section. Organic solvents are highly convenient from the point of view of extractability, due to the large variety of commercially available solvents with different physicochemical properties.

Furthermore, many analytical methods based on conventional liquid-liquid extraction make use of organic solvents, so they can be implemented in SDME methods toward the miniaturisation of those classical procedures. It is worth mentioning, however, that not every organic solvent feasible for liquid-liquid extraction can be directly employed in SDME, since drop dissolution or evaporation can be experienced during the extraction process due to the low extractant phase-to-sample volume ratio used. Besides, the lack of compatibility with some analytical techniques should be kept in mind when selecting certain organic solvents as extractant phases. Furthermore, several organic solvents employed in SDME do not fulfil the requirements to be considered environmental sustainable. In fact, organic solvents such as benzene have been systematically used in different SDME modes, in spite of being banned in labs and industry some decades ago. Thus, in the last years, some organisations have established solvent selection guides that can be consulted for the replacement of organic solvents that present safety and health issues with greener alternatives [25 - 27].

ILs have been employed in SDME as greener alternatives to organic solvents for extraction and preconcentration of target compounds. ILs are organic salts with unique properties, namely, negligible vapour pressure, good thermal stability, non-volatility, non-flammability, and tunable polarity, viscosity, and miscibility with solvents and water. The first application of ILs in SDME was reported in 2003 [28], and from this pioneering work, several dozens of articles involving the application of ILs in a variety of SDME modes and related techniques have been published. 1-alkyl-3-methylimidazolium-based ILs have been the extractant phases of choice in most of applications. It is worth noting that, in spite of being commonly considered as green solvents, certain ILs have been reported to be toxic [29], and most of common ILs are obtained by using chemicals derived from non-renewable resources [30].

Surfactant-containing extractant phases have also been used as alternatives to organic solvents in SDME and related techniques. Aqueous solutions of surfactants were firstly employed to prepare soap bubbles for extraction of sulphur dioxide from gaseous samples [31]. An appealing application of surfactants is the preparation of water immiscible supramolecular solvents. These nano-structured liquid systems, obtained by dispersing a coacervating agent and an alkanoic acid in a continuous phase, -mainly water-, can show a variety of configurations (aqueous micelles, reverse micelles and vesicles) that have a high impact on the solubilisation of target compounds [32]. Coacervates have recently demonstrated their high versatility as extractant phases in SDME and related techniques. Thus, a large variety of compounds showing different polarity can potentially be extracted by coacervates through interactions of different nature, namely hydrogen bond, Van der Waals, ionic and π-cation interactions [33]. Due to their aqueous nature (water comprises approximately 85% of supramolecular systems), coacervates are being considered as green alternatives to volatile organic solvents. Surfactants have also been employed in combination with ILs under the denomination of micellar ILs to increase the extractability of target compounds [34]. Slightly increased enrichment factors were obtained for aromatic compounds when traditional and IL-based surfactants dissolved in 1-butyl-3-methylimidazolium chloride were used as extractant phases in HS-SDME.

Aqueous-based extractant phases have also been proposed in SDME and related techniques. In fact, renewable aqueous droplets were reported in SDME as collectors for soluble gases as early as 1995 [1]. An aqueous microdrop has been used as the acceptor phase in SDME and related techniques that involve three phases, namely, liquid-gas-liquid and liquid-liquid-liquid based SDME and related techniques. In addition, water was also proposed as extractant phase in reversed DI-SDME for extraction of polar compounds from water immiscible samples. Aqueous drops containing a pH modifier were proposed for the extraction of ionisable compounds. Thus, a slightly acidic microdrop was firstly proposed for the enrichment of ammonia from gaseous samples [1], while a 0.1 M NaOH microdrop was used for the enrichment of atmospheric hydrogen sulphide [35]. The use of aqueous soluble reagents that selectively react with the target compounds have also been reported in the literature. For instance, a tetramethylbenzidine-containing aqueous droplet was used for the colorimetric determination of gaseous chlorine [17]. In addition, noble metals, such as palladium, platinum and rhodium, have been employed in a variety of HS-SDME procedures for the trapping of hydride-forming elements prior to their determination by atomic spectrometric techniques [36, 37]. Water is the greenest option when selecting the extractant phase in SDME and related techniques, although it cannot be obviated that the use of high concentrations of these modifiers can make methodologies less green in accordance with the National Environmental Methods Index criteria [38].

11.4. COUPLING OF SDME WITH DIFFERENT DETECTION TECHNIQUES

SDME is used for a wide range of analytes both organic and inorganic, and it involves the use of a broad spectrum of analytical techniques. Nevertheless, the coupling of SDME with detection techniques is not always easy, and the transfer of the drop into the instrument may require certain considerations. Different aspects have to be taken into account, in special the compatibility of the volume, the nature of solvent used in the drop and the means and ways of incorporating it into the instrument. In this respect, SDME is especially suitable in combination with some techniques such as gas chromatography (GC) and normal phase liquid chromatography (LC) when an organic non-polar solvent is used in the drop. On the contrary, the number of organic solvents that can be successfully used in atomic absorption spectrometry (AAS) is limited, and even they can be unsuitable when using reverse-phase LC or capillary electrophoresis. In these cases, the evaporation of the organic solvent and re-dissolution with an alternative compatible solvent (not without problems), or the use of aqueous drops in some SDME modalities such as HS-SDME or LLLME (not always possible), are practical ways to solve these problems. In addition, the direct injection of the drop in certain instruments designed to admit large sample volumes in their normal operation is not suitable without a previous drop dilution. Some effective solutions have been proposed, *e.g.,* the use of an electrothermal vaporisation (ETV) unit for sample introduction in inductively coupled plasma-mass spectrometry (ICP-MS). In this section the coupling of SDME with different analytical techniques is discussed.

11.4.1. Separation Techniques

Use of separation techniques is usual after SDME, in special GC. Organic solvents, volumes and even the syringe used are crossroads of SDME and GC. Therefore, it is not surprising that GC has been the technique of choice for final analysis in many cases. For the coupling of SDME with other separation techniques such as CE or reversed-phase HPLC, different strategies need to be designed in order to make organic solvents used in SDME compatible with these tecnhiques. In general, volumes used in SDME and separation techniques are attuned.

11.4.1.1. Gas Chromatography

A large number of applications using the coupling SDME-GC have been published since the first report published by Jeannot and Cantwell [3] in 1996. These authors used a Teflon rod with a spherical recess to hold a drop of octane. After extraction, the rod was removed and a GC syringe was used for sampling and injecting the drop into a chromatograph. In a second paper, these same workers proposed the use of the GC syringe for both drop suspension and injection [39]. Since then, the standard GC syringe continues being the preferred system for the development of most of SDME methods.

Direct coupling is possible since most of organic solvent drops are well suited for GC (*e.g.,* 1-octanol, toluene, dodecanol or undecanol). Therefore, it is not surprising that initial applications of SDME would appear in combination with GC. Thus, He and Lee [16] developed dynamic LPME using the syringe needle as a microseparatory funnel for determining chlorobenzene by GC with electron capture detector (ECD). Liu and Lee [11] proposed the CFME technique for determination of nitroaromatic compounds and chlorinated benzenes by GC-ECD. Also, HS-SDME was first developed for GC analysis [40]. While all usual detectors (*e.g.*, ECD, flame ionisation detector, FID) have been used, gas chromatography-mass spectrometry (GC-MS) is the most popular approach after SDME due to its sensitivity and selectivity.

Although manual injection of the enriched drop into the chromatograph injection port is carried out with adequate precision, SDME-GC can be employed in an automated way. Complete automation can be achieved using a programmable autosampler such as the CCT Combi PAL from Agilent. For instance, a fully automated dynamic in-syringe LPME procedure coupled with GC-MS using this autosampler has been reported by Lee and Lee [41] for determining carbamate pesticides.

Usually, analytes are derivatisation for their conversion into volatile compounds for GC analysis [42]. SDME facilitates *in situ* derivatisation through some options such as in-syringe derivatisation and in-drop derivatisation. For example, in-syringe silylation has been used for phenols. After extraction, the derivatisation reagent is taken with the syringe where it is mixed with the sample by successive movement of the plunger. Then, the syringe is sealed and heated for a certain time [43]. Extraction and derivatisation can be also carried out simultaneously in the drop. For instance, Fiamegos and Stalikas [44] proposed a one-step SDME-derivatisation with an ion-pairing agent under mild conditions for determining endocrine-disrupting by GC-FID.

SFODME has also been used along with GC since the usual organic solvents can be directly injected in the chromatograph. For instance, 1-dodecanol was used for extracting organochlorine pesticides from water samples. The solidified solvent was transferred into a vial and melted immediately prior to injection [45]. On the contrary, ILs are not compatible with direct injection in a gas chromatograph due to their negligible vapour pressure, so their use as extractant in SDME-GC is limited and special coupling systems need to be used. A removable interface developed by Valcarcel´s group prevents the IL from passing to the column. It must be replaced by a clean one every five injections [46].

11.4.1.2. Capillary Electrophoresis

Nowadays, CE is a separation technique widely used in fields such as bioanalysis, thereby allowing the application of SDME to the determination of amino acids, peptides and chiral analysis, yet not without difficulty [47]. High separation efficiency in short times and with low sample volumes (usually microliters of sample are accommodated in the capillary) are inherent to CE. Despite this, its sensitivity is lower than that of other separation techniques such as HPLC, and therefore, analyte preconcentration is usual before CE analysis. In addition, some matrices make separations difficult and even some risk of clogging of the capillary occurs. A prior clean-up step is usually needed in CE. SDME is able to solve both problems, yet its use along with CE can be limited by the organic nature of the extractant phase. In order to make the running buffer in CE compatible with the drop, the three-phase modality of SDME, *i.e.*, LLLME, is usually employed prior to CE. In this case, an enriched aqueous extractant acceptor phase is obtained, thus being compatible with CE. For instance, fluoroquinolones in urine have been extracted from the aqueous sample solution to a mixture of dichloromethane and toluene on the top of sample and from there to a 0.1 M NaOH drop immersed in the organic phase [48].

SDME and CE can be coupled off- and on-line [47]. In the first case, a syringe is used and the drop is transferred to a sample vial with the corresponding diluent (when several tens of microliters are required). On-line couplings are possible in different ways, the use of the capillary inlet tip for hanging the drop being the most common procedure. The capillary is filled with the suitable background electrolyte (BGE) and an aqueous drop covered with a film of the organic phase is formed by adjusting the backward pressure. The hydrophobicity of the capillary tip is enhanced with octadecyl trimethoxysilane to improve adhesion of the drop. The enriched extractant is then injected in the capillary using an adequate forward pressure [49]. Alternatively, Fang *et al.* [50] proposed the direct analysis of water-immiscible solvent samples by capillary zone electrophoresis (CZE) using SDME with an on-line back-extraction field-amplified sample-injection method. First, the analytes (cocaine and thebaine) are extracted by DI-SDME with chloroform. Then, the drop is transferred to a sample vial and sealed with a water plug for direct electroinjection placing the tip of the capillary carefully in the aqueous phase. When a high voltage is applied, the positively charged analytes migrate into the capillary from the chloroform to the aqueous phase.

The disadvantage of using an organic solvent in the drop can be eliminated if HS-SDME mode with an aqueous acceptor phase is used. For example, Zhang *et al.* [51] determined volatile phenolic compounds from an acidified sample using a basic aqueous drop (a diluted solution of NaOH). After microextraction, the drop was introduced into a CE vial for direct analysis. BGE can also be used in the aqueous drop, being fully compatible with CE [52].

ILs have also been used as solvents for SDME-CE since its direct injection into the capillary is possible without the current instability that occurs with classical-non polar solvents. The on-line coupling of IL-SDME with CE has been proposed by Wang *et al.* [53] for determination of phenols. The IL is suspended at the tip of the capillary, thus providing an increased stability as compared to classical organic solvents. The chosen IL should have the ability to extract analytes and dissolubility in the CE buffer.

11.4.1.3. Liquid Chromatography

The selection of an extraction phase compatible with the mobile phase is essential in the coupling SDME-LC. Some compatible organic solvents have been used directly (*e.g.*, n-hexane, o-dibutyl phthalate, cyclohexane, iso-octane, chloroform, toluene or xylene or even mixtures such as n-octanol and chloroform) [54]. In spite of this, evaporation of the organic phase to dryness and dissolution of the residue in an adequate medium seems to be the most usual strategy for analyte injection in the chromatograph. For example, He and Lee [55] used CFME-HPLC-UV to determine five pesticides in water after evaporation of the solvent drop (carbon tetrachloride) and reconstituting this volume with the mobile phase. This procedure is less convenient than the direct injection of the drop into the instrument, yet it has facilitated the application of SDME-LC for analytes unsuitable by other way.

If analytes are ionisable, back-extraction into a second aqueous phase is a suitable option for LC analysis. This is an appropriate solution for reversed-phase HPLC, which was first proposed by Ma and Cantwell [21]. In this work, a membrane with n-octane was located between the aqueous sample (with neutral analytes) and the aqueous drop. The acid pH of the drop is the key for extraction of basic analytes from the organic phase by protonation of methamphetamine, mephentermine, methoxyphenamine and 2-phenylethylamine.

In general, ILs can be directly analysed by HPLC, therefore they have been used as extractants for SDME-LC. For example, 1-hexyl-3-methylimidazolium hexafluorophosphate [C_6MIM][PF_6] and DI-SDME-HPLC have been used for preconcentration of different mercury species after the formation of dithizonates. The drop was directly injected into the HPLC system using tetrahydrofuran and methanol as mobile phase at pH 4 [56]. SFODME has been also used along with

HPLC without problems. For instance, 1-undecanol allows the extraction and determination of some fat-soluble vitamins in aqueous samples. The solidified solvent was immediately melted and then directly injected into the chromatograph in an isocratic mode with methanol as mobile phase [57].

Derivatisation could be required for certain compounds. Like in GC, in-syringe derivatisation and in-drop derivatisation have been proposed for this coupling. For example, to determine ammonium in aqueous samples by HPLC with a fluorescence detector, the derivatisation reagent (o-phthaldialdehyde) forms a thin film on the inner syringe wall to achieve derivatisation [58]. Volatile aldehydes can be extracted by HS-SDME and simultaneously derivatised using 2,4-dinitrophenylhydrazine in the drop for their determination by HPLC-UV [59].

11.4.2. Atomic Spectrometry

In 2003, Chamsaz *et al.* [60] used for first time electrothermal atomic absorption spectrometry (ETAAS) with HS-SDME for arsenic determination following hydride generation. Since then, the coupling of different atomic detectors with SDME has undergone a remarkable growth, although ETAAS continues being the most popular with this purpose. The most used tecniques with SDME, *i.e.*, ETAAS and ETV-ICP-MS, have in common that the volumes of acceptor phase and measurement fit well. On the contrary, the combination of flame atomic absorption spectrometry (FAAS), atomic fluorescence spectrometry (AFS), inductively coupled plasma optical emission spectrometry (ICP-OES) or ICP-MS with SDME is more troublesome, since dilution of the enriched extract should be avoided. Speciation is also possible through sample derivatisation and/or in-drop derivatisation.

11.4.2.1. Electrothermal Atomic Absorption Spectrometry

Electrothermal atomisation systems are especially well suited for combination with the sample microvolumes used in SDME. With a few exceptions, manual injection of the drop into the graphite furnace *via* the dosing hole is carried out. Sequential injection (SI) systems have also been proposed for automation of sample introduction in the atomiser after SDME. Home-made flow-through extraction cells and commercial multiposition valves are used in these systems

[61, 62].

Classical organic solvents are tipically used to extract metal complexes or ionic pairs. Some characteristics of these solvents should be taken into account for injection into the graphite furnace. For instance, chlorine-containing organic solvents can form volatile metal chlorides, resulting in a significant loss of some analytes. In addition, organic solvents can require higher ashing temperatures and times in comparison with aqueous solutions due to a deep penetration of solvent into the graphite. In order to minimise this problem, permanent modifiers such as Rh or Ir have been proposed. Nevertheless, drawbacks associated with this issue should be highlighted, *i.e.*, irreproducible coatings, lower efficiency, relatively short tube lifetimes, limitations on temperature programs and applicability to simple matrices [63].

Aqueous drops have been used along with HS-SDME, thereby eliminating the above-mentioned problems. Typically, an aqueous solution of Pd(II) allows the sequestration of volatile compounds of As(III), Se(IV), Sb(III) or methylmercury hydrides obtaining very good limits of detection (LODs) [36, 64 - 66]. *In situ* hydride chemical generation or *in situ* photogeneration of Se vapours in the presence of organic acids have been also used with this purpose [67]. The hydrogen generated as by-product reduces Pd(II) forming Pd(0) finely dispersed in the drop that causes the catalytic decomposition of the hydride and its sequestration. No additional matrix modifier is necessary in these cases. In addition, interferences from the sample matrix and extractant are less likely to occur.

ILs such as 1-alkyl-3-methylimidazolium hexafluorophosphates ([$C_n$mim][PF_6], n 4, 6, 8), have been also used in SDME-ETAAS as an alternative to volatile organic solvents for extraction of metal ions as chelates or ion pairs with direct injection into the graphite furnace [68, 69]. The low volatility and immiscibility with water and other organic solvents make them suitable for DI-SDME and HS-SDME. In spite of this, IL properties such as thermal stability even at high temperatures, high viscosity and inmiscibility with aqueous solutions (*i.e.*, matrix modifiers) may cause problems in ETAAS. Tungsten coils (W-coil) have been proposed as an alternative atomiser to the graphite tube for the direct determination of Cd using $[C_4mim][PF_6]$ as extractant. This atomiser has a heating rate about 10 times faster than that of a graphite tube. Higher temperatures and longer times in pyrolysis and cleaning steps were required in order to eliminate matrix interferences and carbon deposition from the IL [70]. SFODME has been also used with ETAAS. A hot injection (about 80 °C) is recommended in order to avoid problems caused by solidification of the organic solvent [71].

11.4.2.2. Flame Atomic Absorption Spectrometry

FAAS is a technique widely used in routine analysis with a relatively low sensitivity that can be circumvented by analyte preconcentration. In general, a sample volume of 1-4 mL is required to carry out a single measurement, which appears incompatible with the usual volumes of SDME. In most cases, SFODME with drop dilution previously to FAAS analysis has been reported, *e.g*, 1-undecanol containing 1,5-diphenyl carbazide as complexing agent has been used for Cu(II) determination by FI-FAAS in water [72].

Microsample introduction systems could solve those problems of volume incompatibility without resorting to drop dilution. For example, Mitani and Anthemidis [73] proposed the on-line LPME based on drop-in-plug sequential injection lab-at-valve platform coupled to FAAS. This system includes a flow-through microextraction chamber that is assembled at one port of the selection valve. A complex of Pb with ammonium pyrrolidine dithiocarbamate is formed on-line and then extracted into a plug of chloroform and micro-droplets of aqueous phase are formed into the plug of the extractant.

11.4.2.3. Electrothermal Vaporisation-inductively Coupled Plasma-mass Spectrometry

An ETV unit requires few microliters of sample in opposition to the usual pneumatic sample introduction systems, thus facilitating the sample introduction into the plasma. Besides enhancing sensitivity, the coupling SDME-ETV-ICP-MS has relevant advantages, namely: i) simultaneous multi-elemental determinations are feasible; ii) introduction of organic solvents that can destabilise or even extinguish the plasma is facilitated; iii) polyatomic interferences due to C are decreased, and iv) the efficiency of transport to the plasma is improved.

Xia *et al.* [12] published for first time a cycle-flow SDME system with ETV-IC--MS for determining Be, Co, Pd, and Cd in biological tissues with benzoylacetone as both extractant and chemical modifier. Like in ETAAS, HS-SDME and ETV-ICP-MS has been proposed for simultaneous determination of As, Sb, Bi, Pb, Sn and Hg after chemical vapour generation in a closed vial and trapping of the generated volatiles in an aqueous drop containing Pd(II) [37].

The [C_4MIM][PF_6] IL has been used as extractant using cycle-flow SDME in combination with ETV-ICP-MS for determining Co, Hg and Pb complexed with 1-(2-Pyridylazo)-2-naphthol (PAN) [74]. Recent research shows the use of SFOME in combination with ETV-ICP-MS. The main difficulty lies in the high boiling points of solvents, thus requiring high drying temperature for their removal. So, analyte losses can occur in this step, *e.g.,* with 1-dodecanol (b.p. 259 °C) losses of Co, Pd, Cd, Hg, Pb and Bi have been observed. In this sense, a mixture of 1-dodecanol with p-xylene (b.p. 138 °C) facilitates the solvent evaporation at 200 °C, hence avoiding losses for these elements [75].

11.4.2.4. Gas Chromatography-inductively Coupled Plasma-mass Spectrometry

GC can provide another efficient tool for sample introduction in the plasma due to the nature of the mobile phase. The high resolution inherent to GC can be added to the high sensitivity of ICP-MS. In spite of these advantages, the use of GC-ICP-MS is limited to volatile organometallic compounds. For example, HS-SDME has been coupled with GC-ICP-MS for speciation of butyltin compounds in environmental and biological samples. Given the low volatility of the organotins, sodium tetraethylborate ($NaBEt_4$) and sodium tetrahydroborate ($NaBH_4$) were used as derivatising reagents [76].

11.4.2.5. Atomic Fluorescence Spectrometry

Although AFS has been used after SDME, drop dilution and/or drop treatment have been carried out in all cases. So, preconcentration advantages of SDME were under-exploited. For instance, HS-SDME-AFS with an IL in the drop has been used for determining organomercurials. Then, this drop was treated with an acidic potassium permanganate solution and heated in boiling water for 20 min to oxidize mercury species [77]. Another example may be found in the extraction of

Hg complexed with diethylditiocarbamate (DDTC) by SFODME-AFS using 1-undecanol as extractant. A dilution of the drop with 2 mL of ethanol was neccessary [78].

11.4.3. Mass Spectrometry

In this section, direct coupling of SDME with mass spectrometry, without a previous separation technique or plasma, is considered. Usually SDME is coupled with matrix-assisted laser desorption/ionisation (MALDI) mass spectrometry given that drop volumes fit well with those used in this technique (0.5-2 µL). In these cases, the drop is directly placed on the MALDI target plate. The extractant drop must be compatible with the MALDI matrix solution so that they can be mixed. It should also be relatively volatile or should provide uniform crystal formation, according to the sample-matrix preparation procedure used.

MALDI-MS has become an important analytical tool for a large number of biomacromolecules, thus increasing the scope of SDME applications to proteins, peptides, *etc*. For example, Sudhir *et al.* [79, 80] proposed the use of DI-SDME with atmospheric pressure MALDI-MS (AP-MALDI-MS) for the identification of peptides using gold or silver nanoparticles in toluene with tetraalkylammonium bromide in the drop. Gram negative bacteria can be detected in different solutions using a time-of-flight detector (MALDI-TOF-MS) after microextraction in a drop with platinum nanoparticles prepared in an IL (1-butyl-3-methylimidazolium hexafluorophosphate). The anionic bacteria have a strong electrostatic attraction to the positively-charged head group of IL. In addition, the IL is used as MALDI matrix, hence improving the signal precision [81].

SDME-MALDI-MS has been also used for the characterisation of low-molecular weight compounds. Some examples can be cited: the coupling of HS-SDME with MALDI-Fourier transform (FT)-MS for determining volatile basic components in tobacco [82], the screening of low molecular weight drugs (< 500 Da) from urine by DI-SDME-AP-MALDI-MS following the addition of methyltrioctylammonium as matrix ion signal suppressor in the MALDI target plate [83] or the determination of quinine by dynamic LPME and AP-MALDI-MS using an appropriate internal standard [84].

DI-SDME has been combined with MALDI tandem mass spectrometry (AP-MALDI-MS/MS/MS) for the determination of dopamine (a hydrophilic drug) by hydrogen-bonding interaction with the drop solvent (octanol) [85].

11.4.4. Molecular Spectroscopy

As indicated above, couplings with SDME have been mainly focused on GC, HPLC, GF-AAS and ETV-ICP-MS. In spite of the versatility and simplicity of UV-*vis* spectrophotometry and related techniques, only in the last few years these couplings have become popular. This delay can be explained by the incompatibility of volumes in SDME and in conventional instruments. Although it is possible to dilute the drop and to use a microvolume cuvette (about 50 μL) prior to measurement, it still entails a loss of sensitivity. Initially, special designs of optical systems were used with this purpose, such as liquid droplets, liquid films/droplets and falling drops [86]. For instance, Liu and Dasgupta proposed an optical system based on a dynamically growing and falling liquid drop, which was later adapted for solvent microextraction [87]. More recently, microvolume spectrophotometers and spectrofluorimeters appeared in the market [86], which has largely facilitated the coupling of molecular spectroscopy with SDME.

11.4.4.1. UV-vis Spectrophotometry

The possibility of using measurement volumes at the microliter level is the key to extend the use of SDME-UV-*vis* spectrophotometry. In general, advances in microvolume UV-*vis* spectrometers have extended to the sample compartment through miniaturisation but also in radiation sources and detectors. Different systems and accessories have been developed in recent years, and some of them are already in the market, namely: confined drop-based systems, liquid-core waveguides (LCWs), microcells, UV-transmissive pipette tips and variable path-length systems [86].

Confined drop-based UV-*vis* spectrometers have been extensively used in combination with SDME. Bendicho´s group proposed for the first time the coupling of SDME with this type of instruments for iodine determination [88]. In general, these spectrometers are based on surface tension to hold a sample (0.5-2 μL) in place between two pedestals within the optical path forming the sample-measurement column. This system is marketed by Thermo Fisher Scientific (Nanodrop), though other companies offer similar UV-*vis* spectrometers (*e.g.*, BioSpec nano by Shimadzu). Numerous applications have been published since then [89].

The properties of the extractant drop and its compatibility with the pedestal (to avoid possible chemical attacks) should be taken into account in order to form a liquid column. In addition, evaporation can occur during measurements. Some examples of solvents used with this purpose are: N,N′-dimethylformamide [88], toluene [90], xylene [91] or methyl isobutyl ketone [92]. HS-SDME [88], DI-SDME [90] and other modalities such as DSDME [92] have been proposed in combination with UV-*vis* spectrophotometry. Aqueous drops have been widely used in conjunction with the HS-SDME mode, *e.g.*, Griess reagents were used in an acetic acid-containing drop for determining nitrite at low concentrations [93]. In most cases, in-drop derivatisation with the corresponding reagents for forming coloured compounds has been carried out.

The disadvantage of using these miniaturised systems lies in the decreased path length, *e.g.*, Nanodrop has a path length 10 times lesser than conventional spectrophotometers. Nevertheless, this decrease in sensitivity is offset by the high preconcentration factors reached with SDME, *e.g.*, a preconcentration factor of 623 is obtained for iodine, thus compensating the 10-fold loss in sensitivity caused by the optical path, which results in improved detection limit (62.3 times) [88].

11.4.4.2. Luminescence

Microvolume fluorospectrometry and chemiluminescence have been also used in combination with SDME. As in the previous case, for detection with a conventional fluorospectrometer, the drop can be transferred to a semi-micro cuvette and diluted, *e.g.*, the determination of volatile trimethylamine has been carried out using IL-HS-SDME and quantum dots (QDs)-based fluorimetric detection, a dilution of a 20 μL drop with 300 μL of acetonitrile being necessary [94]. However, the use of a commercial microvolume fluorospectrometer (*i.e.*, a confined drop-based fluorospectrometer, Nanodrop) equipped with light-emitting diodes (LEDs) as excitation sources and a charge coupled device (CCD) array allows the direct coupling of fluorescence measurements and SDME. Bendicho's group [95] reported for first time a methodology based on HS-SDME and this microvolume fluorospectrometer for determining formaldehyde in textile samples. Analyte was extracted in an aqueous drop containing Hantzsch reagents for in-drop derivatisation.

So far, only the HS-SDME mode has been used in combination with fluorescence detection, both with aqueous and organic drops. For instance, an aqueous drop containing bovine serum albumin and fluorescein has been proposed for terpene screening in fragrance-free cosmetics based on fluorescein quenching [96], and the enzyme alcohol dehydrogenase and the cofactor β-NAD has been used in a drop for determining ethanol in alcohol free cosmetics [97]. Organic solvents have been used in a drop containing QDs as fluorophores. The quenching on QD fluorescence was seen for Se(IV) and CH_3Hg^+, thus allowing their determination [98, 99]. As mentioned above, ILs have also been used in the drop in order to contain CdSe/ZnS QDs. Thus, volatile trimethylamine was quantified through the enhancing effect produced on the fluorescence of the drop [94].

The LLLME mode has been combined with chemiluminescence in order to reach a sensitive, selective and simple assay for Sb(III) and total Sb. It was based on immobilisation of CdSe QDs and inhibition of the chemiluminescence from CdSe QDs/H_2O_2 system in presence of antimony species. A tube luminometer from Berthold Technologies was used. It allows measuring chemiluminescence using the drop (1.5 μL) and 50 μL of 1.5 M H_2O_2 [100].

11.4.4.3. Turbidimetry

HS-SDME combined with in-drop precipitation and turbidity measurement (*i.e.*, microvolume turbidimetry using a microvolume UV-*vis* spectrometer) provides several advantages over classical turbidimetry: i) usual interferences in turbidimetry (absorbance, turbidity and fluorescence from matrix sample) do not occur; ii) measurements can be made at shorter wavelengths with the consequent sensitivity enhancement; iii) analyte is preconcentrated; iv) the cuvetteless system used solves problems such as scattering and background fluorescence, both associated with optical windows. Notwithstanding, precipitation is a complex process since several processes are involved (namely, nucleation, crystal growth, and eventually, aggregation and break-up), some of them being undesirable in turbidimetric analysis. Therefore, an exhaustive control of variables should be carried out for in-drop precipitation [101].

To our knowledge, only two works using HS-SDME and microvolume turbidimetry have been published. Precipitation of sulfide as ZnS has been used for preconcentration and determination of the anion in waters [101]. Chemical oxygen demand has been measured using UV-induced oxidation with hydrogen peroxide and HS-SDME with in-drop precipitation. A precipitate of $BaCO_3$ is formed in the drop as a result of the generation of CO_2 after organic matter oxidation [102].

11.4.4.4. Infrared Spectroscopy

Infrared (IR) spectroscopy is a versatile technique in analytical labs but with two constraints derived from the low sensitivity and the interference caused by water. SDME can solve these problems since analyte preconcentration can occur in an organic solvent. The coupling of SDME with IR has been proposed by Verma *et al.* [103]. These authors used the diffuse reflectance Fourier transform infrared spectroscopy (DRS-FTIR) for determining chromium(VI) [103] and vanadate [104] in different environmental and biological fluids. The drop containing the respective complexes with N1-hydroxy-N1, N2-diphenylbenzamidine (HOA) and the extractant (dichloromethane) was deposited over 0.1 g of KBr (pre-weighted and finely ground). It was dried over a water bath at 60 °C for 2 min and then this matrix was mixed before filling the sample holder.

11.4.4.5. Ion-Mobility Spectrometry

Ion-mobility spectrometry (IMS) is a low-cost and fast analytical technique used for determining molecules in a gas phase by their ion mobility in a weak electric field. However, IMS is not sufficiently selective and sensitive when dealing with complex matrices, so different preconcentration procedures have been applied before the determination step, SDME among them. The coupling of SDME with IMS was first proposed by the Valcárcel´s group [105]. The main difficulties arise from the high signals that most of the organic solvents provide in this technique and the impossibility to inject directly a liquid in the conventional system. So, analytes should be previously volatilised and solvent should be removed. These workers suggested the use of ILs as extractants combined with a specially designed injection unit for solving those problems. ILs are trapped because of their high viscosity, and no signal is obtained due to their low volatility. IL-H--SDME-IMS has been used for determining halocompounds in water [106]. When more complex matrices such as wine are analysed, a greater clean-up must be carried out, *e.g.*, using solid-phase extraction before IL-HS-SDME-IMS [106] or IL-HS-SDME coupled with multicapillary column (MC) separation and IMS [107]. Analytical characteristics achieved with these couplings are comparable to those from other analytical techniques such as MS.

11.5. SPECIFIC APPLICATIONS

SDME has been applied to almost every area of analysis (*e.g*, environment, foods, biology, cosmetics, textile, pharmaceutical, clinical, *etc*). Most applications have used the direct immersion and the headspace modes. For analytes with enhanced volatility, HS-SDME is the preferred mode, since fast extractions are generally achieved along with a suitable clean-up. Selected applications of the most employed SDME modes (*i.e.*, HS, DI, SFO, DS) are included in Table **2** along with the analytical technique used for detection and LODs achieved.

Table 2. Selected applications of SDME mainly for the determination of anions, metal ions, organometals and organic compounds.

Analyte	**Matrix**	**SDME Mode**	**Analytical technique**	**LOD (ng/mL)**	**Reference**
Acrylamide	Potato crips	DI-SDME	GC	57	[8]
Al	Water, drinks	SDME (cycle flow)	ETV-ICP-MS	3.3 ppt	[108]
Alcohols	Beer	HS-SDME	GC	--	[18]
Aldehydes	Water	HS-SDME	GC-MS	0.08-0.32	[109]
Aliphatic amines	Water	HS-SDME	GC-MS	0.6-1.1	[110]
Aliphatic amines	Seafood	HS-SDME	CE	<400	[111]
Alkylamines	Wastewater	HS-SDME	IC	0.96-8.2	[112]

(Table 2) contd.....

Analyte	Matrix	SDME Mode	Analytical technique	LOD (ng/mL)	Reference
Amine drugs	Urine	LLLME	CE	0.5	[113]
Aminoacids, peptides	Water	SDME	CE	--	[114]
Antifungal drugs	Biological	SFODME	HPLC	0.01-0.1	[115]
Aromatic Amines	Waters	HS-SDME	HPLC-UV	0.09-0.38	[116]
As, Se, Sb	Water	HS-SDME	ETAAS	0.1-0.2	[36]
As species	Water	SDME	CE	0.1-07	[117]
As species	Water	SFODME	ETAAS	9.2 ppt	[118]
As, Sb, Bi, Pb, Sn, Hg	Water	HS-SDME	ETV-ICP-MS	0.01-0.8	[37]
Be, Co, Pd, Cd	Biological	DI-SDME	ETV-ICP-MS	0.12-1.5 ppt	[12]
Benzophenone-3	Urine	DI-SDME	HPLC	1.3	[119]
Bisphenol A	Printing paper	DI-SDME	HPLC	0.03 mg/g	[120]
Bromate, iodate,	Seawater	DI-SDME	GC-MS	10-20 ppt	[121]
BTEX	Waters	HS-SDME	GC-MS	20 ppt benzene	[122]
BTEX	Waters	SFODME	GC-FID	2.3-54 ppt	[123]
Bu-Sn compounds	shellfish, water	HS-SDME	GC-ICP-MS	0.8 ppt TBT	[76]
Carbamate pestides	Water	DI-SDME	GC-MS	30-80 ppt	[124]
Carbonyl comp.	Water	HS-SDME	HPLC-DAD	1.7-24.1	[125]
Carbonyl compounds	Biological, oil	HS-SDME	GC	1-3	[126]
Cationic surfactants	Wastewater	DI-SDME	AP-MALDI-MS	10	[127]
Cd	Rize, water	DI-SDME	UV-Vis Spectrophotometry	0.0005	[128]
Cd, Pb	Water	DI-SDME	ETAAS	2, (Cd) 90 (Pb) ppt	[129]
Cd, Pb	Biological, environmental	DI-SDME	ETV-ICP-MS	4.6 (Cd), 2.9 (Pb)	[130]
Cd, Pb	Water	SFODME	ETAAS	10 (Pb), 0.5 (Cd) ppt	[131]
Chem. Warfare agents	Water	DI-SDME	GC-MS	10-75	[132]
Chloride	Water	HS-SDME	μ-spectrophotometry	2.8	[133]
Chlorobenzenes	Wastewater	DI-SDME	GC	-	[134]

(Table 2) contd.....

Analyte	Matrix	SDME Mode	Analytical technique	LOD (ng/mL)	Reference
Cianide	Water	HS-SDME	μ-spectrophotometry	4.3	[135]
Co,Hg,Pb	Serum, water	SDME (cycle-flow)	ETV-ICP-MS)	1.5-6.7 ppt	[74]
Co, Ni	Water	SFODME	ETAAS	0.4 (Co), 0.3 (Ni) ppt	[136]
Cr(VI)	Waters	DI-SDME	ETAAS	0.02	[61]
Dialkyl phthalate esters	Food simulants	DI-SDME	GC-FID	--	[137]
Diclorofenac	Water	DSDME	HPLC	0.1	[138]
Dopamine	Urine	DI-SDME	AP-MALDI-MS	40 ppm	[85]
DQO	Waters	HS-SDME	μ-turbidimetry	1.2 ppm O_2	[102]
Endosulfan	Water	DI-SDME	GC-ECD	0.01	[139]
Essential oils	Leaves	HS-SDME	GC-FID	3.3-20.5 μg/100 g	[140]
Ethanol	Cosmetics	HS-SDME	μ-fluorospectrometry	0.04	[97]
Fluoride	Toothpaste and waters	HS-SDME	GC-FID	4.4	[141]
Fluoride	Milk	HS-SDME	IC	3.8	[142]
Formaldehyde	Textiles	HS-SDME	μ-fluorospectrometry	26	[95]
Fungicides	Water	SDME	GC-ECD	6-10 ppt	[143]
Hexanal, heptanal	Blood	HS-SDME	GC-MS	0.12-0.16 nM	[144]
Heavy metals	Water	SFODME	ETV-ICP-MS	4-17 ppt	[75]
Hg	Water	DI-SDME	ETV-AAS	0.01	[145]
Hg	Saliva	SFODME	CV-AFS	2.5 ppt	[78]
Hg species	Water	DI-SDME	HPLC-PAD	1-22.8	[56]
Hg(II), MeHg	Water, fish	HS-SDME	ETAAS	5 (MeHg)	[146]
Iodine	Pharmaceutical, salt, milk powder	DI-SDME	GC-MS	25 ppt	[147]
Iodine	Infant formulas	HS-SDME	GC-ECD	0.25	[148]
Iodine	Food, pharmaceutical, water	HS-SDME	μ-spectrophotometry	0.69	[88]
Labile sulphide	Water	HS-SDME	μ-turbidimetry	0.5	[101]
Methyl-Hg	Fish	HS-SDME	ETAAS	5	[66]
MMT	Water, gasoline	HS-SDME	ETAAS	0.01	[149]
MMT	Water	HS-SDME	GC-MS	0.2	[150]

(Table 2) contd.....

Analyte	Matrix	SDME Mode	Analytical technique	LOD (ng/mL)	Reference
Mn	Fish, water	DI-SDME	ETAAS	0.03	[151]
Mn	Water	DI-SDME	ETAAS	0.5	[68]
NH_3	Environmental, Biological	HS-SDME	CE	1.5 μM	[152]
Nitrite	Water	HS-SDME	μ-spectrophotometry	1.5	[93]
3-Nitroaniline	Water	DSDME	HPLC	1	[153]
Nitrotoluene comp.	Water	SFODME	GC-FID	0.3-0.5	[154]
OCPs	Fish	SDME	GC-MS	0.5 ng/g	[155]
OCPs	Vegetables	DI-SDME	GC-MS	0.05-0.2	[156]
OCPs	Water	SFODME	GC-ECD	7-19 ppt	[45]
OP pesticides	Water	DI-SDME	GC-NPD	0.012-0.02	[157]
OPP	Rat liver	HS-SDME	GC-ECD	1 ng/g	[158]
OPPs	Fruit juices	SDME	GC-MS	< 5	[159]
OPPs	Water, fruit	SDME (DI, cycle flow)	GC-FPD	0.21-0.56	[160]
OPPs	Water	DI-SDME	GC-MS	10-73 ppt	[161]
OPPs	Water	DI-SDME	GC-NPD	12-20 ppt	[162]
OPPs, carbamate	Water	DI-SDME	GC-MS	0.02-0.05 ppt	[163]
OPPs	Water	SFODME	GC-FPD	0.01-0.04	[164]
PAHs	Wastewater	DI-SDME	GC-FID	-	[165]
PAHs	Environmental	HS-SDME	HPLC-fluorescence	4-25 ppt	[166]
Parabens	Water, cosmetic	DI-SDME	GC-MS	0.001-0.015	[167]
Pb	Water, foods	SDME	ETAAS	0.015	[168]
PCBs	Water	HS-SDME	GC-MS	0.005	[169]
Peptides	Biological	DI-SDME	AP-MALDI-MS	160-210 nM	[80]
Pesticides	Honey	HS-SDME, DI-SDME	-	0.09-19.3	[170]
Pesticides	Grapes, apples	HS-SDME	GC-MS	0.0003	[171]
Pesticides	Vegetables	HS-SDME	GC-ECD, GC-NPD	Sub-ppb	[172]
Pesticides	Coconut water	DI-SDME	GC-MS	0.1-0.88	[173]
Phenolic comp.	Cigarette smoke	DI-SDME	HPLC-MS	0.05-0.3	[174]
Phenols	Waters	DI-SDME	CE	<50	[53]
Phenols	Waters	HS-SDME	GC	0.1-04	[175]
Phenols	Water	DI-SDME	GC-MS	4-61 ppt	[43]

(Table 2) contd.....

Analyte	Matrix	SDME Mode	Analytical technique	LOD (ng/mL)	Reference
Phosphate	Water	DSDME	μ -spectrophotometry	6.1 nM	[92]
Phtahlate esters	Water	SFODME	GC-MS	0.02-0.05	[176]
Polybrominated diphenyl ethers	Water, urine	SFODME	HPLC	0.01-0.04	[177]
Polyphenols	Herbal infusions, fruits, functional foods	SFODME	GC-MS	0.011-0.13	[178]
R-Sn compounds	Water, solids	HS-SDME	GC-MS	1 ppt	[179]
Sb(III), total Sb	Water	HS-SDME	ETAAS	0.025	[65]
Se	Water	HS-SDME	ETAAS	0.15	[64]
Se	Water	HS-SDME	μ -fluorospectrometry	0.08	[98]
Several species	Water	HS-SDME	μ -fluorospectrometry	6.3 nM	[180]
Sufentanil and Alfentanil	Urine	DI-SDME	GC-FID	5-6	[181]
Sulphite preservatives	Foods	HS-SDME	μ -spectrophotometry	0.06 μg/g	[182]
TBT, TPhT	Water	HS-SDME	HPLC-fluorimetry	0.62-0.95	[183]
Terpenes	Cosmetics	HS-SDME	μ -fluorospectrometry	0.49-0.87 μg/g	[96]
Terpenoids	Leaves	HS-SDME	GC-FID	-	[184]
Tl	Water	DI-SDME	ETAAS	0.7	[185]
TMA	Fish	HS-SDME	fluorimetry	14	[94]
TMA	Fish	HS-SDME	μ -spectrophotometry	24	[91]
Trihalomethanes	Water	HS-SDME	GC-IMS		[105]
Trihalomethanes	Water	HS-SDME	GC	0.1-0.9	[186]
Vanadium	Biological fluids	DI-SDME	IR spectroscopy	40	[104]
VOCs	Tobacco	HS-SDME	GC-MS	0.1-3.4 μg/g	[187]
Volatile Aldehydes	Cucumbers	HS-SDME	GC-MS	0.05 mg/kg	[188]
Weak acid dissociable cianyde	Wastewater, river water	HS-SDME	CE	0.15 μM	[189]

ABBREVIATIONS

AAS	atomic absorption spectrometry
AFS	atomic fluorescence spectrometry
AP	atmospheric pressure

BGE	background electrolyte
CCD	charge coupled device
CE	capillary electrophoresis
CFME	continuous flow microextraction
CZE	capillary zone electrophoresis
DDME	drop-to-drop microextraction
DDTC	diethylditiocarbamate
DI-DSDME	direct immersion single-drop microextraction
DRS	diffuse reflectance spectroscopy
DSDME	directly suspended drop microextraction
dynamic HS-LPME	dynamic headspace liquid-phase microextraction
dynamic-LPME	dynamic liquid-phase microextraction
ECD	electron capture detector
ETAAS	electrothermal atomic absorption spectrometry
ETV	electrothermal vaporisation
FAAS	flame atomic absorption spectrometry
FID	flame ionisation detector
FT	fourier transform
GC	gas chromatography
HOA	N1-hydroxy-N1,N2-diphenylbenzamidine
HPLC	high performance liquid chromatography
HS-SDME	headspace single-drop microextraction
ICP-MS	inductively coupled plasma-mass spectrometry
ICP-OES	inductively coupled plasma optical emission spectrometry
IL	ionic liquid
IMS	Ion-mobility spectrometry
IR	infrared spectroscopy
LC	liquid chromatography
LCWs	liquid-core waveguides
LED	light-emitting diode
LLLME	liquid-liquid-liquid microextraction
LOD	limit of detection
LPME	liquid phase microextraction
MALDI	matrix-assisted laser desorption/ionisation

MC	multicapillary column
PAN	1-(2-Pyridylazo)-2-naphthol
QD	quantum dot
SDME	single-drop microextraction
SDS	sodium dodecyl sulphate
SFODME	solidification floating organic drop microextraction
SI	sequential injection
TOF	time-of-flight detector

SCOPE OF THE CHAPTER

The chapter entitled: 'Single-drop microextraction and related techniques' addresses the miniaturisation of solvent extraction using drops of a few microliter volume. A report on their main operation modes, type of extractants, detection systems adapted for measuring the analyte/s in microvolumes, as well as main applications of the most extended SDME modes is included.

CONFLICT OF INTEREST

The authors confirm that they have no conflict of interest to declare for this publication.

ACKNOWLEDGEMENTS

Financial support from the Spanish Ministry of Economy and Competitiveness (Projects CTQ2012-32788 and CTQ2015-68146-P) (MINECO/FEDER) is gratefully acknowledged. F. Pena-Pereira thanks Xunta de Galicia for financial support as a post-doctoral researcher of the I2C program.

REFERENCES

[1] Liu, S.; Dasgupta, P.K. Liquid droplet. A renewable gas sampling interface. *Anal. Chem.,* **1995**, *67*, 2042-2049.
[http://dx.doi.org/10.1021/ac00109a023]

[2] Liu, H.; Dasgupta, P.K. Analytical chemistry in a drop. Solvent extraction in a microdrop. *Anal. Chem.,* **1996**, *68*(11), 1817-1821.
[http://dx.doi.org/10.1021/ac960145h] [PMID: 21619093]

[3] Jeannot, M.A.; Cantwell, F.F. Solvent microextraction into a single drop. *Anal. Chem.,* **1996**, *68*(13),

2236-2240.
[http://dx.doi.org/10.1021/ac960042z] [PMID: 21619310]

[4] Jeannot, M.A.; Przyjazny, A.; Kokosa, J.M. Single drop microextraction--development, applications and future trends. *J. Chromatogr. A,* **2010**, *1217*(16), 2326-2336.
[http://dx.doi.org/10.1016/j.chroma.2009.10.089] [PMID: 19932482]

[5] Pena-Pereira, F.; Lavilla, I.; Bendicho, C. Liquid-phase microextraction techniques within the framework of green chemistry. *TrAC. Trends Analyt. Chem.,* **2010**, *29*, 617-628.
[http://dx.doi.org/10.1016/j.trac.2010.02.016]

[6] Pena-Pereira, F.; Lavilla, I.; Bendicho, C. Liquid-phase microextraction approaches combined with atomic detection: a critical review. *Anal. Chim. Acta,* **2010**, *669*(1-2), 1-16.
[http://dx.doi.org/10.1016/j.aca.2010.04.050] [PMID: 20510897]

[7] Jain, A.; Verma, K.K. Recent advances in applications of single-drop microextraction: a review. *Anal. Chim. Acta,* **2011**, *706*(1), 37-65.
[http://dx.doi.org/10.1016/j.aca.2011.08.022] [PMID: 21995911]

[8] Kaykhaii, M.; Abdi, A. Rapid and sensitive determination of acrylamide in potato crisps using reversed-phase direct immersion single drop microextraction-gas chromatography. *Anal. Methods,* **2013**, *5*, 1289-1293.
[http://dx.doi.org/10.1039/c2ay26560e]

[9] Wu, H.F.; Yen, J.H.; Chin, C.C. Combining drop-to-drop solvent microextraction with gas chromatography/mass spectrometry using electronic ionization and self-ion/molecule reaction method to determine methoxyacetophenone isomers in one drop of water. *Anal. Chem.,* **2006**, *78*(5), 1707-1712.
[http://dx.doi.org/10.1021/ac052076p] [PMID: 16503626]

[10] Zhang, A.L.; Zha, Y. Rapid drop-to-drop liquid–liquid microextraction by help of surface acoustic wave. *Chem. Eng. Process. Process Intensif.,* **2012**, *62*, 145-149.
[http://dx.doi.org/10.1016/j.cep.2012.07.014]

[11] Liu, W.; Lee, H.K. Continuous-flow microextraction exceeding 1000-fold concentration of dilute analytes. *Anal. Chem.,* **2000**, *72*(18), 4462-4467.
[http://dx.doi.org/10.1021/ac000240x] [PMID: 11008784]

[12] Xia, L.; Hu, B.; Jiang, Z.; Wu, Y.; Liang, Y. Single-drop microextraction combined with low-temperature electrothermal vaporization ICPMS for the determination of trace Be, Co, Pd, and Cd in biological samples. *Anal. Chem.,* **2004**, *76*(10), 2910-2915.
[http://dx.doi.org/10.1021/ac035437e] [PMID: 15144204]

[13] Yangcheng, L.; Quan, L.; Guangsheng, L.; Youyuan, D. Directly suspended droplet microextraction. *Anal. Chim. Acta,* **2006**, *566*, 259-264.
[http://dx.doi.org/10.1016/j.aca.2006.02.072]

[14] Khalili Zanjani, M.R.; Yamini, Y.; Shariati, S.; Jönsson, J.A. A new liquid-phase microextraction method based on solidification of floating organic drop. *Anal. Chim. Acta,* **2007**, *585*(2), 286-293.
[http://dx.doi.org/10.1016/j.aca.2006.12.049] [PMID: 17386676]

[15] Farajzadeh, M.A.; Djozan, D.; Khorram, P. Development of a new microextraction method based on a

dynamic single drop in a narrow-bore tube: application in extraction and preconcentration of some organic pollutants in well water and grape juice samples. *Talanta,* **2011**, *85*(2), 1135-1142. [http://dx.doi.org/10.1016/j.talanta.2011.05.044] [PMID: 21726749]

[16] He, Y.; Lee, H.K. Liquid-phase microextraction in a single drop of organic solvent by using a conventional microsyringe. *Anal. Chem.,* **1997**, *69*, 4634-4640. [http://dx.doi.org/10.1021/ac970242q]

[17] Liu, H.; Dasgupta, P.K. A renewable liquid droplet as a sampler and a windowless optical cell. Automated sensor for gaseous chlorine. *Anal. Chem.,* **1995**, *67*, 4221-4228. [http://dx.doi.org/10.1021/ac00119a003]

[18] Tankeviciute, A.; Kazlauskas, R.; Vickackaite, V. Headspace extraction of alcohols into a single drop. *Analyst (Lond.),* **2001**, *126*, 1674-1677. [http://dx.doi.org/10.1039/b103493f]

[19] Shen, G.; Lee, H.K. Headspace liquid-phase microextraction of chlorobenzenes in soil with gas chromatography-electron capture detection. *Anal. Chem.,* **2003**, *75*(1), 98-103. [http://dx.doi.org/10.1021/ac020428b] [PMID: 12530824]

[20] Ma, M.; Cantwell, F.F. Solvent microextraction with simultaneous back-extraction for sample cleanup and preconcentration : Quantitative extraction. *Anal. Chem.,* **1998**, *70*, 3912-3919. [http://dx.doi.org/10.1021/ac980174n]

[21] Ma, M.; Cantwell, F.F. Solvent microextraction with simultaneous back-extraction for sample cleanup and preconcentration: Preconcentration into a single microdrop. *Anal. Chem.,* **1999**, *71*, 388-393. [http://dx.doi.org/10.1021/ac9805899]

[22] Gao, W.; Chen, G.; Chen, T.; Zhang, X.; Chen, Y.; Hu, Z. Directly suspended droplet microextraction combined with single drop back-extraction as a new approach for sample preparation compatible with capillary electrophoresis. *Talanta,* **2011**, *83*(5), 1673-1679. [http://dx.doi.org/10.1016/j.talanta.2010.11.066] [PMID: 21238767]

[23] Kubáň, P.; Boček, P. Micro-electromembrane extraction across free liquid membranes. Instrumentation and basic principles. *J. Chromatogr. A,* **2014**, *1346*, 25-33. [http://dx.doi.org/10.1016/j.chroma.2014.04.047] [PMID: 24792701]

[24] Raterink, R.J.; Lindenburg, P.W.; Vreeken, R.J.; Hankemeier, T. Three-phase electroextraction: a new (online) sample purification and enrichment method for bioanalysis. *Anal. Chem.,* **2013**, *85*(16), 7762-7768. [http://dx.doi.org/10.1021/ac4010716] [PMID: 23724849]

[25] Henderson, R.K.; Jiménez-González, C.; Constable, D.J.; Alston, S.R.; Inglis, G.G.; Fisher, G.; Sherwood, J.; Binks, S.P.; Curzons, A.D. Expanding GSK's solvent selection guide - embedding sustainability into solvent selection starting at medicinal chemistry. *Green Chem.,* **2011**, *13*, 854-862. [http://dx.doi.org/10.1039/c0gc00918k]

[26] Curzons, A.D.; Constable, D.C.; Cunningham, V.L. Solvent selection guide: a guide to the integration of environmental, health and safety criteria into the selection of solvents. *Clean Technol. Environ. Policy,* **1999**, *1*, 82-90. [http://dx.doi.org/10.1007/s100980050014]

[27] Jiménez-González, C.; Curzons, A.D.; Constable, D.J.; Cunningham, V.L. Expanding GSK's solvent selection guide - application of life cycle assessment to enhance solvent selections. *Clean Technol. Environ. Policy,* **2005**, *7*, 42-50.
[http://dx.doi.org/10.1007/s10098-004-0245-z]

[28] Liu, J.F.; Jiang, G.B.; Chi, Y.G.; Cai, Y.Q.; Zhou, Q.X.; Hu, J.T. Use of ionic liquids for liquid-phase microextraction of polycyclic aromatic hydrocarbons. *Anal. Chem.,* **2003**, *75*(21), 5870-5876.
[http://dx.doi.org/10.1021/ac034506m] [PMID: 14588028]

[29] Zhao, D.; Liao, Y.; Zhang, Z. Toxicity of ionic liquids. *CLEAN - Soil, Air. Water,* **2007**, *35*, 42-48.

[30] Imperato, G.; König, B.; Chiappe, C. Ionic green solvents from renewable resources. *Eur. J. Chem.,* **2006**, *2007*(7), 1049-1058.

[31] Kanyanee, T.; Borst, W.L.; Jakmunee, J.; Grudpan, K.; Li, J.; Dasgupta, P.K. Soap bubbles in analytical chemistry. Conductometric determination of sub-parts per million levels of sulfur dioxide with a soap bubble. *Anal. Chem.,* **2006**, *78*(8), 2786-2793.
[http://dx.doi.org/10.1021/ac052198h] [PMID: 16615794]

[32] Ballesteros-Gómez, A.; Sicilia, M.D.; Rubio, S. Supramolecular solvents in the extraction of organic compounds. A review. *Anal. Chim. Acta,* **2010**, *677*(2), 108-130.
[http://dx.doi.org/10.1016/j.aca.2010.07.027] [PMID: 20837178]

[33] López-Jiménez, F.J.; Rubio, S.; Pérez-Bendito, D. Single-drop coacervative microextraction of organic compounds prior to liquid chromatography. Theoretical and practical considerations. *J. Chromatogr. A,* **2008**, *1195*(1-2), 25-33.
[http://dx.doi.org/10.1016/j.chroma.2008.05.002] [PMID: 18501913]

[34] Yao, C.; Twu, P.; Anderson, J.L. Headspace single drop microextraction using micellar ionic liquid extraction solvents. *Chromatographia,* **2010**, *72*, 393-402.
[http://dx.doi.org/10.1365/s10337-010-1675-x]

[35] Cardoso, A.A.; Liu, H.; Dasgupta, P.K. Fluorometric fiber optic drop sensor for atmospheric hydrogen sulfide. *Talanta,* **1997**, *44*(6), 1099-1106.
[http://dx.doi.org/10.1016/S0039-9140(96)02202-3] [PMID: 18966842]

[36] Fragueiro, S.; Lavilla, I.; Bendicho, C. Headspace sequestration of arsine onto a Pd(II)-containing aqueous drop as a preconcentration method for electrothermal atomic absorption spectrometry. *Spectrochim. Acta B At. Spectrosc.,* **2004**, *59*, 851-855.
[http://dx.doi.org/10.1016/j.sab.2004.03.002]

[37] Gil, S.; de Loos-Vollebregt, M.T.; Bendicho, C. Optimization of a single-drop microextraction method for multielemental determination by electrothermal vaporization inductively coupled plasma mass spectrometry following *in situ* vapor generation. *Spectrochim. Acta B At. Spectrosc.,* **2009**, *64*, 208-214.
[http://dx.doi.org/10.1016/j.sab.2008.12.002]

[38] Keith, L.H.; Gron, L.U.; Young, J.L. Green analytical methodologies. *Chem. Rev.,* **2007**, *107*(6), 2695-2708.
[http://dx.doi.org/10.1021/cr068359e] [PMID: 17521200]

[39] Jeannot, M.A.; Cantwell, F.F. Solvent microextraction as a speciation tool: determination of free

progesterone in a protein solution. *Anal. Chem.,* **1997**, *69*(15), 2935-2940.
[http://dx.doi.org/10.1021/ac970207j] [PMID: 9253247]

[40] Przyjazny, A.; Austin, J.F.; Essenmacher, A.T. Headspace liquid-phase microextraction: a novel preconcentration technique for volatile organic pollutants. *Proceedings of U.S. Environmental Protection Agency Office of Water the 6th Polish Conference on Analytical Chemistry,* **2000** Gliwice, Poland, pp. 135-136.

[41] Lee, J.; Lee, H.K. Fully automated dynamic in-syringe liquid-phase microextraction and on-column derivatisation of carbamate pesticides with gas chromatography/mass spectrometric analysis. *Anal. Chem.,* **2011**, *83*(17), 6856-6861.
[http://dx.doi.org/10.1021/ac200807d] [PMID: 21761858]

[42] Hyötyläinen, T.; Riekkola, M.L. Sorbent- and liquid-phase microextraction techniques and membrane-assisted extraction in combination with gas chromatographic analysis: A review. *Anal. Chim. Acta,* **2008**, *614*(1), 27-37.
[http://dx.doi.org/10.1016/j.aca.2008.03.003] [PMID: 18405678]

[43] Saraji, M.; Bakhshi, M. Determination of phenols in water samples by single-drop microextraction followed by in-syringe derivatisation and gas chromatography-mass spectrometric detection. *J. Chromatogr. A,* **2005**, *1098*(1-2), 30-36.
[http://dx.doi.org/10.1016/j.chroma.2005.08.063] [PMID: 16314158]

[44] Fiamegos, Y.C.; Stalikas, C.D. In-drop derivatisation liquid-phase microextraction assisted by ion-pairing transfer for the gas chromatographic determination of phenolic endocrine disruptors. *Anal. Chim. Acta,* **2007**, *597*(1), 32-40.
[http://dx.doi.org/10.1016/j.aca.2007.06.021] [PMID: 17658310]

[45] Farahani, H.; Yamini, Y.; Shariati, S.; Khalili-Zanjani, M.R.; Mansour-Baghahi, S. Development of liquid phase microextraction method based on solidification of floated organic drop for extraction and preconcentration of organochlorine pesticides in water samples. *Anal. Chim. Acta,* **2008**, *626*(2), 166-173.
[http://dx.doi.org/10.1016/j.aca.2008.08.001] [PMID: 18790117]

[46] Aguilera-Herrador, E.; Lucena, R.; Cardenas, S.; Valcárcel, M. Direct coupling of ionic liquid based single-drop microextraction and GC/MS. *Anal. Chem.,* **2008**, *80*(3), 793-800.
[http://dx.doi.org/10.1021/ac071555g] [PMID: 18154274]

[47] AlOthman, Z.A.; Dawod, M.; Kim, J.; Chung, D.S. Single-drop microextraction as a powerful pretreatment tool for capillary electrophoresis: A review. *Anal. Chim. Acta,* **2012**, *739*, 14-24.
[http://dx.doi.org/10.1016/j.aca.2012.06.005] [PMID: 22819045]

[48] Gao, W.; Chen, G.; Chen, Y.; Zhang, X.; Yin, Y.; Hu, Z. Application of single drop liquid-liqui--liquid microextraction for the determination of fluoroquinolones in human urine by capillary electrophoresis. *J. Chromatogr. B Analyt. Technol. Biomed. Life Sci.,* **2011**, *879*(3-4), 291-295.
[http://dx.doi.org/10.1016/j.jchromb.2010.11.040] [PMID: 21176888]

[49] Choi, K.; Kim, S.J.; Jin, Y.G.; Jang, Y.O.; Kim, J.S.; Chung, D.S. Single drop microextraction using commercial capillary electrophoresis instruments. *Anal. Chem.,* **2009**, *81*(1), 225-230.
[http://dx.doi.org/10.1021/ac801886b] [PMID: 19117452]

[50] Fang, H.; Zeng, Z.; Liu, L.; Pang, D. On-line back-extraction field-amplified sample injection method

for directly analyzing cocaine and thebaine in the extractants by solvent microextraction. *Anal. Chem.,* **2006**, *78*(4), 1257-1263.
[http://dx.doi.org/10.1021/ac0516537] [PMID: 16478120]

[51] Zhang, J.; Su, T.; Lee, H.K. Headspace water-based liquid-phase microextraction. *Anal. Chem.,* **2005**, *77*(7), 1988-1992.
[http://dx.doi.org/10.1021/ac040129h] [PMID: 15801728]

[52] Xie, H.Y.; He, Y.Z.; Gan, W.E.; Yu, C.Z.; Han, F.; Ling, D.S. In-line coupling headspace liquid-phase microextraction with capillary electrophoresis. *J. Chromatogr. A,* **2010**, *1217*(8), 1203-1207.
[http://dx.doi.org/10.1016/j.chroma.2009.12.028] [PMID: 20034631]

[53] Wang, Q.; Qiu, H.; Li, J.; Liu, X.; Jiang, S. On-line coupling of ionic liquid-based single-drop microextraction with capillary electrophoresis for sensitive detection of phenols. *J. Chromatogr. A,* **2010**, *1217*(33), 5434-5439.
[http://dx.doi.org/10.1016/j.chroma.2010.06.059] [PMID: 20621300]

[54] Xu, L.; Basheer, C.; Lee, H.K. Developments in single-drop microextraction. *J. Chromatogr. A,* **2007**, *1152*(1-2), 184-192.
[http://dx.doi.org/10.1016/j.chroma.2006.10.073] [PMID: 17097670]

[55] He, Y.; Lee, H.K. Continuous flow microextraction combined with high-performance liquid chromatography for the analysis of pesticides in natural waters. *J. Chromatogr. A,* **2006**, *1122*(1-2), 7-12.
[http://dx.doi.org/10.1016/j.chroma.2006.04.078] [PMID: 16716335]

[56] Pena-Pereira, F.; Lavilla, I.; Bendicho, C.; Vidal, L.; Canals, A. Speciation of mercury by ionic liquid-based single-drop microextraction combined with high-performance liquid chromatography-photodiode array detection. *Talanta,* **2009**, *78*(2), 537-541.
[http://dx.doi.org/10.1016/j.talanta.2008.12.003] [PMID: 19203620]

[57] Sobhi, H.R.; Yamini, Y.; Esrafili, A.; Abadi, R.H. Suitable conditions for liquid-phase microextraction using solidification of a floating drop for extraction of fat-soluble vitamins established using an orthogonal array experimental design. *J. Chromatogr. A,* **2008**, *1196-1197*, 28-32.
[http://dx.doi.org/10.1016/j.chroma.2008.05.005] [PMID: 18502432]

[58] Muniraj, S.; Yan, C.T.; Shih, H.K.; Ponnusamy, V.K.; Jen, J.F. Determination of ammonium in aqueous samples using new headspace dynamic in-syringe liquid-phase microextraction with *in situ* derivitazation coupled with liquid chromatography-fluorescence detection. *Anal. Chim. Acta,* **2012**, *754*, 54-60.
[http://dx.doi.org/10.1016/j.aca.2012.10.001] [PMID: 23140954]

[59] Xu, H.; Lv, L.; Hu, S.; Song, D. High-performance liquid chromatographic determination of hexanal and heptanal in human blood by ultrasound-assisted headspace liquid-phase microextraction with in-drop derivatisation. *J. Chromatogr. A,* **2010**, *1217*(16), 2371-2375.
[http://dx.doi.org/10.1016/j.chroma.2009.09.068] [PMID: 19819460]

[60] Chamsaz, M.; Arbab-Zavar, M.H.; Nazari, S. Determination of arsenic by electrothermal atomic absorption spectrometry using headspace liquid phase microextraction after *in situ* hydride generation. *J. Anal. At. Spectrom.,* **2003**, *18*, 1279-1282.
[http://dx.doi.org/10.1039/b303169a]

[61] Pena, F.; Lavilla, I.; Bendicho, C. Immersed single-drop microextraction interfaced with sequential injection analysis for determination of Cr(VI) in natural waters by electrothermal-atomic absorption spectrometry. *Spectrochim. Acta B At. Spectrosc.,* **2008**, *63*, 498-503. [http://dx.doi.org/10.1016/j.sab.2008.01.005]

[62] Anthemidis, A.N.; Adam, I.S. Development of on-line single-drop micro-extraction sequential injection system for electrothermal atomic absorption spectrometric determination of trace metals. *Anal. Chim. Acta,* **2009**, *632*(2), 216-220. [http://dx.doi.org/10.1016/j.aca.2008.10.078] [PMID: 19110096]

[63] Pena-Pereira, F.; Lavilla, I.; Bendicho, C. Miniaturized preconcentration methods based on liquid–liquid extraction and their application in inorganic ultratrace analysis and speciation: a review. *Spectrochim. Acta B At. Spectrosc.,* **2009**, *64*, 1-15. [http://dx.doi.org/10.1016/j.sab.2008.10.042]

[64] Fragueiro, S.; Lavilla, I.; Bendicho, C. Hydride generation-headspace single-drop microextraction-electrothermal atomic absorption spectrometry method for determination of selenium in waters after photoassisted prereduction. *Talanta,* **2006**, *68*(4), 1096-1101. [http://dx.doi.org/10.1016/j.talanta.2005.07.011] [PMID: 18970437]

[65] Pena-Pereira, F.; Lavilla, I.; Bendicho, C. Headspace single-drop microextraction with *in situ* stibine generation for the determination of antimony (III) and total antimony by electrothermal-atomic absorption spectrometry. *Mikrochim. Acta,* **2009**, *164*, 77-83. [http://dx.doi.org/10.1007/s00604-008-0036-z]

[66] Gil, S.; Fragueiro, S.; Lavilla, I.; Bendicho, C. Determination of methylmercury by electrothermal atomic absorption spectrometry using headspace single-drop microextraction with *in situ* hydride generation. *Spectrochim. Acta B At. Spectrosc.,* **2005**, *60*, 145-150. [http://dx.doi.org/10.1016/j.sab.2004.10.008]

[67] Figueroa, R.; Garcia, M.; Lavilla, I.; Bendicho, C. Photoassisted vapor generation in the presence of organic acids for ultrasensitive determination of Se by electrothermal-atomic absorption spectrometry following headspace single-drop microextraction. *Spectrochim. Acta B At. Spectrosc.,* **2005**, *60*, 1556-1563. [http://dx.doi.org/10.1016/j.sab.2005.10.009]

[68] Manzoori, J.L.; Amjadi, M.; Abulhassani, J. Ionic liquid-based single drop microextraction combined with electrothermal atomic absorption spectrometry for the determination of manganese in water samples. *Talanta,* **2009**, *77*(4), 1539-1544. [http://dx.doi.org/10.1016/j.talanta.2008.09.045] [PMID: 19084676]

[69] Martinis, E.M.; Wuilloud, R.G. Cold vapor ionic liquid-assisted headspace single-drop microextraction: a novel preconcentration technique for mercury species determination in complex matrix samples. *J. Anal. At. Spectrom.,* **2010**, *25*, 1432-1439. [http://dx.doi.org/10.1039/c004678g]

[70] Wen, X.; Deng, Q.; Wang, J.; Yang, S.; Zhao, X. A new coupling of ionic liquid based-single drop microextraction with tungsten coil electrothermal atomic absorption spectrometry. *Spectrochim. Acta A Mol. Biomol. Spectrosc.,* **2013**, *105*, 320-325. [http://dx.doi.org/10.1016/j.saa.2012.12.040] [PMID: 23318776]

[71] López-García, I.; Rivas, R.E.; Hernández-Córdoba, M. Liquid-phase microextraction with solidification of the organic floating drop for the preconcentration and determination of mercury traces by electrothermal atomic absorption spectrometry. *Anal. Bioanal. Chem.,* **2010**, *396*(8), 3097-3102. [http://dx.doi.org/10.1007/s00216-010-3500-7] [PMID: 20174984]

[72] Şahin, C.A.; Tokgöz, I. A novel solidified floating organic drop microextraction method for preconcentration and determination of copper ions by flow injection flame atomic absorption spectrometry. *Anal. Chim. Acta,* **2010**, *667*(1-2), 83-87. [http://dx.doi.org/10.1016/j.aca.2010.04.012] [PMID: 20441870]

[73] Mitani, C.; Anthemidis, A.N. On-line liquid phase micro-extraction based on drop-in-plug sequential injection lab-at-valve platform for metal determination. *Anal. Chim. Acta,* **2013**, *771*, 50-55. [http://dx.doi.org/10.1016/j.aca.2013.02.003] [PMID: 23522112]

[74] Xia, L.; Li, X.; Wu, Y.; Hu, B.; Chen, R. Ionic liquids based single drop microextraction combined with electrothermal vaporization inductively coupledplasma mass spectrometry for determination of Co, Hg and Pb in biological and environmental samples. *Spectrochim. Acta B At. Spectrosc.,* **2008**, *63*, 1290-1296. [http://dx.doi.org/10.1016/j.sab.2008.09.018]

[75] Guo, X.; He, M.; Chen, B.; Hu, B. Solidified floating organic drop microextraction combined with ETV-ICP-MS for the determination of trace heavy metals in environmental water samples. *Talanta,* **2012**, *94*, 70-76. [http://dx.doi.org/10.1016/j.talanta.2012.02.053] [PMID: 22608416]

[76] Xiao, Q.; Hu, B.; He, M. Speciation of butyltin compounds in environmental and biological samples using headspace single drop microextraction coupled with gas chromatography-inductively coupled plasma mass spectrometry. *J. Chromatogr. A,* **2008**, *1211*(1-2), 135-141. [http://dx.doi.org/10.1016/j.chroma.2008.09.089] [PMID: 18922539]

[77] Liu, J.F.; Chi, Y.G.; Jiang, G.B. Screening the extractability of some typical environmental pollutants by ionic liquids in liquid-phase microextraction. *J. Sep. Sci.,* **2005**, *28*(1), 87-91. [http://dx.doi.org/10.1002/jssc.200401805] [PMID: 15688636]

[78] Yuan, C.G.; Wang, J.; Jin, Y. Ultrasensitive determination of mercury in human saliva by atomic fluorescence spectrometry based on solidified floating organic drop microextraction. *Mikrochim. Acta,* **2012**, *177*, 153-158. [http://dx.doi.org/10.1007/s00604-012-0768-7]

[79] Sudhir, P.R.; Wu, H.F.; Zhou, Z.C. Identification of peptides using gold nanoparticle-assisted single-drop microextraction coupled with AP-MALDI mass spectrometry. *Anal. Chem.,* **2005**, *77*(22), 7380-7385. [http://dx.doi.org/10.1021/ac051162m] [PMID: 16285689]

[80] Sudhir, P.R.; Shrivas, K.; Zhou, Z.C.; Wu, H.F. Single drop microextraction using silver nanoparticles as electrostatic probes for peptide analysis in atmospheric pressure matrix-assisted laser desorption/ionization mass spectrometry and comparison with gold electrostatic probes and silver hydrophobic probes. *Rapid Commun. Mass Spectrom.,* **2008**, *22*(19), 3076-3086. [http://dx.doi.org/10.1002/rcm.3710] [PMID: 18777509]

[81] Ahmad, F.; Wu, H.F. Characterization of pathogenic bacteria using ionic liquid *via* single drop

microextraction combined with MALDI-TOF MS. *Analyst (Lond.),* **2011**, *136*(19), 4020-4027. [http://dx.doi.org/10.1039/c1an15350a] [PMID: 21804984]

[82] Sun, S.; Cheng, Z.; Xie, J.; Zhang, J.; Liao, Y.; Wang, H.; Guo, Y. Identification of volatile basic components in tobacco by headspace liquid-phase microextraction coupled to matrix-assisted laser desorption/ionization with Fourier transform mass spectrometry. *Rapid Commun. Mass Spectrom.,* **2005**, *19*(8), 1025-1030. [http://dx.doi.org/10.1002/rcm.1884] [PMID: 15768451]

[83] Shrivas, K.; Wu, H.F. Single drop microextraction as a concentrating probe for rapid screening of low molecular weight drugs from human urine in atmospheric-pressure matrix-assisted laser desorption/ionization mass spectrometry. *Rapid Commun. Mass Spectrom.,* **2007**, *21*(18), 3103-3108. [http://dx.doi.org/10.1002/rcm.3192] [PMID: 17708597]

[84] Shrivas, K.; Wu, H.F. Quantitative bioanalysis of quinine by atmospheric pressure-matrix assisted laser desorption/ionization mass spectrometry combined with dynamic drop-to-drop solvent microextraction. *Anal. Chim. Acta,* **2007**, *605*(2), 153-158. [http://dx.doi.org/10.1016/j.aca.2007.10.032] [PMID: 18036378]

[85] Wu, H.F.; Lin, C.H. Direct combination of immersed single-drop microextraction with atmospheric pressure matrix-assisted laser desorption/ionization tandem mass spectrometry for rapid analysis of a hydrophilic drug *via* hydrogen-bonding interaction and comparison with liquid-liquid extraction and liquid-phase microextraction using a dual gauge microsyringe with a hollow fiber. *Rapid Commun. Mass Spectrom.,* **2006**, *20*(16), 2511-2515. [http://dx.doi.org/10.1002/rcm.2612] [PMID: 16862620]

[86] Pena-Pereira, F.; Costas-Mora, I.; Romero, V.; Lavilla, I.; Bendicho, C. Advances in miniaturized UV-*vis* spectrometric systems. *TrAC. Trends Analyt. Chem.,* **2011**, *30*, 1637-1648. [http://dx.doi.org/10.1016/j.trac.2011.04.018]

[87] Liu, H.; Dasgupta, P.K. A liquid drop: What is it good for? *Microchem. J.,* **1997**, *57*, 127-136. [http://dx.doi.org/10.1006/mchj.1997.1498]

[88] Pena-Pereira, F.; Lavilla, I.; Bendicho, C. Headspace single-drop microextraction coupled to microvolume UV-*vis* spectrophotometry for iodine determination. *Anal. Chim. Acta,* **2009**, *631*(2), 223-228. [http://dx.doi.org/10.1016/j.aca.2008.10.048] [PMID: 19084630]

[89] Andruch, V.; Kocúrová, L.; Balogh, I.S.; Škrlíková, J. Recent advances in coupling single-drop and dispersive liquid–liquid microextraction with UV–vis spectrophotometry and related detection techniques. *Microchem. J.,* **2012**, *102*, 1-10. [http://dx.doi.org/10.1016/j.microc.2011.10.006]

[90] Sharma, N.; Pillai, A.K.; Pathak, N.; Jain, A.; Verma, K.K. Liquid-phase microextraction and fibre-optics-based cuvetteless CCD-array micro-spectrophotometry for trace analysis. *Anal. Chim. Acta,* **2009**, *648*(2), 183-193. [http://dx.doi.org/10.1016/j.aca.2009.07.006] [PMID: 19646583]

[91] Pena-Pereira, F.; Lavilla, I.; Bendicho, C. Colorimetric assay for determination of trimethylamine-nitrogen (TMA-N) in fish by combining headspace-single-drop microextraction and microvolume UV-vis spectrophotometry. *Food Chem.,* **2010**, *119*, 402-407.

[http://dx.doi.org/10.1016/j.foodchem.2009.07.038]

[92] Pena-Pereira, F.; Cabaleiro, N.; de la Calle, I.; Costas, M.; Gil, S.; Lavilla, I.; Bendicho, C. Directly suspended droplet microextraction in combination with microvolume UV-*vis* spectrophotometry for determination of phosphate. *Talanta,* **2011**, *85*(2), 1100-1104. [http://dx.doi.org/10.1016/j.talanta.2011.05.032] [PMID: 21726744]

[93] Senra-Ferreiro, S.; Pena-Pereira, F.; Lavilla, I.; Bendicho, C. Griess micro-assay for the determination of nitrite by combining fibre optics-based cuvetteless UV-*vis* micro-spectrophotometry with liquid-phase microextraction. *Anal. Chim. Acta,* **2010**, *668*(2), 195-200. [http://dx.doi.org/10.1016/j.aca.2010.04.038] [PMID: 20493298]

[94] Carrillo-Carrión, C.; Simonet, B.M.; Valcárcel, M. (CdSe/ZnS QDs)-ionic liquid-based headspace single drop microextraction for the fluorimetric determination of trimethylamine in fish. *Analyst (Lond.),* **2012**, *137*(5), 1152-1159. [http://dx.doi.org/10.1039/c2an15914g] [PMID: 22268066]

[95] Sáenz, M.; Alvarado, J.; Pena-Pereira, F.; Senra-Ferreiro, S.; Lavilla, I.; Bendicho, C. Liquid-phase microextraction with in-drop derivatisation combined with microvolume fluorospectrometry for free and hydrolyzed formaldehyde determination in textile samples. *Anal. Chim. Acta,* **2011**, *687*(1), 50-55. [http://dx.doi.org/10.1016/j.aca.2010.12.006] [PMID: 21241845]

[96] Cabaleiro, N.; de la Calle, I.; Bendicho, C.; Lavilla, I. Fast screening of terpenes in fragrance-free cosmetics by fluorescence quenching on a fluorescein-bovine serum albumin probe confined in a drop. *Anal. Chim. Acta,* **2012**, *719*, 61-67. [http://dx.doi.org/10.1016/j.aca.2011.12.061] [PMID: 22340532]

[97] Cabaleiro, N.; de la Calle, I.; Bendicho, C.; Lavilla, I. Enzymatic single-drop microextraction for the assay of ethanol in alcohol-free cosmetics using microvolume fluorospectrometry detection. *Anal. Chim. Acta,* **2012**, *733*, 28-33. [http://dx.doi.org/10.1016/j.aca.2012.04.039] [PMID: 22704372]

[98] Costas-Mora, I.; Romero, V.; Pena-Pereira, F.; Lavilla, I.; Bendicho, C. Quantum dots confined in an organic drop as luminescent probes for detection of selenium by microfluorospectrometry after hydridation: study of the quenching mechanism and analytical performance. *Anal. Chem.,* **2012**, *84*(10), 4452-4459. [http://dx.doi.org/10.1021/ac300221s] [PMID: 22506873]

[99] Costas-Mora, I.; Romero, V.; Lavilla, I.; Bendicho, C. *In situ* building of a nanoprobe based on fluorescent carbon dots for methylmercury detection. *Anal. Chem.,* **2014**, *86*(9), 4536-4543. [http://dx.doi.org/10.1021/ac500517h] [PMID: 24678836]

[100] Costas-Mora, I.; Romero, V.; Lavilla, I.; Bendicho, C. Solid-state chemiluminescence assay for ultrasensitive detection of antimony using on-vial immobilization of CdSe quantum dots combined with liquid-liquid-liquid microextraction. *Anal. Chim. Acta,* **2013**, *788*, 114-121. [http://dx.doi.org/10.1016/j.aca.2013.06.007] [PMID: 23845489]

[101] Lavilla, I.; Pena-Pereira, F.; Gil, S.; Costas, M.; Bendicho, C. Microvolume turbidimetry for rapid and sensitive determination of the acid labile sulfide fraction in waters after headspace single-drop microextraction with *in situ* generation of volatile hydrogen sulfide. *Anal. Chim. Acta,* **2009**, *647*(1),

112-116.
[http://dx.doi.org/10.1016/j.aca.2009.05.035] [PMID: 19576394]

[102] Akhoundzadeh, J.; Costas, M.; Lavilla, I.; Chamsaz, M.; Bendicho, C. Miniaturized and green method for determination of chemical oxygen demand using UV-induced oxidation with hydrogen peroxide and single drop microextraction. *Mikrochim. Acta,* **2013**, *180*, 1029-1036.
[http://dx.doi.org/10.1007/s00604-013-1024-5]

[103] Verma, D.; Verma, S.K.; Deb, M.K. Single-drop micro-extraction and diffuse reflectance Fourier transform infrared spectroscopic determination of chromium in biological fluids. *Talanta,* **2009**, *78*(1), 270-277.
[http://dx.doi.org/10.1016/j.talanta.2008.11.020] [PMID: 19174237]

[104] Verma, D.; Deb, M.K. Determination of vanadium(V) employing a new combined single drop microextraction and diffuse reflectance Fourier transform infrared spectroscopy technique. *Int. J. Environ. Anal. Chem.,* **2012**, *92*, 59-75.
[http://dx.doi.org/10.1080/03067310903410998]

[105] Aguilera-Herrador, E.; Lucena, R.; Cárdenas, S.; Valcárcel, M. Ionic liquid-based single drop microextraction and room-temperature gas chromatography for on-site ion mobility spectrometric analysis. *J. Chromatogr. A,* **2009**, *1216*(29), 5580-5587.
[http://dx.doi.org/10.1016/j.chroma.2009.05.071] [PMID: 19523638]

[106] Márquez-Sillero, I.; Aguilera-Herrador, E.; Cárdenas, S.; Valcárcel, M. Determination of 2,4,6-tricholoroanisole in water and wine samples by ionic liquid-based single-drop microextraction and ion mobility spectrometry. *Anal. Chim. Acta,* **2011**, *702*(2), 199-204.
[http://dx.doi.org/10.1016/j.aca.2011.06.046] [PMID: 21839198]

[107] Márquez-Sillero, I.; Cárdenas, S.; Valcárcel, M. Direct determination of 2,4,6-tricholoroanisole in wines by single-drop ionic liquid microextraction coupled with multicapillary column separation and ion mobility spectrometry detection. *J. Chromatogr. A,* **2011**, *1218*(42), 7574-7580.
[http://dx.doi.org/10.1016/j.chroma.2011.06.032] [PMID: 21737082]

[108] Xia, L.; Hu, B.; Jiang, Z.; Wu, Y.; Li, L.; Chen, R. 8-Hydroxyquinoline-chloroform single drop microextraction and electrothermal vaporization ICP-MS for the fractionation of aluminium in natural waters and drinks. *J. Anal. At. Spectrom.,* **2005**, *20*, 441-446.
[http://dx.doi.org/10.1039/B416938G]

[109] Deng, C.; Yao, N.; Li, N.; Zhang, X. Headspace single-drop microextraction with in-drop derivatisation for aldehyde analysis. *J. Sep. Sci.,* **2005**, *28*(17), 2301-2305.
[http://dx.doi.org/10.1002/jssc.200500085] [PMID: 16342795]

[110] Deng, C.; Li, N.; Wang, L.; Zhang, X. Development of gas chromatography-mass spectrometry following headspace single-drop microextraction and simultaneous derivatisation for fast determination of short-chain aliphatic amines in water samples. *J. Chromatogr. A,* **2006**, *1131*(1-2), 45-50.
[http://dx.doi.org/10.1016/j.chroma.2006.07.084] [PMID: 16920125]

[111] Mark, J.J.; Kumar, A.; Demattio, H.; Hoffmann, W.; Malik, A.; Matysik, F.M. Single-drop microextraction, microchip electrophoresis and contactless conductivity detection for the determination of aliphatic amines in the biodegradation process of seafood samples. *Electroanalysis,*

2011, *23*, 161-168.
[http://dx.doi.org/10.1002/elan.201000445]

[112] Xun, Y.; Hang, Y.; Luan, G. Ion chromatography for rapid and sensitive determination of three alkylamines in wastewater after headspace single-drop microextraction. *Anal. Lett.,* **2014**, *47*, 25-33.
[http://dx.doi.org/10.1080/00032719.2013.832268]

[113] Choi, K.; Kim, J.; Jang, Y.O.; Chung, D.S. Direct chiral analysis of primary amine drugs in human urine by single drop microextraction in-line coupled to CE. *Electrophoresis,* **2009**, *30*(16), 2905-2911.
[http://dx.doi.org/10.1002/elps.200800829] [PMID: 19691052]

[114] Park, Y.K.; Choi, K.; Ahmed, A.Y.; AlOthman, Z.A.; Chung, D.S. Selective preconcentration of amino acids and peptides using single drop microextraction in-line coupled with capillary electrophoresis. *J. Chromatogr. A,* **2010**, *1217*(20), 3357-3361.
[http://dx.doi.org/10.1016/j.chroma.2010.03.029] [PMID: 20381809]

[115] Adlnasab, L.; Ebrahimzadeh, H.; Yamini, Y.; Mirzajani, F. Optimization of a novel method based on solidification of floating organic droplet by high-performance liquid chromatography for evaluation of antifungal drugs in biological samples. *Talanta,* **2010**, *83*(2), 370-378.
[http://dx.doi.org/10.1016/j.talanta.2010.09.031] [PMID: 21111148]

[116] Zhou, Q.; Ye, C. Ionic liquid for improved single-drop microextraction of aromatic amines in water samples. *Mikrochim. Acta,* **2008**, *162*, 153-159.
[http://dx.doi.org/10.1007/s00604-007-0857-1]

[117] Cheng, K.; Choi, K.; Kim, J.; Sung, I.H.; Chung, D.S. Sensitive arsenic analysis by carrier-mediated counter-transport single drop microextraction coupled with capillary electrophoresis. *Microchem. J.,* **2013**, *106*, 220-225.
[http://dx.doi.org/10.1016/j.microc.2012.07.005]

[118] Ghambarian, M.; Khalili-Zanjani, M.R.; Yamini, Y.; Esrafili, A.; Yazdanfar, N. Preconcentration and speciation of arsenic in water specimens by the combination of solidification of floating drop microextraction and electrothermal atomic absorption spectrometry. *Talanta,* **2010**, *81*(1-2), 197-201.
[http://dx.doi.org/10.1016/j.talanta.2009.11.056] [PMID: 20188908]

[119] Vidal, L.; Chisvert, A.; Canals, A.; Salvador, A. Sensitive determination of free benzophenone-3 in human urine samples based on an ionic liquid as extractant phase in single-drop microextraction prior to liquid chromatography analysis. *J. Chromatogr. A,* **2007**, *1174*(1-2), 95-103.
[http://dx.doi.org/10.1016/j.chroma.2007.07.077] [PMID: 17720175]

[120] Gao, L.; Zou, J.; Liu, H.; Zeng, J.; Wang, Y.; Chen, X. Determination of bisphenol A in thermal printing papers treated by alkaline aqueous solution using the combination of single-drop microextraction and HPLC. *J. Sep. Sci.,* **2013**, *36*(7), 1298-1303.
[http://dx.doi.org/10.1002/jssc.201201060] [PMID: 23471677]

[121] Reddy-Noone, K.; Jain, A.; Verma, K.K. Liquid-phase microextraction-gas chromatography-mass spectrometry for the determination of bromate, iodate, bromide and iodide in high-chloride matrix. *J. Chromatogr. A,* **2007**, *1148*(2), 145-151.
[http://dx.doi.org/10.1016/j.chroma.2007.03.027] [PMID: 17391685]

[122] Aguilera-Herrador, E.; Lucena, R.; Cárdenas, S.; Valcárcel, M. Ionic liquid-based single-drop microextraction/gas chromatographic/mass spectrometric determination of benzene, toluene,

ethylbenzene and xylene isomers in waters. *J. Chromatogr. A,* **2008**, *1201*(1), 106-111. [http://dx.doi.org/10.1016/j.chroma.2008.06.010] [PMID: 18586254]

[123] Faraji, H.; Tajbakhsh, M.; Helalizadeh, M. Validation and optimization of a liquid-phase microextraction method for trace analysis of BTEX in water samples. *Anal. Methods,* **2012**, *4*, 3372-3380. [http://dx.doi.org/10.1039/c2ay25336d]

[124] Saraji, M.; Esteki, N. Analysis of carbamate pesticides in water samples using single-drop microextraction and gas chromatography-mass spectrometry. *Anal. Bioanal. Chem.,* **2008**, *391*(3), 1091-1100. [http://dx.doi.org/10.1007/s00216-008-2087-8] [PMID: 18415085]

[125] Pillai, A.K.; Gautam, K.; Jain, A.; Verma, K.K. Headspace in-drop derivatisation of carbonyl compounds for their analysis by high-performance liquid chromatography-diode array detection. *Anal. Chim. Acta,* **2009**, *632*(2), 208-215. [http://dx.doi.org/10.1016/j.aca.2008.11.021] [PMID: 19110095]

[126] Fiamegos, Y.C.; Stalikas, C.D. Gas chromatographic determination of carbonyl compounds in biological and oil samples by headspace single-drop microextraction with in-drop derivatisation. *Anal. Chim. Acta,* **2008**, *609*(2), 175-183. [http://dx.doi.org/10.1016/j.aca.2007.12.042] [PMID: 18261512]

[127] Shrivas, K.; Wu, H.F. A rapid, sensitive and effective quantitative method for simultaneous determination of cationic surfactant mixtures from river and municipal wastewater by direct combination of single-drop microextraction with AP-MALDI mass spectrometry. *J. Mass Spectrom.,* **2007**, *42*(12), 1637-1644. [http://dx.doi.org/10.1002/jms.1266] [PMID: 17828802]

[128] Wen, X.; Deng, Q.; Guo, J.; Yang, S. Ultra-sensitive determination of cadmium in rice and water by UV-*vis* spectrophotometry after single drop microextraction. *Spectrochim. Acta A Mol. Biomol. Spectrosc.,* **2011**, *79*(3), 508-512. [http://dx.doi.org/10.1016/j.saa.2011.03.021] [PMID: 21530375]

[129] Jiang, H.; Hu, B. Determination of trace Cd and Pb in natural waters by direct single drop microextraction combined with electrothermal atomic absorption spectrometry. *Mikrochim. Acta,* **2008**, *161*, 101-107. [http://dx.doi.org/10.1007/s00604-007-0785-0]

[130] Li, L.; Hu, B.; Xia, L.; Jiang, Z. Determination of trace Cd and Pb in environmental and biological samples by ETV-ICP-MS after single-drop microextraction. *Talanta,* **2006**, *70*(2), 468-473. [http://dx.doi.org/10.1016/j.talanta.2006.03.006] [PMID: 18970794]

[131] Rivas, R.E.; López-García, I.; Hernández-Córdoba, M. Microextraction based on solidification of a floating organic drop followed by electrothermal atomic absorption spectrometry for the determination of ultratraces of lead and cadmium in waters. *Anal. Methods,* **2010**, *2*, 225-230. [http://dx.doi.org/10.1039/b9ay00237e]

[132] Palit, M.; Pardasani, D.; Gupta, A.K.; Dubey, D.K. Application of single drop microextraction for analysis of chemical warfare agents and related compounds in water by gas chromatography/mass spectrometry. *Anal. Chem.,* **2005**, *77*(2), 711-717.

[http://dx.doi.org/10.1021/ac0486948] [PMID: 15649077]

[133] Pillai, A.K.; Jain, A.; Verma, K.K. Headspace single-drop microextraction and fibre optics-based cuvetteless micro-spectrophotometry for the determination of chloride involving oxidation with permanganate. *Talanta,* **2010**, *80*(5), 1816-1822. [http://dx.doi.org/10.1016/j.talanta.2009.10.027] [PMID: 20152416]

[134] Sun, J.; Qiao, B.; Dai, J. Determination of chlorobenzenes in wastewater by single-drop microextraction coupled to capillary gas chromatography. *Adv. Mat. Res.,* **2014**, *838-841*, 2566-2569.

[135] Jain, A.; Pillai, A.K.; Sharma, N.; Verma, K.K. Headspace single-drop microextraction and cuvetteless microspectrophotometry for the selective determination of free and total cyanide involving reaction with ninhydrin. *Talanta,* **2010**, *82*(2), 758-765. [http://dx.doi.org/10.1016/j.talanta.2010.05.048] [PMID: 20602966]

[136] Bidabadi, M.S.; Dadfarnia, S.; Shabani, A.M. Solidified floating organic drop microextraction (SFODME) for simultaneous separation/preconcentration and determination of cobalt and nickel by graphite furnace atomic absorption spectrometry (GFAAS). *J. Hazard. Mater.,* **2009**, *166*(1), 291-296. [http://dx.doi.org/10.1016/j.jhazmat.2008.11.052] [PMID: 19117672]

[137] Batlle, R.; Nerín, C. Application of single-drop microextraction to the determination of dialkyl phthalate esters in food simulants. *J. Chromatogr. A,* **2004**, *1045*(1-2), 29-35. [http://dx.doi.org/10.1016/j.chroma.2004.06.001] [PMID: 15378876]

[138] Sarafraz-Yazdi, A.; Mofazzeli, F.; Es'haghi, Z. Directly suspended droplet thre liquid phase microextraction of diclofenac prior to LC. *Chromatographia,* **2008**, *67*, 49-53. [http://dx.doi.org/10.1365/s10337-007-0449-6]

[139] López-Blanco, M.C.; Blanco-Cid, S.; Cancho-Grande, B.; Simal-Gándara, J. Application of single-drop microextraction and comparison with solid-phase microextraction and solid-phase extraction for the determination of alpha- and beta-endosulfan in water samples by gas chromatography-electro--capture detection. *J. Chromatogr. A,* **2003**, *984*(2), 245-252. [http://dx.doi.org/10.1016/S0021-9673(02)01873-3] [PMID: 12564696]

[140] Adam, M.; Dobiás, P.; Eisner, A.; Ventura, K. Headspace single-drop microextraction of herbal essential oils. *J. Sep. Sci.,* **2008**, *31*(2), 356-363. [http://dx.doi.org/10.1002/jssc.200700377] [PMID: 18172916]

[141] Kaykhaii, M.; Ghalehno, M.H. Rapid and sensitive determination of fluoride in toothpaste and water samples using headspace single drop microextraction-gas chromatography. *Anal. Methods,* **2013**, *5*, 5622-5626. [http://dx.doi.org/10.1039/c3ay41004h]

[142] Yiping, H.; Caiyun, W. Ion chromatography for rapid and sensitive determination of fluoride in milk after headspace single-drop microextraction with *in situ* generation of volatile hydrogen fluoride. *Anal. Chim. Acta,* **2010**, *661*(2), 161-166. [http://dx.doi.org/10.1016/j.aca.2009.12.018] [PMID: 20113730]

[143] Liu, Y.; Zhao, E.; Zhou, Z. Single-drop microextraction and gas chromatographic determination of fungicide in water and wine samples. *Anal. Lett.,* **2006**, *39*, 2333-2344. [http://dx.doi.org/10.1080/00032710600755843]

[144] Li, N.; Deng, C.; Yin, X.; Yao, N.; Shen, X.; Zhang, X. Gas chromatography-mass spectrometric analysis of hexanal and heptanal in human blood by headspace single-drop microextraction with droplet derivatization. *Anal. Biochem.,* **2005**, *342*(2), 318-326. [http://dx.doi.org/10.1016/j.ab.2005.04.024] [PMID: 15913536]

[145] Bagheri, H.; Naderi, M. Immersed single-drop microextraction-electrothermal vaporization atomic absorption spectroscopy for the trace determination of mercury in water samples. *J. Hazard. Mater.,* **2009**, *165*(1-3), 353-358. [http://dx.doi.org/10.1016/j.jhazmat.2008.09.128] [PMID: 19036515]

[146] Sarica, D.Y.; Tuerker, A.R. Speciation and determination of inorganic mercury and methylmercury by headspace single drop microextraction and electrothermal atomic absorption spectrometry in water and fish. *CLEAN - Soil, Air. Water,* **2012**, *40*, 523-530.

[147] Das, P.; Gupta, M.; Jain, A.; Verma, K.K. Single drop microextraction or solid phase microextraction-gas chromatography-mass spectrometry for the determination of iodine in pharmaceuticals, iodized salt, milk powder and vegetables involving conversion into 4-iodo-N,N-dimethylaniline. *J. Chromatogr. A,* **2004**, *1023*(1), 33-39. [http://dx.doi.org/10.1016/j.chroma.2003.09.056] [PMID: 14760847]

[148] Akhoundzadeh, J.; Chamsaz, M.; Yazdinezhad, S.R.; Arbaz-zavvar, M.H. Head-space single drop microextraction combined with gas chromatography with an electron capture detector for determination of iodine in infant formulas. *Anal. Methods,* **2013**, *5*, 778-783. [http://dx.doi.org/10.1039/C2AY25931A]

[149] Rahmani, M.; Kaykhaii, M. Determination of methylcyclopentadienyl-manganese tricarbonyl in gasoline and water *via* ionic-liquid headspace single drop microextraction and electrothermal atomic absorption spectrometry. *Mikrochim. Acta,* **2011**, *174*, 413-419. [http://dx.doi.org/10.1007/s00604-011-0648-6]

[150] Pereira, F.J.; Bendicho, C.; Kalogerakis, N.; Psillakis, E. Headspace single drop microextraction of methylcyclopentadienyl-manganese tricarbonyl from water samples followed by gas chromatography-mass spectrometry. *Talanta,* **2007**, *74*(1), 47-51. [http://dx.doi.org/10.1016/j.talanta.2007.05.024] [PMID: 18371611]

[151] Lemos, V.A.; Vieira, U.S. Single-drop microextraction for the determination of manganese in seafood and water samples. *Mikrochim. Acta,* **2013**, *180*, 501-507. [http://dx.doi.org/10.1007/s00604-013-0961-3]

[152] Pranaityte, B.; Jermak, S.; Naujalis, E.; Padarauskas, A. Capillary electrophoretic determination of ammonia using headspace single-drop microextraction. *Microchem. J.,* **2007**, *86*, 48-52. [http://dx.doi.org/10.1016/j.microc.2006.10.001]

[153] Yazdi, A.S.; Mofazzeli, F.; Es'haghi, Z. Determination of 3-nitroaniline in water samples by directly suspended droplet three-phase liquid-phase microextraction using 18-crown-6 ether and high-performance liquid chromatography. *J. Chromatogr. A,* **2009**, *1216*(26), 5086-5091. [http://dx.doi.org/10.1016/j.chroma.2009.04.090] [PMID: 19464691]

[154] Adlnasab, L.; Ebrahimzadeh, H.; Yamini, Y. Optimized conditions for liquid-phase microextraction based on solidification of floating organic droplet for extraction of nitrotoluene compounds by using response surface methodology. *Anal. Methods,* **2012**, *4*, 190-195.

[http://dx.doi.org/10.1039/C1AY05449J]

[155] Shrivas, K.; Wu, H.F. Ultrasonication followed by single-drop microextraction combined with GC/MS for rapid determination of organochlorine pesticides from fish. *J. Sep. Sci.,* **2008**, *31*(2), 380-386. [http://dx.doi.org/10.1002/jssc.200700380] [PMID: 18246583]

[156] Zhang, M.; Huang, J.; Wei, C.; Yu, B.; Yang, X.; Chen, X. Mixed liquids for single-drop microextraction of organochlorine pesticides in vegetables. *Talanta,* **2008**, *74*(4), 599-604. [http://dx.doi.org/10.1016/j.talanta.2007.06.041] [PMID: 18371681]

[157] Tian, F.; Liu, W.; Fang, H.; An, M.; Duan, S. Determination of six organophosphorus pesticides in water by single-drop microextraction coupled with GC-NPD. *Chromatographia,* **2014**, *77*, 487-492. [http://dx.doi.org/10.1007/s10337-013-2609-1]

[158] Wielgomas, B.; Czarnowski, W. Headspace single-drop microextraction and GC-ECD determination of chlorpyrifos-ethyl in rat liver. *Anal. Bioanal. Chem.,* **2008**, *390*(7), 1933-1941. [http://dx.doi.org/10.1007/s00216-008-1831-4] [PMID: 18299820]

[159] Rahul, C.B.; Roy, P. Single drop microextraction-gas chromatography mass spectrometry: a method for determination of organophosphorus pesticides in fruit juices and water. *Res. J. Chem. Environ.,* **2006**, *10*, 5-12.

[160] Xiao, Q.; Hu, B.; Yu, C.; Xia, L.; Jiang, Z. Optimization of a single-drop microextraction procedure for the determination of organophosphorus pesticides in water and fruit juice with gas chromatography-flame photometric detection. *Talanta,* **2006**, *69*(4), 848-855. [http://dx.doi.org/10.1016/j.talanta.2005.11.024] [PMID: 18970647]

[161] Lambropoulou, D.A.; Psillakis, E.; Albanis, T.A.; Kalogerakis, N. Single-drop microextraction for the analysis of organophosphorus insecticides in water. *Anal. Chim. Acta,* **2004**, *516*, 205-211. [http://dx.doi.org/10.1016/j.aca.2004.03.055]

[162] Tian, F.; Liu, W.; Fang, H.; An, M.; Duan, S. Determination of six organophosphorus pesticides in water by single-drop microextraction coupled with GC-NPD. *Chromatographia,* **2014**, *77*, 487-492. [http://dx.doi.org/10.1007/s10337-013-2609-1]

[163] Chen, H.; Chen, R.; Feng, R.; Li, S. Simultaneous analysis of carbamate and organophosphorus pesticides in water by single-drop microextraction coupled with GC-MS. *Chromatographia,* **2009**, *70*, 165-172. [http://dx.doi.org/10.1365/s10337-009-1154-4]

[164] Khalili-Zanjani, M.R.; Yamini, Y.; Yazdanfar, N.; Shariati, S. Extraction and determination of organophosphorus pesticides in water samples by a new liquid phase microextraction-gas chromatography-flame photometric detection. *Anal. Chim. Acta,* **2008**, *606*(2), 202-208. [http://dx.doi.org/10.1016/j.aca.2007.11.032] [PMID: 18082651]

[165] Sun, J.; Qiao, B.; Sun, D. Determination of polycyclic aromatic hydrocarbons in wastewater by single-drop microextraction coupled to capillary gas chromatography. *Adv. Mat. Res.,* **2014**, *838-841*, 2582-2585.

[166] Wu, Y.; Xia, L.; Chen, R.; Hu, B. Headspace single drop microextraction combined with HPLC for the determination of trace polycyclic aromatic hydrocarbons in environmental samples. *Talanta,* **2008**, *74*(4), 470-477.

[http://dx.doi.org/10.1016/j.talanta.2007.05.057] [PMID: 18371664]

[167] Saraji, M.; Mirmahdieh, S. Single-drop microextraction followed by in-syringe derivatization and GC-MS detection for the determination of parabens in water and cosmetic products. *J. Sep. Sci.,* **2009**, *32*(7), 988-995. [http://dx.doi.org/10.1002/jssc.200800635] [PMID: 19266549]

[168] Manzoori, J.L.; Amjadi, M.; Abulhassani, J. Ultra-trace determination of lead in water and food samples by using ionic liquid-based single drop microextraction-electrothermal atomic absorption spectrometry. *Anal. Chim. Acta,* **2009**, *644*(1-2), 48-52. [http://dx.doi.org/10.1016/j.aca.2009.04.029] [PMID: 19463561]

[169] He, S.W.; Shen, C.Y.; Wei, X.Q.; Jin, M.C.; Cai, M.Q. Determination of trace PCBs in water by GC-MS with ionic liquid based headspace single-drop microextraction under ultrasound. *Adv. Mater. Res.,* **2013**, *726-731*, 74-80.

[170] Amvrazi, E.G.; Martini, M.A.; Tsiropoulos, N.G. Headspace single-drop microextraction of common pesticide contaminants in honey-method development andcomparison with other extraction methods. *Int. J. Environ. Anal. Chem.,* **2012**, *92*, 450-465. [http://dx.doi.org/10.1080/03067319.2011.585716]

[171] Amvrazi, E.G.; Tsiropoulos, N.G. Chemometric study and optimization of extraction parameters in single-drop microextraction for the determination of multiclass pesticide residues in grapes and apples by gas chromatography mass spectrometry. *J. Chromatogr. A,* **2009**, *1216*(45), 7630-7638. [http://dx.doi.org/10.1016/j.chroma.2009.08.092] [PMID: 19783253]

[172] Amvrazi, E.G.; Tsiropoulos, N.G. Application of single-drop microextraction coupled with gas chromatography for the determination of multiclass pesticides in vegetables with nitrogen phosphorus and electron capture detection. *J. Chromatogr. A,* **2009**, *1216*(14), 2789-2797. [http://dx.doi.org/10.1016/j.chroma.2008.07.070] [PMID: 18692190]

[173] Pereira dos Anjos, J.; de Andrade, J.B. Determination of nineteen pesticides residues (organophosphates, organochlorine, pyrethroids, carbamate, thiocarbamate and strobilurin) in coconut water by SDME/GC-MS. *Microchem. J.,* **2014**, *112*, 119-126. [http://dx.doi.org/10.1016/j.microc.2013.10.001]

[174] Saha, S.; Mistri, R.; Ray, B.C. A rapid and selective method for simultaneous determination of six toxic phenolic compounds in mainstream cigarette smoke using single-drop microextraction followed by liquid chromatography-tandem mass spectrometry. *Anal. Bioanal. Chem.,* **2013**, *405*(28), 9265-9272. [http://dx.doi.org/10.1007/s00216-013-7351-x] [PMID: 24057026]

[175] Zhao, F.Q.; Li, J.; Zeng, B.Z. Coupling of ionic liquid-based headspace single-drop microextraction with GC for sensitive detection of phenols. *J. Sep. Sci.,* **2008**, *31*(16-17), 3045-3049. [http://dx.doi.org/10.1002/jssc.200800308] [PMID: 18704999]

[176] Farahani, H.; Ganjali, M.R.; Dinarvand, R.; Norouzi, P. Screening method for phthalate esters in water using liquid-phase microextraction based on the solidification of a floating organic microdrop combined with gas chromatography-mass spectrometry. *Talanta,* **2008**, *76*(4), 718-723. [http://dx.doi.org/10.1016/j.talanta.2008.03.002] [PMID: 18656647]

[177] Liu, H.; Zhang, M.; Wang, X.; Zou, Y.; Wang, W.; Ma, M.; Li, Y.; Wang, H. Extraction and

determination of polybrominated diphenyl ethers in water and urine samples using solidified floating organic drop microextraction along with high performance liquid chromatography. *Mikrochim. Acta,* **2012**, *176*, 303-309.
[http://dx.doi.org/10.1007/s00604-011-0713-1]

[178] Viñas, P.; Martínez-Castillo, N.; Campillo, N.; Hernández-Córdoba, M. Directly suspended droplet microextraction with in injection-port derivatization coupled to gas chromatography-mass spectrometry for the analysis of polyphenols in herbal infusions, fruits and functional foods. *J. Chromatogr. A,* **2011**, *1218*(5), 639-646.
[http://dx.doi.org/10.1016/j.chroma.2010.12.026] [PMID: 21185565]

[179] Colombini, V.; Bancon-Montigny, C.; Yang, L.; Maxwell, P.; Sturgeon, R.E.; Mester, Z. Headspace single-drop microextration for the detection of organotin compounds. *Talanta,* **2004**, *63*(3), 555-560.
[http://dx.doi.org/10.1016/j.talanta.2003.11.035] [PMID: 18969468]

[180] Costas-Mora, I.; Romero, V.; Pena-Pereira, F.; Lavilla, I.; Bendicho, C. Quantum dot-based headspace single-drop microextraction technique for optical sensing of volatile species. *Anal. Chem.,* **2011**, *83*(6), 2388-2393.
[http://dx.doi.org/10.1021/ac103223e] [PMID: 21344872]

[181] Fakhari, A.R.; Tabani, H.; Nojavan, S. Immersed single-drop microextraction combined with gas chromatography for the determination of sufentanil and alfentanil in urine and wastewater samples. *Anal. Methods,* **2011**, *3*, 951-956.
[http://dx.doi.org/10.1039/c1ay05037k]

[182] Gómez-Otero, E.; Costas, M.; Lavilla, I.; Bendicho, C. Ultrasensitive, simple and solvent-free micro-assay for determining sulphite preservatives (E220-228) in foods by HS-SDME and UV-vis micro-spectrophotometry. *Anal. Bioanal. Chem.,* **2014**, *406*(8), 2133-2140.
[http://dx.doi.org/10.1007/s00216-013-7293-3] [PMID: 23995438]

[183] Sheikhloie, H.; Saber-Tehrani, M.; Abrumand-Azar, P.; Waqif-Husain, S. Analysis of tributyltin and triphenyltin in water by ionic liquid-headspace single-drop microextraction then HPLC with fluorimetric detection. *Acta Chromatogr.,* **2009**, *21*, 577-589.
[http://dx.doi.org/10.1556/AChrom.21.2009.4.5]

[184] Tang, B.; Tian, M.; Row, K.H. Determination of terpenoids in *Chamaecyparis obtusa* leaves by headspace single-drop microextraction with gas chromatography detection. *Anal. Lett.,* **2014**, *47*, 48-57.
[http://dx.doi.org/10.1080/00032719.2013.832278]

[185] Chamsaz, M.; Arbab-Zavar, M.H.; Darroudi, A.; Salehi, T. Preconcentration of thallium (I) by single drop microextraction with electrothermal atomic absorption spectroscopy detection using dicyclohexano-18-crown-6 as extractant system. *J. Hazard. Mater.,* **2009**, *167*(1-3), 597-601.
[http://dx.doi.org/10.1016/j.jhazmat.2009.01.019] [PMID: 19232464]

[186] Yamini, Y.; Hosseini, M.H.; Hojaty, M.; Arab, J. Headspace solvent microextraction of trihalomethane compounds into a single drop. *J. Chromatogr. Sci.,* **2004**, *42*(1), 32-36.
[http://dx.doi.org/10.1093/chromsci/42.1.32] [PMID: 14971396]

[187] Sha, Y.; Meng, J.; Zhang, Y.; Deng, C.; Wu, D. Determination of volatile organic acids in tobacco by single-drop microextraction with in-syringe derivatization followed by GC-MS. *J. Sep. Sci.,* **2010**,

33(2), 212-217.
[http://dx.doi.org/10.1002/jssc.200900523] [PMID: 20039308]

[188] Ligor, T.; Buszewski, B. Single-drop microextraction and gas chromatography-mass spectrometry for the determination of volatile aldehydes in fresh cucumbers. *Anal. Bioanal. Chem.,* **2008**, *391*(6), 2283-2289.
[http://dx.doi.org/10.1007/s00216-008-2098-5] [PMID: 18415084]

[189] Jermak, S.; Pranaityte, B.; Padarauskas, A. Ligand displacement, headspace single-drop microextraction, and capillary electrophoresis for the determination of weak acid dissociable cyanide. *J. Chromatogr. A,* **2007**, *1148*(1), 123-127.
[http://dx.doi.org/10.1016/j.chroma.2007.03.022] [PMID: 17382334]

CHAPTER 12

Membrane-Based Microextraction Techniques with Emphasis on Hollow-fiber Microextraction

Jan Åke Jönsson and **Estelle Larsson***

Center for Analysis and Synthesis, Department of Chemistry, Lund University, P.O. Box 124, SE-221 00 Lund, Sweden

Abstract: Various membrane-based extraction techniques are used in analytical chemistry mainly for pretreatment before analyte determination using chromatographic or other techniques. Membrane extraction can also be applied for the extraction of various metal ions and is then followed by atomic absorption or similar detection techniques. They allow high selectively for a number of analytes from chemically more or less complex samples and high concentration enrichment, easily thousands of times. Currently, the most common format for membrane extraction utilizes hollow-fiber membranes, which permit easy and versatile operation with a minimum of cost. A number of applications have been presented involving determination of polar and medium-polar compounds as acids and bases in samples of environmental and biological origin, usually in combination with liquid chromatography and mass spectrometry. For other analytes in which membrane extractions are various metal ions, the membrane extraction is followed by techniques like atomic absorption and similar. Also applications to non-polar compounds in mainly environmental samples, followed by gas chromatography, are described in the literature.

In this chapter, the basic theoretical principles for the common variants of membrane extraction are described. Guidelines for the selection of operational parameters as well as concrete advice for the practical implementation are provided.

Keywords: Biomedical samples, Chromatography, Determination of acids, Determination of bases, Enrichment, Environmental samples, Extraction, Food samples, Hollow fibers, Mass spectrometry, Membrane extraction, Sample prepa-

* **Corresponding author Estelle Larsson:** Department of Occupational and Environmental Medicine, Skåne University Hospital, SE-221 85 Lund, Sweden; Tel: +46 46 17 69 31; E-mail:estelle.larsson@med.lu.se

M. Valcárcel, S. Cárdenas, R. Lucena (Eds.)

ration, Selectivity, Supported liquid membrane (SLM).

12.1. INTRODUCTION

An important area of current research in analytical chemistry aims to the development of alternative extraction techniques meeting some of the disadvantages of classical techniques, such as solvent extraction or solid phase extraction. These activities are somewhat moderated by the fact that the area of chemical analysis (especially regarding environmental and pharmaceutical applications, but also in other areas) is heavily regulated by various international and local rules, leading to the acceptance of new technology being slow and costly.

The use of membranes for extraction in analytical chemistry has been increasing during recent years. Such techniques have a number of clear advantages over other extraction techniques, especially concerning selectivity, enrichment power and automation potential, but also economy and occupational health aspects. They can selectively extract and enrich analytes from chemically more or less complex samples in very efficient ways [1 - 8].

In membrane extraction, the membrane separates the sample phase (often called donor solution) from the acceptor phase and the analyte molecules pass through the membrane from the donor to the acceptor. This principle gives a clear advantage over other sample preparation techniques, as all compounds that reach the analytical instrument must travel through the membrane, acting like a barrier between the sample and the analytical instrument. There is no direct connection and no possibility for transferring compounds into the analytical instrument in other ways. This leads to a high selectivity, *i.e.* efficient discrimination between analytes and various unwanted matrix compounds.

There are in principle two variants of liquid membrane extraction, three-phase systems and two-phase systems [1, 2].

The three-phase system (aq/org/aq), where analytes are extracted from the aqueous sample into an organic liquid, immobilized in a porous hydrophobic membrane support, and further to a second aqueous phase, is suitable for the extraction of polar compounds (acidic or basic, charged, metals, *etc.*) and it is compatible with reversed phase HPLC.

The two-phase system (aq/org), where analytes are extracted into an organic solvent separated from the aqueous sample by a hydrophobic porous membrane is suitable for more hydrophobic analytes and compatible with gas chromatography.

Both these techniques utilize a supported liquid membrane (SLM) *i.e.* an organic liquid supported in the pores of a hydrophobic porous material. In this paper, the focus is on basic and practical considerations for two- or three-phase supported liquid membrane extraction systems in hollow fibers, techniques which are nowadays usually called as hollow fiber liquid phase microextraction (HF-LPME).

Earlier, similar extractions were mainly performed in flat-sheet format, which is described in detail in the literature [1 - 3] and also in a practical tutorial [4]. The hollow-fiber techniques were developed especially in pharmaceutical applications by Rasmussen and Pedersen-Bjergaard [5, 6] and Lee with coworkers [7].

One main advantage with HF-LPME is that it is not sensitive to the presence of particles in the sample. Clogging (as for instance when using solid phase extraction (SPE)) is no problem, which means that filtration is not required prior to extraction of turbid samples.

It is possible to successfully apply HF-LPME to slurries, as was recently described with sewage sludge [9] and in biological samples [10]. Furthermore, compared to LLE, there is no problem with the formation of emulsions in the water-organic interface, which otherwise can be a significant practical problem.

Using hollow fibers, it is possible to perform extractions for sample clean-up and enrichment with very cheap and simple equipment, leading to efficient enrichment and high enrichment factors (easily thousands of times) for ultra-trace analysis. The current “record” [11] seems to be 28 000 times.

12.2. THEORY

12.2.1. Three-phase SLM Basics

To discuss the theory for membrane extraction in some detail, the best start is to consider three-phase liquid membrane extraction principles with acidic or basic analytes. The principles for other modes can then be derived from this treatment.

Three-phase extraction is a combination of extraction of a compound (analyte) from the sample into an organic solvent (in the membrane) followed by a second extraction (back-extraction) into a second aqueous phase. This sequence of two extractions, usually with different pH in the two aqueous phases, is a common procedure in classic liquid-liquid extraction. In membrane extraction, these two extraction steps occur simultaneously, a process sometimes called as pertraction, so the mass transfer kinetics will be more efficient, compared to the situation when these steps are performed in sequence in separation funnels.

Generally, the rate of mass transfer from donor to acceptor is proportional to the concentration difference, ΔC, of the diffusing species over the membrane:

$$\Delta C = \alpha_{\mathrm{D}} K_{\mathrm{D}} C_{\mathrm{D}} - \alpha_{\mathrm{A}} K_{\mathrm{A}} C_{\mathrm{A}}$$

$$K_{\mathrm{D}} = \frac{C_{\mathrm{M}}}{\alpha_{\mathrm{D}} C_{\mathrm{D}}} \quad K_{\mathrm{A}} = \frac{C_{\mathrm{M}}}{\alpha_{\mathrm{A}} C_{\mathrm{A}}} \tag{1}$$

where C_D, C_M, C_A are the total concentrations (all species) in the donor, membrane and acceptor phases, respectively. α_{D}, *etc.* are the extractable (usually uncharged species) fractions of the total concentrations. Thus, $\alpha_{\mathrm{D}} \cdot C_{\mathrm{D}}$ is the concentration in the donor of the extractable form of the analyte. K_{D} and K_{A} are the partition ratios (partition coefficients) *i.e.* pertaining to the extractable form only, between donor phase and membrane, and between the acceptor phase and membrane, respectively. Usually, K_{D} and K_{A} are very similar, but if ionic strengths, *etc.*, are different in the donor and acceptor, they may differ.

At thermodynamic equilibrium between all three phase, ΔC is zero and from Eq.

(1) we then get the distribution ratio (coefficient) D between the donor and acceptor phases (*i.e.* between two aqueous phases), as:

$$D = \left(\frac{C_{\mathrm{A}}}{C_{\mathrm{D}}} \right)_{equil} = \frac{\alpha_{\mathrm{D}} K_{\mathrm{D}}}{\alpha_{\mathrm{A}} K_{\mathrm{A}}} \quad \textbf{(2)}$$

As $K_{\mathrm{A}} \approx K_{\mathrm{D}}$, D will be mainly determined only by the α-values, which for acids or bases can easily be varied over many orders of magnitude by selecting suitable pH values. For a base, the α-value is given by the following equation, derived from the definition of *pKa* (of the corresponding acid):

$$\alpha = \frac{[\mathrm{B}]}{[\mathrm{B}] + [\mathrm{BH}^{+}]} = \frac{1}{1 + 10^{(pK_a - pH)}} \quad \textbf{(3)}$$

With this, the equilibrium distribution coefficient for acidic or basic analytes in a three-phase membrane extraction system can be written as:

$$D = \left(\frac{C_{\mathrm{A}}}{C_{\mathrm{D}}} \right)_{equil} = \frac{\alpha_{\mathrm{D}} K_{\mathrm{D}}}{\alpha_{\mathrm{A}} K_{\mathrm{A}}} = \frac{\left(1 + 10^{s(pH_{\mathrm{A}} - pKa)} \right)}{\left(1 + 10^{s(pH_{\mathrm{D}} - pKa)} \right)} \frac{K_{\mathrm{D}}}{K_{\mathrm{A}}} \quad \textbf{(4)}$$

where $s = 1$ for acids and $s = -1$ for bases, and, usually, $K_{\mathrm{A}} \approx K_{\mathrm{D}}$. So, when we know the *pKa* of the analyte and the pH of the phases, the alpha-factors and the overall distribution constant can be easily calculated.

For other types of transport than permeation of acids or bases (*e.g.* carrier-mediated), Eqs. (1-2) will be still valid, while the definition of the alpha values will be different. In principle it is possible to derive similar equations as Eqs. (3-4).

Generally, the total number of sample molecules in an SLM system is distributed between all three phases according to the following equation:

$$C_I V_D = C_D V_D + C_M V_M + C_A V_A \quad \textbf{(5)}$$

Here, C_I signifies the initial concentration of the analyte in the sample (donor phase) and the initial concentration in the other phases is zero. C_D, C_M and C_A are the concentrations at any time in the donor, membrane and acceptor phases with volumes V_D, V_M and V_A, respectively.

Combining with Eqs. (1) and (2), we can write for the conditions at equilibrium:

$$C_I V_D = C_A \frac{\alpha_A K_A}{\alpha_D K_D} V_D + C_A \alpha_A K_A V_M + C_A V_A \quad \textbf{(6)}$$

The first term on the right hand side in this equation is the amount of analyte remaining in the donor, the second is the amount which is in the membrane and the third is the amount which is extracted into the acceptor. Obviously, for an efficient extraction, the first and second terms should be negligible compared to the third.

The acceptor concentration C_A is zero from the beginning of the extraction and increases successively, usually to values well over the initial sample (donor) concentration C_I. The kinetics of this will be discussed below. The concentration enrichment factor *Ee* (in other texts sometimes written *EF* or *E*), is the ratio:

$$Ee = \frac{C_A}{C_I} \quad \textbf{(7)}$$

At equilibrium this can be combined with Eq. (6) which gives:

$$Ee_{equil} = \frac{1}{\left(\frac{\alpha_A K_A}{\alpha_D K_D} + \alpha_A K_A \frac{V_M}{V_D} + \frac{V_A}{V_D} \right)} \quad \textbf{(8)}$$

We can also define the extraction efficiency E as the fraction of the initial analyte amount that is transferred to the acceptor:

$$E = \frac{C_{\mathrm{A}} V_{\mathrm{A}}}{C_{\mathrm{I}} V_{\mathrm{D}}} \tag{9}$$

which at equilibrium in combination with Eq. (6) becomes:

$$E_{equil} = \frac{1}{\left(\frac{\alpha_{\mathrm{A}} K_{\mathrm{A}} V_{\mathrm{D}}}{\alpha_{\mathrm{D}} K_{\mathrm{D}} V_{\mathrm{A}}} + \alpha_{\mathrm{A}} K_{\mathrm{A}} \frac{V_{\mathrm{M}}}{V_{\mathrm{A}}} + 1 \right)} \tag{10}$$

In some papers the extraction efficiency is termed recovery, *R*. This is not recommended, as it can be confused with spike recovery or relative recovery, important quantities in the validation of analytical methods for which an ideal value of 100% is expected. The value of *E* can be much less than 100% in many applications without leading to bad quantitation.

12.2.2. Kinetics

To model the increase in the acceptor concentration during the extraction process, we can derive a first-order kinetic equation which can be written:

$$Ee = \frac{C_{\mathrm{A}}}{C_{\mathrm{I}}} = Ee_{equil} \cdot \left(1 - \exp(-k \cdot \frac{1}{Ee_{equil}} \cdot t \right) \tag{11}$$

In this equation, *k* is an overall rate constant for the mass transfer from the donor to the acceptor.

A plot of Eq. (11), see Fig. (**1**), shows the generalized extraction profile, which contains three distinct regions, first a virtually linear rise (called kinetic regime), after that an intermediate region, and finally the equilibrium regime where the enrichment factor approaches its equilibrium value Ee_{equil} according to Eq. (8).

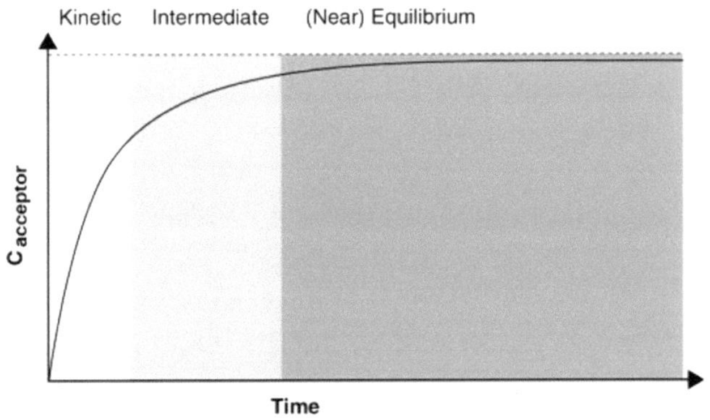

Fig. (1). Generalized extraction profile. (After P. Mayer, J. Tolls, J.L.M. Hermens, D. Mackay. Equilibrium sampling devices. Environ. Sci. Technol., 2003, 37, 184A-191A. Copyright 2003 American Chemical Society).

12.2.3. Complete Trapping

With acidic or basic analytes, trapping in the acceptor is accomplished by setting the acceptor pH conditions so that α_A is a very small number, approaching 0. As a rule of thumb [12], for basic analytes, the acceptor pH should be less than *pKa* - 3.3 and the donor pH larger than *pKa* + 2 (and correspondingly for acidic analytes). From Eq. (3) it is seen that this corresponds to $\alpha_A < 0.0005$ and $\alpha_D >$ 0.99. These conditions are called "complete trapping" or "infinite sink", and lead to an equilibrium distribution coefficient $D > 2000$ (c.f. Eq. (2)).

From Eqs. (8) and (10) it is easily seen that when $\alpha_A \rightarrow 0$:

$$Ee_{equil} \rightarrow \frac{V_D}{V_A} \qquad E_{equil} \rightarrow 1 \tag{12}$$

Thus, the extraction in this idealized case is complete at equilibrium; if the extraction is taken to equilibrium, all analyte is transferred to from the sample to the acceptor and the concentration enrichment factor is simply equal to the ratio between the phase volumes. Note that α_D, and thus the sample pH does not influence Eq. (12), but in practice it is found that the extraction is faster if α_D approaches 1.

Also note that the condition $\alpha_A \approx 0$ refers to the whole extraction process, it is imperative that it is still valid at the end of the extraction experiment. This might sometimes not be the case, due to for example insufficient buffering of the acceptor.

12.2.4. Linear, Exhaustive and Equilibrium Extraction

In many applications, the extraction is often stopped well before equilibrium. In this case, the extraction is performed within the linear part (kinetic regime) of the extraction curve (Fig. **1**). If Eq. (11) is approximated by its McLaurin series for small values for t, it is found that

$$C_A \approx C_I \cdot k \cdot t \qquad \textbf{(13)}$$

So, as a good approximation, the acceptor concentration of the analyte will increase linearly with time, and at any given time it is proportional to the initial concentration of the analyte in the sample. This linear extraction is the basis for much of the work with membrane extraction, where the extraction is incomplete and stopped after a certain time, far before equilibrium is attained. Obviously, if the slope of the extraction curve varies with some experimental parameter, the precision of the extraction is compromised, so a close control of the extraction conditions is needed to obtain a good reproducibility. This should include the sample matrix, as different matrices could lead to different rate constants and thus to large uncertainty. This problem can be overcome by standard addition calibration or careful matrix matching [13, 14]. Furthermore, if the extraction is driven too far, into the intermediate region of (Fig. **1**), the proportionality between the concentration in the sample and that in the acceptor will be uncertain.

With a limited sample volume and complete trapping conditions, it is also possible to conduct the extraction until exhaustive extraction equilibrium according to Eq. (12). The sample will then be depleted with regard to analyte, so a decrease of mass transfer is due to decrease of analyte concentration and the equilibrium occurs only after virtually all analyte has been transferred to the acceptor. Then the only parameters that determine the enrichment factor in this case are the volumes of the sample and the acceptor. As these volumes are possible to know reasonably well, the uncertainty of the extraction process is potentially smaller than for extraction in the linear region. It was experimentally shown that it is in this way possible to obtain better repeatability and reproducibility and less matrix effects than with the linear extraction [15]. It is nevertheless important that the extraction time is so long that equilibrium is reached even if the kinetics for a certain sample might differ. Also it is very important that the complete trapping conditions are valid during the whole course of extraction, necessitating an efficiently buffered acceptor phase.

A third and somewhat different approach, equilibrium extraction, involves selection of a relatively low value of D and a virtually infinite phase ratio ($V_D >> V_A$ and $V_D >> V_M$). With these conditions Eq. (8) reduces to:

$$\begin{aligned} Ee_{equil} &= D \\ E_{equil} &= 0 \end{aligned} \qquad \textbf{(14)}$$

with D as in Eq. (2). Then the concentration of analyte in the sample will not be influenced by the extraction, so this mode of operation permits extraction and sampling without influencing for example absorption, adsorption, redox or other speciation conditions in the sample. This is the basis for a novel sampling technique, called ESTM (Equilibrium sampling through membranes) intended for measurement of freely dissolved fractions of environmental pollutants (as distinguished from total concentrations). This parameter is of environmental interest, as it describes transport processes and biological activities better than total concentrations [16]. The ESTM principle is especially suitable for polar compounds and metals, in contrast to SPME-based methods with similar purposes, which are more suitable for nonpolar compounds. ESTM has been evaluated both with phenolic pollutants [17] and metal ions [18]. An analogous problem in pharmacology is the measurement of drug-protein binding, to which ESTM has been successfully applied [19, 20].

12.2.5. Carrier-Mediated Transport

In contrast to the simple principles for extraction of acids or bases described above, extraction of permanently charged species like metals and certain organic ions has to be performed with the help of various reagents.

To extract metal ions by liquid membranes, ion-pairing or chelating reagents can be added either to the donor phase or to the membrane to help the analytes to be transported through the membrane. Various trapping reagents can also be added to the acceptor phase to prevent analytes to be extracted back into the membrane. There are several reviews in the field of metal extraction using liquid membrane extraction [21 - 23].

Aliquat-336 (methyltrioctylammonium chloride) is a compound which can be added to the membrane [24] for this purpose. This tertiary ammonium ion is permanently positively charged, but as ion pair with chloride it can be added to a suitable membrane solvent.

Another commonly used extraction reagent for metal ions is diethylhexyl phosphoric acid (DEHPA). It has been used for liquid membrane extraction of various metals [25].

Analogous principles have been also used to extract charged organic compounds as anions, for example was Aliquat-336 used for the extraction of amino acids [26], glyphosate [27], peptides [28] and drugs [29]. Using DEHPA, organic compounds such as amino acids [30] and polyamines [31, 32] can be extracted as cations. In this way also very polar compounds can be extracted, even if they are simple acids or bases and should be amenable to direct permeation as described above. Sometimes high polarity and surface active properties prevent them from being extracted into the membrane liquid even if uncharged.

For the extraction of permanently charged anionic surfactants, trihexyl amine was used as a carrier [33]. The carrier is then added to the sample in the donor. In the acidic donor this amine is charged, and can form ion-pairs with the analytes, while in the alkaline acceptor, the amine is neutralized, thus "killing" the ion-pair.

12.2.6. Two-phase Liquid Membrane Extraction

For two-phase LPME, (also called microporous membrane liquid-liquid extraction, MMLLE), the basic principles are simpler than those for the three-phase variant, as there is only one phase boundary involved in the extraction, usually from an aqueous to an organic phase. This is chemically equivalent to a single liquid-liquid extraction in separatory funnels or other devices. In this case, the extraction is driven by the difference in chemical potential of the analytes in organic solvent and in aqueous solution.

In principle and with obvious modifications, Eq. (1) is valid also for 2-phase LPME but the definition of the relevant distribution coefficient is different:

$$\Delta C = \alpha_{\mathrm{D}} K_{\mathrm{D}} C_{\mathrm{D}} - C_{\mathrm{A}}$$

$$K_{\mathrm{D}} = \frac{C_{\mathrm{A}}}{\alpha_{\mathrm{D}} C_{\mathrm{D}}} \tag{15}$$

Here, K_{D} is the partition coefficient between the organic acceptor phase and the aqueous donor (sample) phase. For basic compounds α_{D} is given by Eq. (3), for acids correspondingly, while for non-chargeable compounds $\alpha_{\mathrm{D}} = 1$. Also here, at equilibrium $\Delta C = 0$, which leads to the equilibrium distribution coefficient D:

$$D = \left(\frac{C_{\mathrm{A}}}{C_{\mathrm{D}}}\right)_{equil} = \alpha_{\mathrm{D}} \cdot K_{\mathrm{D}} \tag{16}$$

Thus, in two-phase LPME the organic/aqueous partition coefficient directly governs the extraction, in contrast to three phase mode where the trapping conditions (*i.e.* α_{A}) have the highest influence. The technique works best if log K_{D} >3, *i.e.* for relatively nonpolar compounds.

In many cases the octanol-water partition coefficient (log K_{OW}) is used as a general measure of hydrophobicity and an estimate for the partition coefficient, even if the organic solvents used usually are other than octanol. As a rule of thumb, and if the actual value for K_{D} is not known, it can be said that compounds with log $K_{\mathrm{OW}} > 3$ will be possible to extract with two-phase LPME.

The enrichment factor and extraction efficiency at equilibrium, as described for three-phase membrane extraction in Eqs. (8) and (10) take the following, simpler form in the two-phase case:

$$Ee_{equil} = \frac{1}{\left(\frac{1}{\alpha_D K_D} + \frac{V_A}{V_D}\right)} \quad \textbf{(17)}$$

$$E_{equil} = \frac{1}{\left(\frac{V_D}{\alpha_D K_D V_A} + 1\right)} \quad \textbf{(18)}$$

The kinetic Eq. (11) is valid in the same (slightly simplified way) also for the two-phase case, as well as the considerations around (Fig. **1**).

However, the concept of complete trapping as described above (2.3) is not applicable here, but still the three different end-points for extraction (2.3-2.4) are valid, and all of them are described in the literature. For linear extraction, the principle is exactly the same as described above, but for exhaustive extraction, Eqs. (17-18) is modified to:

$$Ee_{equil} \rightarrow \frac{V_D}{V_A} \qquad \text{when } K_D \rightarrow \infty \quad \textbf{(19)}$$

$$E \rightarrow 1$$

With enough high partition coefficients, it is possible to approach the exhaustive (depletive) conditions, leading to close to 100% extraction efficiency. A work regarding the extraction of polybrominated diphenyl ethers (which have very high organic/water partition coefficients) [34] is an example where 100% extraction efficiency is obtained for the most hydrophobic compounds. Here it is necessary to include the volume of organic phase in the membrane pores in the acceptor volume in the calculation.

Also a nondepletive extraction can be made in the two-phase mode. If the organic/water phase ratio is small enough, so the concentration in the aqueous sample is not significantly changed at equilibrium, then Eq. (14) applies. There are examples in the literature (*e.g* [35].) of this mode of extraction, but for such measurements, it is usually better to apply SPME [16], which will follow the same principles, but with a bonded polymer organic phase instead of a liquid.

The two-phase technique can be seen as a complement to the three-phase membrane extraction, permitting membrane based extraction to be extended to further classes of compounds. Compared with three-phase, the two-phase membrane extraction techniques has the following characteristics:

- It is applicable to hydrophobic, preferably uncharged compounds, *i.e.*, those that cannot be extracted with three-phase systems.
- The maximum concentration enrichment possible is limited by the partition coefficient (Eq. 16), whereas in three-phase extraction it is dependent on the degree of trapping (Eq. 3), which can be influenced by controlling the pH of the different phases. Therefore three-phase extraction provides more degrees of freedom by which the conditions of extraction can be tuned.
- The extract ends up in an organic solvent, not in water. Thus it is more easily interfaced to gas chromatography and normal-phase HPLC than three-phase extraction, which is most compatible with reversed phase HPLC.
- The hardware is identical or similar, so practical procedures should be similar, considering the point above.

12.3. PRACTICAL CONSIDERATIONS

12.3.1. Choice of Fiber

The hollow fibers (HF) most commonly applied are Accurel® PP polypropylene fibers available from Membrana GmbH (Wuppertal, Germany), The most common versions are either 50/280 with a wall thickness of 50 μm, a pore size 0.1 μm and an i.d. of 0.3 mm or Q3/2 with a wall thickness of 200 μm, a pore size of 0.2 μm and an i.d. of 0.6 mm. The thicker wall of the Q3/2 fibers could lead to easier handling but slower mass transfer kinetics, but usually the mass transfer is limited by convection in the donor phase. Polypropylene fibers have in several studies been shown to be resistant to extreme pH in both the basic and acidic region of the pH scale. In older practice, often flat porous PTFE sheets were used [4].

12.3.2. Choice of Membrane Liquid

The membrane liquid for three-phase extraction should possess a number of properties. Since it is supposed to constitute a barrier between the aqueous donor and acceptor it is important that it is hydrophobic enough to not be miscible with water and that it will not dissolve in water during the time of the experiment (up to several hours). On the other hand, for more polar compounds it should have enough polarity to readily dissolve these compounds. Viscosity is another important parameter; the membrane liquid should preferably have as low viscosity as possible. Environmental and health aspects should also be taken into account, *i.e.* chlorinated or other highly toxic solvents are best avoided. Usually di-*n*-hexyl ether is a good choice for more polar analytes and *n*-undecane for less polar ones. In order to increase mass transfer for polar compounds, some percent of TOPO (trioctyl phosphine oxide) added to the membrane liquid can be effective [36]. For carrier mediated extraction (see section 2.5 above) or for other purposes, the membrane liquid could contain various additives.

12.3.3. Sample Pretreatment

For efficient extraction, the analytes should be present as uncharged species in the sample. For acidic analytes this is achieved by lowering the pH of the sample by addition of an acid, usually sulfuric acid. For basic compounds, in an analogous way, a base, such as sodium hydroxide, is added. It is important to be aware of that changing the sample pH might also impact the adsorption of analytes to particulate or dissolved organic matter if present in the sample, both directly (uncharged analytes are subject to stronger hydrophobic interactions) and indirectly (changing the pH will also change the overall charge of the organic matter, such as humic acids, and thereby its sorption properties). If equilibrium extraction in order to determine free dissolved fractions of analytes in environmental [17, 18] or pharmacological contexts [19, 20] should be performed, all manipulation of the sample should be avoided.

As was mentioned above, HF-LPME is not sensitive to the presence of particles in the sample, which means that filtration is usually not required prior to extraction of turbid samples. Rather it could influence the result by removing particle-bound analyte, why filtration is not recommended.

The sample volume is usually not critical, except for ESTM where it has to be large enough to avoid depletion of the freely dissolved fraction of the analyte. Otherwise, for environmental samples, it is mainly based on what is practical to handle and 50 or 100 mL are typical sample volumes used. For biological fluids, *e.g.* blood plasma, the sample volume is typically much smaller and limited by the volume that is possible to obtain. The small acceptor volumes which are possible to obtain with hollow fibers, permit high enrichment factors also with relatively small sample volumes.

12.3.4. Choice of Acceptor for Three-phase Extraction

10-20 μL is usually a suitable acceptor volume; it yields high enrichment, but still enough for injection of for example 5 μL on HPLC. Suitable fiber lengths are 20 cm of the 0.3 mm HF and 3.7 cm of the 0.6 mm HF.

For complete trapping (see section 2.3) the pH of the acceptor should be at least 3.3 units above the pKa values for acidic compounds and below for basic compounds. It is necessary that the acceptor has a sufficient buffer capacity, so changes in the acceptor pH are avoided. In some cases protons might be transported through the membrane, leading to pH changes.

Three-phase membrane extraction is usually employed in trace analysis, especially for complex matrices and is often followed by LC-MS analysis. Therefore, MS compatibility of the acceptor is an important aspect, *i.e.* volatile buffers such as ammonium formate, ammonium carbonate or ammonium acetate are preferred over for example phosphate buffers.

12.3.5. Choice of Membrane and Acceptor Liquid for Two-phase Extraction

For two-phase extraction, the choice of organic phase is quite wide. Most organic solvents, except the most volatile ones are possible to use. As the acceptor solvent is usually directly injected into a GC, too high or too low volatility can be impractical. Good results have been obtained with toluene, undecane, 1-octanol and di-*n*-hexyl ether among other. For other suggestions see a recent comprehensive review [37].

12.3.6. HF Preparation

The fibers are cut to desired length with a scalpel. To avoid contamination they are then preferably washed in acetone. It is usually enough to put the fibers in a bottle or a Petri dish with some acetone, but the process can be speeded up by sonication. The fibers are then allowed to air dry and thereafter stored in a Petri dish with lid until used.

12.3.6.1. Three Phase Extraction Procedure with 0.3 mm Fiber

This version of HF extraction was first described by Liu and coworkers [17]. See Fig. (**2**). It is probably the cheapest and most efficient way of membrane extraction, but demands some practical skill. Insulin syringes with 0.3 mm needle diameter are cheap and can be purchased in a pharmacy. A small amount of acceptor solution is drawn into the syringe and one end of the fiber is then gently put on the syringe needle and the fiber is filled. The fiber is then immersed into the membrane liquid for approx. 1 min. Afterwards it is quickly dipped into water to wash of surplus solvent on the outside of the HF. A further small volume of acceptor is then pushed through the fiber which is thereafter gently removed from the syringe. The ends are then sealed, either with a piece of aluminum foil or PTFE tape or by heating with a soldering iron. Since the fiber is light and will float on the surface, a copper wire can be coiled around it as a weight, whereafter the fiber is placed in the sample bottle on a magnetic stirrer and extracted for a suitable amount of time. After extraction, the fiber is removed from the sample with a pair of tweezers, and the ends cut off with a scalpel. A syringe is then filled with air, put into one end of the fiber and the acceptor liquid is pushed into a vial (with conical insert) for HPLC injection.

Sometimes it is needed to dilute or neutralize the acceptor liquid after extraction. By weighing the vial before and after adding the acceptor, the accurate acceptor volume for each extract can be determined, and optimal choice of neutralizer volume can be chosen and the dilution calculated and corrected for.

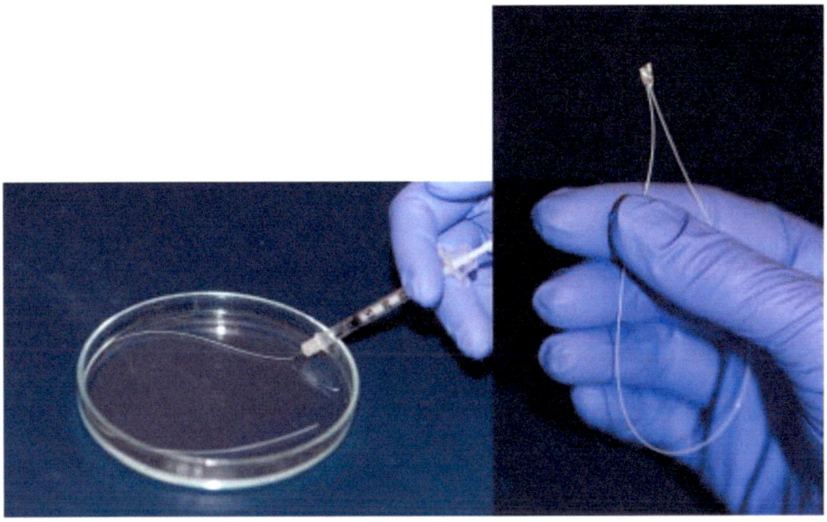

Fig. (2). (**a-b**) Hollow fiber extraction with 0.3 mm fiber. (Copyright 2014 Göran Larsson, Lund).

12.3.6.2. Three Phase Extraction Procedure with 0.6 mm Fiber

This early procedure was published by H.K. Lee and coworkers [38]. It was originally termed Hollow fiber-protected LPME and it is commonly used. It is easily performed, demanding a minimum of practice, but it requires one HPLC syringe for each sample. For 0.6 mm HF one end of the fiber is firstly squeezed with a pair of tweezers with flat ends and then sealed by heating using a hot metal device, *e.g.* a soldering iron, heat pistol or similar. An HPLC syringe with a suitable needle diameter is then filled with acceptor buffer, the fiber is gently put onto the syringe and the acceptor is pushed into the fiber until it is seen exiting through the pores. This is a way to investigate that the fiber is completely sealed, if not, the acceptor will only exit through the end. The fiber is then dipped into a bottle with membrane liquid for a few minutes and then quickly immersed into a beaker with water to wash of surplus solvent on the outside. After that, the syringe is left in the fiber, the fiber is lowered into the sample and the syringe secured with a clamp. See (Fig. **3**). After extraction, the syringe is removed from the sample, the end of the fiber cut off and the acceptor is withdrawn into the syringe and transferred to a weighted HPLC vial. If one syringe for each sample is not available, the syringe can be replaced with a metal rod of suitable diameter during the extraction. Also here, it might be necessary to dilute or neutralize the extract as discussed above in section 3.7.

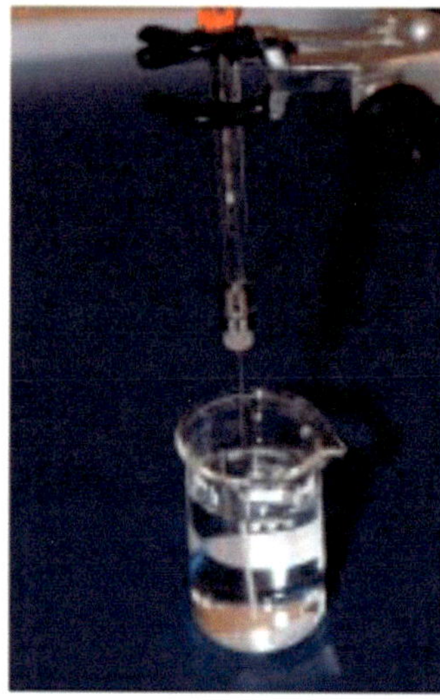

Fig. (3). Hollow fiber extraction with 0.6 mm fiber. (Copyright 2014 Göran Larsson, Lund).

12.3.6.3. Solvent Bar

The simplest version of HF-LPME is probably the "Solvent bar" device, first described by H. K. Lee and coworkers [39]. It can be arranged either as a 2-phase or a 3-phase system and is simply a piece of hollow fiber, usually with i.d. 0.6 mm, and closed in both ends by heating. It is filled with a syringe in a similar way as described in section 3.8. The syringe is then removed before the remaining end is closed. To harvest the extract one end is cut and a syringe is used to remove an aliquot of the contents.

12.3.6.4. Two-Phase Extraction Procedures

All of the HF membrane devices described above can be applied to two-phase extraction. Here, no aqueous acceptor solution is used; instead the organic solvent constitutes both the acceptor and the membrane liquid, making the procedure less complicated. Also, a hydrophobic solvent penetrates the pores of the hydrophobic fiber much easier than an aqueous acceptor. This means that the solvent does not need to be injected into the lumen of each fiber using a syringe, even if this is probably the most common practice. Instead a number of fibers can be prepared by sealing them in both ends with either heat or aluminum foil in an empty state. They can then be filled with organic solvent by placing them in a vessel with solvent and sonicating for some minutes. The fibers are then quickly dipped into water to wash off surplus solvent on the surface and then immersed into the samples. After extraction, the fiber is removed from the sample, the ends cut open and the acceptor transferred to a vial with a suitable syringe before GC analysis. This is possible with all formats, perhaps most easily with solvent bars.

12.4. OPTIMIZATION

12.4.1. Acceptor and Sample pH

In order to obtain as efficient extraction as possible, conditions for complete trapping should be attempted. As mentioned above (section 2.3), in three-phase extraction the acceptor pH should be at least 3.3 units below the pKa value of a basic analyte to obtain total protonation (correspondingly for acidic analytes). However, using a more extreme pH might not be advisable. There are observations [36] that the efficiency can suffer in such cases, probably for kinetic reasons related to the mass transfer at the membrane-acceptor interface but this is insufficiently studied.

The sample pH is much less critical. The general recommendation that it should be more than 2 units higher than pKa (for bases; see section 2.3) is not critical. Obviously, these recommendations are not valid in the case of equilibrium extraction, see section 2.4.

12.4.2. Other Parameters

If using a magnetic stirrer, the stirring speed (rpm) is usually optimized. However, a good guideline is that it should be as high as possible without the fiber moving too vigorously, which can cause the membrane liquid/acceptor to be washed out.

Sometimes addition of salt (usually NaCl) to the donor can enhance enrichment by salting out, *i.e.* increasing the partitioning of the analyte into the organic membrane liquid. However, the results are inconsistent; in some studies salt addition does increase enrichment while in others it actually causes a decrease, probably because the increased viscosity decreases mass transfer. The effect of salt addition has to be investigated in each individual study.

If modifiers such as ion pair reagents, carriers *etc.* are employed, the amounts of these need to be optimized. For more information and guidelines about such procedures, see the literature on specific studies.

12.5. MATRIX EFFECTS

The quantification of the initial analyte concentration in the sample is typically calculated by dividing the measured concentration in the acceptor by the enrichment factor, determined by calibration using standard solutions. However, it is important to note that enrichment factors determined by extracting spiked samples in reagent water might not be applicable to real samples. For instance, the presence of particulate matter or dissolved organic matter (DOM) in the samples might lead to adsorption of the analyte, especially if the sample pH is adjusted to make the analyte uncharged and thereby also more susceptible to hydrophobic interactions. If a fraction of the analyte is adsorbed and thereby non-extractable, the mass transfer of analyte from donor to acceptor will be smaller, giving an apparently lower enrichment factor. It has been shown that the enrichment factor can be more than halved for the same analyte in an environmental sample compared to reagent water. To obtain the relevant enrichment factor for environmental or biological samples, a standard addition procedure can be employed where original samples are spiked with different amounts of the analyte. The measured concentration in the acceptor (C_A) is then plotted against the added concentration to the sample and since $Ee = C_A/C_S$, Ee will be the slope of the regression line fitted to the data points. This Ee can be used for calculations of the initial concentration in unknown samples [9].

12.6. APPLICATIONS

During the years, a number of applications of membrane-based extractions (especially HF-LPME) in various fields have been presented and it is beyond the scope of this paper to attempt a review of this. There are several recent publications that summarize such applications [5, 7, 40, 41]. Many applications are found in the biomedical field, involving determination of *e.g.* drug compounds in blood plasma, urine, *etc.* There are also a large number of environmental applications relating to herbicides, phenols and other pollutants in different types of environmental waters [40]. "Emerging pollutants" as drug residues or personal care products, usually being rather polar compounds have also been successfully extracted with three-phase HF-LPME. A recent development is application of HF-LPME also to solid or semi-solid samples such as sewage sludge or meat [9, 10, 42]. In this procedure, the solid sample is first suspended into water and the analyte is allowed to reach partitioning equilibrium between the solid and the aqueous phase. The fiber is then placed into the suspension and extraction performed from the aqueous phase. By using a standard addition approach [9], then the initial concentration in the solid phase can be calculated.

DISCLOSURE

Part of this chapter has been previously published in Membrane extraction: General overview and basic techniques. Comprehensive sampling and sample preparation. Volume 2: Theory of extraction techniques, 2012 (461-474).

CONFLICT OF INTEREST

The authors confirm that they have no conflict of interest to declare for this publication.

ACKNOWLEDGEMENTS

Many students, post-docs and other colleagues, too many to be listed, have been involved in this work, by providing ideas and input for the development of membrane based extraction techniques.

REFERENCES

[1] Jönsson, J.Å.; Mathiasson, L. Membrane-based techniques for sample enrichment. *J. Chromatogr. A,* **2000**, *902*(1), 205-225. [http://dx.doi.org/10.1016/S0021-9673(00)00922-5] [PMID: 11192156]

[2] Jönsson, J.Å.; Mathiasson, L. Membrane extraction in analytical chemistry. *J. Sep. Sci.,* **2001**, *24*, 495-507. [http://dx.doi.org/10.1002/1615-9314(20010801)24:7<495::AID-JSSC495>3.0.CO;2-B]

[3] Cordero, B.M.; Pérez Pavón, J.L.; García Pinto, C.; Fernández Laespada, M.E.; Carabias Martínez, R.; Rodríguez Gonzalo, E. Analytical applications of membrane extraction in chromatography and electrophoresis. *J. Chromatogr. A,* **2000**, *902*(1), 195-204. [http://dx.doi.org/10.1016/S0021-9673(00)00835-9] [PMID: 11192155]

[4] Jönsson, J.Å. Membrane extraction for sample preparation - a practical guide. *Chromatographia.,* **2003**, *57* Suppl. 1, S-317-S-324.

[5] Pedersen-Bjergaard, S.; Rasmussen, K.E. Bioanalysis of drugs by liquid-phase microextraction coupled to separation techniques. *J. Chromatogr. B Analyt. Technol. Biomed. Life Sci.,* **2005**, *817*(1), 3-12.
[http://dx.doi.org/10.1016/j.jchromb.2004.08.034] [PMID: 15680784]

[6] Bårdstu, K.F.; Ho, T.S.; Rasmussen, K.E.; Pedersen-Bjergaard, S.; Jönsson, J.Å. Supported liquid membranes in hollow fiber liquid-phase microextraction (LPME)-practical considerations in the three-phase mode. *J. Sep. Sci.,* **2007**, *30*(9), 1364-1370.
[http://dx.doi.org/10.1002/jssc.200600486] [PMID: 17623480]

[7] Lee, J.; Lee, H.K.; Rasmussen, K.E.; Pedersen-Bjergaard, S. Environmental and bioanalytical applications of hollow fiber membrane liquid-phase microextraction: a review. *Anal. Chim. Acta,* **2008**, *624*(2), 253-268.
[http://dx.doi.org/10.1016/j.aca.2008.06.050] [PMID: 18706332]

[8] Pedersen-Bjergaard, S.; Rasmussen, K.E.; Jönsson, J.Å. Liquid-phase microextraction (LPME) utilizing porous hollow fibers. In: *Handbook of Sample Preparation*; Pawliszyn, J.; Lord, H., Eds.; John Wiley & Sons: Hoboken, N.J., **2010**; pp. 125-148.

[9] Sagrista, E.; Larsson, E.; Ezoddin, M.; Hidalgo, M.; Salvado, V.; Jönsson, J.Å. Determination of non-steroid anti inflammatory drugs (NSAIDs) in sewage sludge by a simple hollow fiber supported liquid membrane extraction and LC-MS. *J. Chromatogr. A,* **2010**, *1217*, 6153-6158.
[http://dx.doi.org/10.1016/j.chroma.2010.08.005] [PMID: 20810117]

[10] Ghaffarzadegan, T.; Nyman, M.; Jönsson, J.Å.; Sandahl, M. Determination of bile acids by hollow fibre liquid-phase microextraction coupled with gas chromatography. *J. Chromatogr. B Analyt. Technol. Biomed. Life Sci.,* **2014**, *944*, 69-74.
[http://dx.doi.org/10.1016/j.jchromb.2013.11.012] [PMID: 24295906]

[11] Hyder, M.; Genberg, J.; Jönsson, J.Å. Application of hollow fiber liquid phase microextraction for pinic acid and pinonic acid analysis from organic aerosols. *Anal. Chim. Acta,* **2012**, *713*, 79-85.
[http://dx.doi.org/10.1016/j.aca.2011.11.046] [PMID: 22200311]

[12] Jönsson, J.Å.; Lövkvist, P.; Audunsson, G.; Nilvé, G. Mass transfer kinetics for analytical enrichment and sample preparation using supported liquid membranes in a flow system with stagnant acceptor liquid. *Anal. Chim. Acta,* **1993**, *227*, 9-24.
[http://dx.doi.org/10.1016/0003-2670(93)85084-W]

[13] Pedersen-Bjergaard, S.; Rasmussen, K.E.; Brekke, A.; Ho, T.S.; Halvorsen, T.G. Liquid-phase microextraction of basic drugs-selection of extraction mode based on computer calculated solubility data. *J. Sep. Sci.,* **2005**, *28*(11), 1195-1203.
[http://dx.doi.org/10.1002/jssc.200401935] [PMID: 16116997]

[14] Berhanu, T.; Liu, J.F.; Romero, R.; Megersa, N.; Jönsson, J.Å. Determination of trace levels of dinitrophenolic compounds in environmental water samples using hollow fiber supported liquid membrane extraction and high performance liquid chromatography. *J. Chromatogr. A,* **2006**, *1103*(1), 1-8.
[http://dx.doi.org/10.1016/j.chroma.2005.10.082] [PMID: 16376356]

[15] Lezamiz, J.; Jönsson, J.Å. Development of a simple hollow fibre supported liquid membrane extraction method to extract and preconcentrate dinitrophenols in environmental samples at ng L(-1) level by liquid chromatography. *J. Chromatogr. A,* **2007**, *1152*(1-2), 226-233. [http://dx.doi.org/10.1016/j.chroma.2006.11.104] [PMID: 17182046]

[16] Mayer, P.; Tolls, J.; Hermens, J.L.; Mackay, D. Equilibrium sampling devices. *Environ. Sci. Technol.,* **2003**, *37*(9), 184A-191A. [http://dx.doi.org/10.1021/es032433i] [PMID: 12775037]

[17] Liu, J.F.; Jönsson, J.Å.; Mayer, P. Equilibrium sampling through membranes of freely dissolved chlorophenols in water samples with hollow fiber supported liquid membrane. *Anal. Chem.,* **2005**, *77*(15), 4800-4809. [http://dx.doi.org/10.1021/ac0503512] [PMID: 16053291]

[18] Romero, R.; Liu, J.F.; Mayer, P.; Jönsson, J.Å. Equilibrium sampling through membranes of freely dissolved copper concentrations with selective hollow fiber membranes and the spectrophotometric detection of a metal stripping agent. *Anal. Chem.,* **2005**, *77*(23), 7605-7611. [http://dx.doi.org/10.1021/ac050763a] [PMID: 16316167]

[19] Barri, T.; Trtic-Petrovic, T.; Karlsson, M.; Jönsson, J.Å. Characterization of drug binding to pure human blood proteins by employing equilibrium sampling through hollow fiber supported liquid membrane and bjerrum and scatchard plots. *J. Pharm. Biomed. Anal.,* **2008**, *48*, 49-56. [http://dx.doi.org/10.1016/j.jpba.2008.04.030] [PMID: 18565712]

[20] Trtić-Petrović, T.; Liu, J.F.; Jönsson, J.Å. Equilibrium sampling through membrane based on a single hollow fibre for determination of drug-protein binding and free drug concentration in plasma. *J. Chromatogr. B Analyt. Technol. Biomed. Life Sci.,* **2005**, *826*(1-2), 169-176. [http://dx.doi.org/10.1016/j.jchromb.2005.08.020] [PMID: 16181816]

[21] de Gyves, J.; Rodrigues de San Miguel, E. Metal ion separations by supported liquid membranes. *Ind. Eng. Chem. Res.,* **1999**, *38*, 2182-2202. [http://dx.doi.org/10.1021/ie980374p]

[22] Sastre, A.M.; Kumar, A.; Shukla, J.P.; Singh, R.K. Improved techniques in liquid membrane separations: an overview. *Separ. Purif. Methods,* **1998**, *27*(2), 213-298. [http://dx.doi.org/10.1080/03602549809351641]

[23] Lopez-Lopez, J.A.; Mendiguchia, C.; Pinto, J.J.; Moreno, C. Liquid membranes for quantification and speciation of trace metals in natural waters. *Trends Analyt. Chem.,* **2010**, *29*(7), 645-653. [http://dx.doi.org/10.1016/j.trac.2010.01.007]

[24] Papantoni, M.; Djane, N-K.; Ndung'u, K.; Jönsson, J.Å.; Mathiasson, L. Trace enrichment of metals using a supported liquid membrane technique. *Analyst (Lond.),* **1995**, *120*, 1471-1477. [http://dx.doi.org/10.1039/AN9952001471]

[25] Djane, N-K.; Ndung'u, K.; Malcus, F.; Johansson, G.; Mathiasson, L. Supported liquid membrane enrichment using an organophosphororus extractant for analytical trace metal determinations in river waters. *Fresenius J. Anal. Chem.,* **1997**, *358*, 822-827. [http://dx.doi.org/10.1007/s002160050516]

[26] Dzygiel, P.; Wieczorek, P.; Mathiasson, L.; Jönsson, J.Å. Enrichment of amino acids by supported

liquid membrane extraction using Aliquat 336 as a carrier. *Anal. Lett.,* **1998**, *31*(7), 1261-1274. [http://dx.doi.org/10.1080/00032719808002861]

[27] Dzygiel, P.; Wieczorek, P. Extraction of glyphosate by a supported liquid membrane technique. *J. Chromatogr. A,* **2000**, *889*(1-2), 93-98. [http://dx.doi.org/10.1016/S0021-9673(00)00120-5] [PMID: 10985540]

[28] Drapala, A.; Jönsson, J.Å.; Wieczorek, P. Peptides analysis in blood plasma using on-line system of SLM and HPLC. *Anal. Chim. Acta,* **2005**, *553*, 9-14. [http://dx.doi.org/10.1016/j.aca.2005.08.008]

[29] Yamini, Y.; Reimann, C.T.; Vatanara, A.; Jönsson, J.Å. Extraction and preconcentration of salbutamol and terbutaline from aqueous samples using hollow fiber supported liquid membrane containing anionic carrier. *J. Chromatogr. A,* **2006**, *1124*(1-2), 57-67. [http://dx.doi.org/10.1016/j.chroma.2006.05.001] [PMID: 16716341]

[30] Wieczorek, P.; Jönsson, J.Å.; Mathiasson, L. Concentration of amino acids using supported liquid membranes with di-2-ethylhexyl phosphoric acid as a carrier. *Anal. Chim. Acta,* **1997**, *346*(2), 191-197. [http://dx.doi.org/10.1016/S0003-2670(97)00123-2]

[31] Romero, R.; Jönsson, J.Å.; Gázquez, D.; Bagur, M.G.; Sánchez-Viñas, M. Multivariate optimization of supported liquid membrane extraction of biogenic amines prior to liquid chromatography determination as dabsyl derivatives. *J. Sep. Sci.,* **2002**, *25*, 584-592. [http://dx.doi.org/10.1002/1615-9314(20020601)25:9<584::AID-JSSC584>3.0.CO;2-F]

[32] Dziarkowska, K.; Jönsson, J.Å.; Wieczorek, P.P. Single hollow fiber SLM extraction of polyamines followed by tosyl chloride derivatization and HPLC determination. *Anal. Chim. Acta,* **2008**, *606*(2), 184-193. [http://dx.doi.org/10.1016/j.aca.2007.11.014] [PMID: 18082649]

[33] Miliotis, T.; Knutsson, M.; Jönsson, J.Å.; Mathiasson, L. Determination of aromatic anionic surfactants using supported liquid membrane technique. *Int. J. Environ. Anal. Chem.,* **1996**, *64*, 35-45. [http://dx.doi.org/10.1080/03067319608028333]

[34] Fontanals, N.; Barri, T.; Bergström, S.; Jönsson, J.Å. Determination of polybrominated diphenyl ethers at trace levels in environmental waters using hollow-fiber microporous membrane liquid-liquid extraction and gas chromatography-mass spectrometry. *J. Chromatogr. A,* **2006**, *1133*(1-2), 41-48. [http://dx.doi.org/10.1016/j.chroma.2006.08.026] [PMID: 16965779]

[35] Liu, J.F.; Hu, X.L.; Peng, J.F.; Jönsson, J.Å.; Mayer, P.; Jiang, G.B. Equilibrium sampling of freely dissolved alkylphenols into a thin film of 1-octanol supported on a hollow fiber membrane. *Anal. Chem.,* **2006**, *78*(24), 8526-8534. [http://dx.doi.org/10.1021/ac0615145] [PMID: 17165849]

[36] Shen, Y.; Grönberg, L.; Jönsson, J.Å. Experimental studies on the enrichment of carboxylic acids with tri-n-octyphosphine oxide as extractant in a supported liquid membrane. *Anal. Chim. Acta,* **1994**, *292*, 31-39. [http://dx.doi.org/10.1016/0003-2670(94)00041-7]

[37] Barri, T.; Jönsson, J.Å. Advances and developments in membrane extraction for gas chromatography: Techniques and applications. *J. Chromatogr. A,* **2008**, *1186*(1-2), 16-38.

[http://dx.doi.org/10.1016/j.chroma.2008.02.002] [PMID: 18314129]

[38] Shen, G.; Lee, H.K. Hollow fiber-protected liquid-phase microextraction of triazine herbicides. *Anal. Chem.,* **2002**, *74*(3), 648-654. [http://dx.doi.org/10.1021/ac010561o] [PMID: 11838687]

[39] Jiang, X.; Lee, H.K. Solvent bar microextraction. *Anal. Chem.,* **2004**, *76*(18), 5591-5596.

[40] Jönsson, J.Å. Membrane-based extraction for environmental analysis. In: *Comprehensive Sampling and Sample Preparation*; Pawliszyn, J.; Le, X.C.; Li, X-F.; Lee, H.K., Eds.; Elsevier, Academic Press: Oxford, U K, **2012**; Vol. 3, pp. 591-602. [http://dx.doi.org/10.1016/B978-0-12-381373-2.00104-6]

[41] Bello Lopez, M.A.; Ramos Payan, M.; Ocana-Gonzalez, J.A.; Fernandez-Torres, R.; Callejon Mochon, M. Analytical applications of hollow fiber liquid phase microextraction (HF-LPME): A review. *Anal. Lett.,* **2012**, *45*, 804-830. [http://dx.doi.org/10.1080/00032719.2012.655676]

[42] Boström, M.L.; Huang, C.; Engström, H.; Larsson, E.; Berglund, O.; Jönsson, J.A. A specific, highly enriching and "green" method for hollow fiber liquid phase microextraction of ionizable pharmaceuticals from fish tissue. *Anal. Methods,* **2014**, *6*, 6031-6037. [http://dx.doi.org/10.1039/C4AY00470A]

SUBJECT INDEX

A

M. Valcárcel, S. Cárdenas & R. Lucena (Eds.)

E

F

G

H

I

L

M

N

O

P

T

U

V

W

Z

Made in the USA
Las Vegas, NV
03 November 2026

57424455R00257